ENVIRONMENTAL RESEARCH ADVANCES SERIES

A TRUE TALE OF SCIENCE AND DISCOVERY

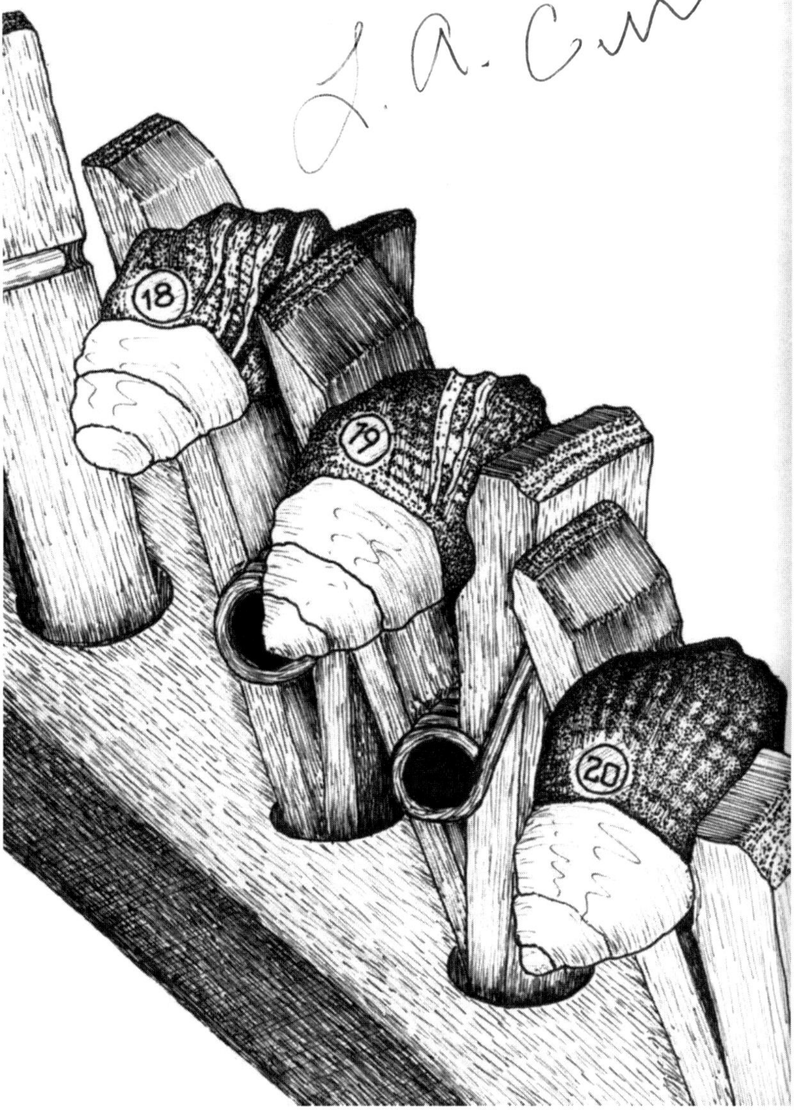

ENVIRONMENTAL RESEARCH ADVANCES SERIES

Environmental Research Advances. Volume 1
Harold J. Benson (Editor)
2008. ISBN: 978-1-60456-314-6

Handbook on Environmental Quality
Evan K. Drury and Tylor S. Pridgen (Editors)
2009. ISBN: 978-1-60741-420-9

Estimating Future Recreational Demand
Peter T. Yao (Editor)
2009. ISBN: 978-1-60692-472-3

Bioengineering for Pollution Prevention
Dianne Ahmann and John R. Dorgan (Editors)
2009. ISBN: 978-1-60692-900-1

Bioengineering for Pollution Prevention
Dianne Ahmann and John R. Dorgan (Editors)
2009. ISBN: 978-1-60876-574-4 (Online book)

Sorbents: Properties, Materials and Applications
Thomas P. Willis (Editor)
2009. ISBN: 978-1-60741-851-1

Bioengineering for Pollution Prevention
Dianne Ahmann and John R. Dorgan
2009. ISBN: 978-1-60876-574-4

Environmental Cost Management
Randi Taylor Mancuso (Editor)
2009. ISBN: 978-1-60741-815-3

Biogeography
Mihails Gailis and Stefans Kalniòð (Editors)
2010. ISBN: 978-1-60741-494-0

A True Tale of Science and Discovery
Lawrence A. Curtis
2010. ISBN: 978-1-60876-595-9

Biological and Environmental Applications of Gas Discharge Plasmas
Graciela Brelles-Mariño (Editor)
2010. ISBN: 978-1-60741-945-7

River Sediments
Greig Ramsey and Seoras McHugh (Editors)
2010. ISBN: 978-1-60741-437-7

A True Tale of Science and Discovery
Lawrence A. Curtis
2010. ISBN: 978-1-60876-595-9

ENVIRONMENTAL RESEARCH ADVANCES SERIES

A TRUE TALE OF SCIENCE AND DISCOVERY

LAWRENCE A. CURTIS

Illustrated by Nathan Lee Tanner

Nova Science Publishers, Inc.
New York

For permission to use material from this book please contact us:
Telephone 631-231-7269; Fax 631-231-8175
Web Site: http://www.novapublishers.com

LIBRARY OF CONGRESS CATALOGING-IN-PUBLICATION DATA

Curtis, Lawrence A.
A true tale of science and discovery / Lawrence A. Curtis.
p. cm.
Includes bibliographical references and index.
ISBN 978-1-60876-595-9 (hardcover)
1. Snails. 2. Discoveries in science. I. Title.
QL430.4.C87 2009
594'.32--dc22

Published by Nova Science Publishers, Inc. ✦ New York

CONTENTS

PREFACE

An attempt is made herein to portray one example of how human knowledge was increased. It concerns marine biology, but is presented in a way that makes it available to a reader not versed in that area of knowledge or science in general. My purpose has been to illustrate how one piece of science was done for those who have never been involved in doing scientific work. I hope that I have succeeded in doing that. I thank all those who advised me in various ways during the process of writing this. In particular I would like to acknowledge my sister, Tina Sculerati, and two friends who helped me a great deal, Karen Hubbard and Anne Colwell.

Chapter 1

THE START

This is an account of people, a place, a time, and the pursuit of knowledge. You are about to embark on one tale of how an increase in the knowledge of the natural world happened. In one guise or another it is an oft repeated tale. This time it concerns me, some of my students, and an obscure, but gloriously complex, marine organism known unofficially as the Eastern Mud Whelk. A drawing of it is presented as a frontispiece to this chapter. This snail is currently officially known as *Ilyanassa obsoleta*. (It's pronounced Il-ee-ah-nassa ob-so-lee-ta.) As recently as the 1960's its scientific name was *Nassarius obsoletus*. Before that its moniker was *Nassa obsoleta* and historically other designations have been applied. People who applied the names were impressed with the fact that the snail lived in the mud and often had a highly deteriorated shell. The word roots, "ilya" (mud), "nassa" (basket), and "obsoletus" (worn out, decayed, old) tell it all: older snails look like eroded mud baskets. Let me provide a word about scientific names. They are used to be as sure as possible exactly which species you are talking about. Common names may be used and they are perhaps easier and more comfortable, but they can refer to more than one organism. For example, around the world there are many different species known as "mud snails," but there is only one *Ilyanassa obsoleta*. [The word "species" is unusual in that it is used in both the singular (a species) and plural (several species) senses.] The first part is the generic name and there can be many species within a single genus. The second part is the specific name. Used together, these names signal only one kind of organism. You are not supposed to refer to a species with its generic name, but I guess I can get away with it here. There is only one species in the genus *Ilyanassa,* and that species is the star of this tale.

I first became aware of this beast while working toward my Masters Degree at the University of New Hampshire in the 1960's. The Piscataqua River meets the

Atlantic Ocean near Portsmouth, New Hampshire and extends inland about 12 miles to terminate in Great Bay. This estuary witnessed much of my work and play while I was there. The snail was not my central scientific interest at the time, but when the tide was out it littered the shoreline.

At the age of 31 in 1973, I got my Ph.D. in Biological Sciences from the University of Delaware. It is important to say here that about a year later I came down with insulin-dependent diabetes. That was a major surprise! It wasn't much of a limitation in the 1970's and early 1980's, but later on it made my work and play much more challenging. My dissertation was about the life cycle of a marine worm, *Sabellaria vulgaris*, which, needless to say, almost nobody cared the first thing about. It seems to be my fate to take up with organisms that do not claim global attention. Nevertheless, life cycles of organisms are central to understanding biology and I found the business of investigating them interesting and a challenge.

In 1975 I managed to obtain a position as Assistant Professor at a branch of the University of Delaware in the southern part of the state. At the time this academic unit was called the Parallel Program. The idea was that, with emphasis on the first two years of University education, course work parallel in quality and rigor to that provided on the main campus would be made available to local students. I could get behind that. If I had just settled into my instructive duties and forgotten about research, that would have pleased administrators of my academic unit, but my nature and perceived duties dictated otherwise. A university professor is supposed to fulfill three basic responsibilities, teaching, research, and service. I considered the third, which basically amounts to committee work, to be a somewhat necessary evil. University administrators have essentially one strategy. If they want something from the faculty, they get the faculty to form a committee to figure out how to give it to them. If faculty want something from them, they get faculty to form a committee to dash and dissipate their energies. So the request won't have to be honored, not anytime soon anyway. Either way, a faculty member on or running a committee is likely (not always) to be involved in a waste of time and whenever possible he or she would do well to avoid this activity.

I wanted to pursue especially teaching and research. Teaching was rewarding to me and I put much effort into it. I taught what was requested of me, proposed and developed new courses, and did such service as I could not avoid, but what to do about research? For a time I tried to go on with studying my dissertation worm, but my interest at the time required taking plankton samples and that was a prohibitively expensive activity.

Let me take a side-step here to make a pledge to readers, similar to one that the late Stephen J. Gould made to his readers in one of his many essays on science. It will be necessary to deal with matters of science, but I will not lie to you. That is true for any purpose, but especially I will not do it in the name of simplification. In fact, there is nothing all that complicated to report here. Above all, I want this to be assessable to persons that are interested, but not versed in the science involved. The biology of animals without backbones (invertebrates) and estuarine ecology are the main fields with which I deal. A reader not acquainted with these disciplines should be able to get along fine. Close attention will sometimes be needed, but I will attempt to explain things so that someone not familiar with the terms and concepts involved can easily follow. Years of teaching have taught me that scientific lingo often prevents people from understanding science. There is no value in weighing a non-scientific reader down with unnecessary terms, but if used wisely terms are useful tools of explanation. Otherwise they just obfuscate.

Perhaps the term plankton is not too necessary to explain, but let me begin with that. It refers to organisms that do not swim strongly and drift at the mercy of currents. These organisms need not be microscopic (tiny), but in this case they were. I was interested in the planktonic developmental stages (larvae) involved in the worm's life cycle. To sample these larvae and pursue this line of research required boat time and a variety of other expensive items. My resources were very limited (non-existent) and I soon recognized that this line was not available to me. I decided an intertidal animal that could be sampled simply by walking out when the tide was down was much more realistic, and so I ultimately focused in on the mud snail as an object of research. Little did I know at the time what a wondrous journey I was starting.

Before they knew what a pain in the neck to them my research endeavors would become, administrators agreed to setting me up with some laboratory space in the Cape Henlopen Laboratory of the University's College of Marine Studies complex in Lewes, Delaware. I had worked there during my dissertation research. It was not much, only one room and no equipment, not even a microscope, but it was my place to work and get started. My lab space was located in an old military building on Cape Henlopen at the mouth of Delaware Bay. The building had been used to store and dispense mines during WWII to prevent enemy submarines from entering the Bay. It was big, a simple rectangle with hoists to move heavy mines around. The roof and asbestos sides were not tight. Some rain and lots of wind would seep in around the edges. It sat maybe a quarter of a mile from the Bay. The first time I experienced it most of the rooms inside did not have walls that reached all the way to the floor, strictly a place to work in summer. Now, inside there were posh, mostly watertight rooms that could be air conditioned and heated. My Ph.D. advisor, the late Dr. Franklin C. Daiber, stepped forward and made the microscopes I had used during my dissertation research available to me. They were very good microscopes, one dissecting (a Wild M5) and one compound (a Wild M20). Now I had some basic equipment and could get started. On the Delaware Bay side of Cape Henlopen there is a sandflat that harbored a huge population of *Ilyanassa*. They occurred along shore in a band about 1,500-2,000 meters long and anywhere from about 25 to 200 meters wide. Serendipity clicked. My lab was there, I had a little equipment, I wanted an intertidal animal to work on, and there was a handy snail population. I did not know much about this snail, but I had some small familiarity from my New Hampshire days, i.e., I knew what its name was, *Nassarius* something. I had a lot to learn, but the juxtaposition of circumstances was pronounced. I began to ponder: what was going on with these snails; why was the species so abundant and successful?

Some characterization of our protagonist is appropriate. Research seeking new information should always start with what is already known and my first job was to look into the literature on this snail. It was first described by T. Say in an 1822 publication. It has been the subject of study on and off at least since then. It is a native and, if for no other reason than its abundance, an ecologically important animal along the Atlantic coast of North America from the Bay of Fundy in Canada to about northern Florida in the U.S.A.. About a century ago it was accidentally introduced on the Pacific coast and it now also occurs there. Along shores in estuarine environments there can be sometimes 1,000's per square meter. The snail is black and the largest ones are generally about an inch (25-30 mm) long (high, in snail anatomy). So this is not a big, dynamic creature. It has neither plume nor pelage. It is not repulsive, but not lovable either. It does not lend itself to film documentaries. *Ilyanassa* is mainly a deposit feeder, an organism that consumes bottom sediments and digests out contained nutrients, but it also eats macroalgae (algae visible to the naked eye) and, most importantly, it conspicuously aggregates on and consumes dead meat (carrion). It is probably best described as a scavenger. That this snail can eat both plant and animal material is unusual because snails tend to be either herbivore or carnivore. This is because the enzymes that aid in digestion of plant material cannot coexist with enzymes that digest animal material. The omnivore capability was the snail trait that initially interested me. Could it be a key to the snail's numerical abundance? Off we go.

Here are some instructions for dealing with this tale. The telling of this saga necessitates that it will be somewhat autobiographical in nature, but that is not my point. It is the tale of a search for biological knowledge. Its central feature is a sort of a down-in-the-trenches rendition of what was found out, how it was found out, and how understanding of the natural world was enhanced. Consequently, considerable attention will have to be devoted to the circumstances under which discoveries were made. My goal in writing this is to give an interested non-scientist an insight into how one contribution to ecological science proceeded. To take you into this world of field biology it will be necessary to take you into the problems and concerns that were of interest to a me as I worked various research projects. Things that interest a person tell you much about that person and what he or she is trying to do. As indicated above it is not so much the person I seek to have you to follow, but rather the process of scientific discovery. A grasp on that process will probably be of greatest value to my intended reader. Taking you on a trip into my world of discovery will require that I maintain a balance. On the one hand I will need to broach problems of science that you likely have not thought about. I suppose I won't totally avoid getting wonkish about certain issues, but it is necessary to tell my intended tale. So have a little patience with me if I err in places on that score. Some over indulgence in information would happen whatever area you might decide to look into, whether it be Appalachian quilts, golf, a martial arts discipline, or perhaps the manufacture of cardboard boxes. We all, certainly including me, have stuff to learn. On the other hand I recognize that I do not need to take you into every last bit of relevant detail to make my point(s). I spoke above about the terms used in science. They all have some concept behind them and if they are not explained one can very quickly get lost. Their overuse can put anyone off that is not used to dealing with them. Some terms will have to be used, but I shall strive to keep their meanings and pertinence to the issues at hand clear. I will also keep their number under control. That also goes for the concepts I employ and any arithmetic I might have to use. By keeping things abundantly clear I shall always be keeping my intended reader in mind.

As perhaps you have already gathered, this will not be a tale of what might be called "high profile" research and discovery. It was not well funded and did not have much moral support from the powers that be. It certainly was not carried out with the aid of plush research grants. Taxpayers went virtually unscathed. It was a bootstrap operation from start to finish. Nevertheless, I like to think that some good science was produced and our knowledge of the natural world enhanced. Research often proceeds in this manner. In the history of biology, more often than not discovery has begun with curiosity driven, lonesome activity. This is a useful lesson for a person that has not (yet) been involved with science. You too could or can do it. It mostly takes interest and desire.

One more thing is useful to include here. I include this bit of reality because scientific endeavors are often portrayed as if they were perfectly done by supremely competent beings. How many times have you heard a conclusion of some sort denounced because it is *non-scientific*? Perhaps it deserves denouncement, but had it been described as a *scientific* conclusion it would automatically be given more respect. Science as a means to find things out is deserving of great respect, but its conclusions do not arise out of flawlessness. Nobody can be a perfect, all-knowing scientist. After all, no one in any venue can be so perfect. It would be inconceivably boring in any case. Did I always do what I did in the best way possible? Was I always correct in every aspect? Not likely, but no scientist can expect to always be immersed in such correctitude. It's just not in the game. As is true in any quest of discovery, I have not always been so competent. There was always trial, error, and outright mistake involved. As a consequence, any conclusion of science is open to question on the basis of new interpretation or new evidence. That very definitely is in the game. What is needed in a scientist is dedication and cleverness in the search for the truth and the keeping of careful records so you can retrace your steps and correct the inevitable mistakes. This is more in keeping with the nature of scientific endeavors than is some kind of flawlessness.

Science is a product of people, a time, and a place all enmeshed and intertwined. Were it someone else driving this research and different colleagues involved, that alone would alter the product. Some of the same phenomena would undoubtedly have been discovered, but the emphasis would almost surely have been different. If time were shifted, either more or less knowledge about the snail's biology would have been available. This too would surely have altered what was discovered. Variability is a cornerstone of biology. Organisms on planet earth are different from place to place (and time to time) because of genes possessed and environmental circumstances. The interdependence of these variables, biological variability, people, time, and place, make it impossible for other circumstances to produce the same result. Science in general is not often thought of this way. It is supposed to consist of universal truths, independent of the circumstances under which they were discovered. This would perhaps be more true in the physical sciences. There variation is considered "error," some sort of mistake in taking measurements. For example, all electrons are expected to weigh the same. If you find differences among them, you are dealing with measurement error. In biology, variability is caused by a combination of measurement error and real differences in organisms caused by different genes and environments. For example, if you are studying the amount of oxygen consumed by snails, there will be differences among snails, partly from measurement error and partly from real differences among the individuals you test. Because of these dual sources of variability, the results of a biological study are particularly contingent on circumstances. This is especially true of a study done outside the laboratory in the field. All this is not to say that no general truths can emerge from biological field studies. While variables are more profuse in the field and firm predictions are harder to come by, we can nevertheless come to understand generally what the important factors are and how they affect individual organisms and populations. Science can certainly be done in the laboratory, but it is my belief that, if it is at all possible, science done out in nature is the way to go and I have done most of my work there. Field research is often very challenging and arriving at an explanation of results more difficult (more variables to deal with), but out in the field is where organisms live and where they function normally.

Science is supposed to be hypothesis-driven. That is, you make some observations of the natural world, form an hypothesis (an educated guess) to explain them, devise a test of the hypothesis, test it, and then see if results of the test support or refute the hypothesis. In brief, these are the steps of the scientific method, at least as usually portrayed. A fairly worn analogy, not my creation, may help illuminate how science is done, some of the issues, and my general approach. Doing science as outlined just above is akin to observing a barn wall, defining a circle (the hypothesis) on the wall, firing an arrow at the circle (the test), and seeing if the arrow finds (results, data) the circle. If it does, the hypothesis is supported and tentatively accepted. Tentatively always, as there could be other explanations for the outcome. This is a perfectly valid way of doing science, but not always a possible approach. When we are exploring the unknown, often no hypothesis can be formed before some data are in hand. We don't know where to draw the circle. So we fire the arrow at the wall (take observations), see where it lands, and then go up and draw a circle around it. This would be an after-the-data-are-in hypothesis, as opposed to one formed before a test is run. Some say this is not a valid way to do science because no prediction is made and you don't know where the arrow is supposed to land. It just lands and after the fact you try to justify why it landed there. If the arrow had for some reason been pulled to another place on the wall, if observations had been otherwise, the circle (the after-the-fact hypothesis to explain observations) would have been drawn in another place. In the end it seems that both ways of doing science are valid and needed. Sometimes an hypothesis can be formed; other times no expectation as to what is going to happen is available and we must be led to our tentative explanation of observations (the hypothesis) by data in hand. I and my colleagues have often (not always) found ourselves in the latter position, taking field observations in a systematic way, not really knowing what to expect, and trying to make sense of where the arrow landed. You have to be honest, no fudging. Data acquisition must not be designed, the arrow must not be aimed, to favor a certain outcome. If not fudged, this too is a perfectly valid way of doing science and I have often used it. It is my hope that understanding of the natural world was increased a bit by what I and my colleagues have found out and what I have to tell.

Experimental setup for the laboratory testing of an *Ilyanassa obsoleta* diet

So, we undertake this journey. As I have said, I will not lie to you in the name of simplification, but at the same time things will have to be explained. Much was found out over this 30+-year tale and that is my basic point here, not my personal tale. There is no mystery to what goes on in science. It is all available and understandable and it is my faith that it can be comprehended by an interested reader not used to dealing with scientific issues. The goal is to empower and interest such a reader. My challenge will be to make the issues understandable without loading my explanations with all the terms and jargon usually employed when explaining for other scientists. There may be times when I explain more deeply than you necessarily find of interest, but almost always you can avoid getting bogged down in such spots if you choose. The key ideas will likely be summarized more briefly further on. So you won't lose the thread. If I have explained less than you would like, which is my hope, references to places where you can look into things further have been provided.

Chapter 2

THE OMNIVOROUS SNAIL

In Chapter 1 I raised the issue of *Ilyanassa obsoleta* being an omnivore, feeding on both plant (used here to refer to any organism capable of photosynthesis) and animal material. In fact, it uses quite a variety of food items. The question is: how can it do this when most snails are either herbivores or carnivores? Among the gastropods *Ilyanassa* is a neogastropod. Snails in this group exist as males or females (sexes in separate individuals) and feature a notch in the shell opening (the aperture) through which a siphon protrudes and takes water into a mantle cavity for respiration (oxygen uptake and carbon dioxide elimination). The neogastropods are almost all strictly carnivorous, but *Ilyanassa* can also use plant food, making it a rarity within the group. How can this be? A paper, written by Charles Jenner in the 1950's based on work done around Woods Hole in Massachusetts, instructed me that, surprisingly, *Ilyanassa* possesses a crystalline style. This is a structure that exists within the snail's stomach. It is a glass-like (crystalline) rod composed partly of protein that spins and grinds against a gastric shield located on the inside of the stomach wall. Think of a broomstick (the style) spinning like an axil inside a sock open on both ends (the stomach). In the process it releases enzymes (e.g., amylase) that digest plant sugars and starch. A crystalline style is found in herbivorous mollusks, for example in bivalves such as clams and mussels. So the ability of this snail to utilize plant food is explained, but questions arise. They always do.

What is the role of each general food type in the snails diet? Can snails get along on just animal or just plant food or do they require both for survival and reproduction? It was 1976 and I was involved in developing and teaching my courses in southern Delaware, trying to come up with an avenue for research, and also making weekly trips to the main campus in Newark, about 90 miles away, to visit people and go to seminars in the Department of Biological Sciences. I had recently gotten my degree in that department and had taught some courses there. At the time there were several ecologically-minded faculty with whom I was friendly. Among us we talked, often over beer, about a wide range of issues in science, rare books, the issues of the day, and less exalted topics. A conversation relative to the question posed above developed between myself and one of the faculty. Lawrence E. Hurd and I decided that we would try to answer it. Because we both went by the name Larry, we came to be known as L raised to the power of 2, L^2.

Our conversations led us to perform a simple (and inadequate by itself) experiment. Two aquaria were maintained, which gave us the idea for a later, more refined experiment. These were gang tanks where the snails were not identifiable as individuals. The object was to follow growth, in terms of changes in size, and survivorship on meat versus vegetable diets. At the end of April 1977 we collected 147 snails (mean shell height = 16.6 mm) from Cape Henlopen. These were divided randomly between the two tanks and maintained until late June 1978, snails in one tank being fed on spinach, in the other on shrimp. They were fed *ad libitum*, i.e., snails had more food available than they consumed.

In neither tank did appreciable growth in size occur, but there was a difference in the proportion of snails surviving. Shell height is measured from the lip of the siphonal notch to the apex (the pointy end) of the shell. The 72 snails fed on spinach started out at an average shell height 16.6 mm. After about 13 months 66 had survived and they had an average size of 16.8 mm. The 75 snails fed on shrimp started out with an average (mean) shell height of 16.6 mm. After the same 13 months only 31 had survived and those 31 had a mean size of 15.6 mm. Thus there was essentially very little or negative growth among snails fed just spinach or shrimp, but there was better survivorship on the vegetable diet (92%) versus the shrimp diet (41%). This exercise told us that *Ilyanassa* could be maintained in the laboratory on frozen spinach or shrimp for quite a long period of time, the vegetable diet was better in terms of survival, but on neither diet did snails do very well. More questions arose. Was this a consequence of laboratory conditions? Was our growth criterion too crude? What would happen on a mixed diet?

With these clues and questions in mind we decided we would expose snails to four different treatments, diets of meat, or vegetable, or both, and a starvation control. The last was needed in case snails could maintain themselves without food for long experimental periods, a not unheard of possibility. On each treatment snail growth would be measured as changes in weight over time. We would also keep track of how many on each treatment survived and for how long. Several snails would be maintained on each diet so statistical comparisons could be made. For example, average snail growth in each of the four treatments might be compared. Given what the two-tank experiment had shown us, we were testing the hypothesis that snails would do better on a mixed diet.

The basic idea was simple enough, but testing an hypothesis, even in the laboratory, is complicated and to make this point I choose to devote some space here on the details of what went into making the experiment doable and effective. It's always a challenge. Fasten your seatbelt. The experiment was to be set up and run in my laboratory space in southern Delaware, close to a source of live snails. From the resources of the Department Biological of Sciences we gathered up four additional 20 gallon aquaria and the associated pumps and filters. We acquired a sensitive electronic balance to weigh snails. That was after trying and failing to resurrect an old, very fine mechanical balance we found under a cover in the basement. It was an instrument of the past, but I was fascinated by its workings. It could not be brought back to function, but even now I still have it, under a cover again, decorating my laboratory. In nature snails live in moderately reduced salinity relative to full strength seawater, which in the Atlantic Ocean is 35 ppt salts (ppt = parts per thousand, in a 1,000 grams of seawater there are 35 grams of salts). Aquaria were filled with artificial seawater made up to and maintained at a salinity of 25 ppt. Snails were collected on 22 April 1978 from the Cape Henlopen sandflat. Their size ranged from 10.8 to 14.4 mm shell height. Snails of this small size would be younger and we didn't want them dying of old age during the experiment, an unfounded worry as it much later turned out. Weight changes in 12 individual snails on each diet would be followed. This meant that each snail had to be maintained separately so we could be sure of its identity at each weighing. We developed a system to do this. Each of the 48 snails was kept within its treatment aquarium in a section of plexiglass tubing (diameter 5 cm, length 10 cm) screened off at both ends and suspended from a crosspiece spanning the aquarium from side to side. There is an illustration of a setup in the frontispiece for this chapter. In each aquarium there were three ranks of four snails. It took us more than a month to develop this system. Should you choose to look, there is a simply wonderful drawing of the setup in the paper we ultimately published on this work (Curtis and Hurd, 1979). [I shall keep the citing of literature to a minimum, but this method of reference directs the reader to the bibliography at the end where the full reference is given. Unspecified references may be found in the papers that I do specifically cite. If for some reason a reader wishes to find a paper, most university libraries carry the journal and there's always the internet.] After allowing some time for snails to get used to the experimental conditions, during which snails were not fed, each snail was blotted to remove as much water as possible and weighed to the nearest hundredth of a milligram (0.01 mg). [A milligram is 1/1,000th of a gram (a gram = 0.035 of an ounce). So the balance was indeed very sensitive. We later rounded snail weights to the nearest milligram.] This was the starting weight for each individual. On the chance that position of a treatment tank on the laboratory bench could affect results, at this time (18 May) we randomly selected which tank would house which treatment. Feeding was begun. Snails were either not fed (starvation tank, a control to be sure that snails needed food during the experiment), or fed frozen shrimp (animal diet tank), or frozen spinach (vegetable diet tank), or both shrimp and spinach (mixed diet tank). Shrimp was a natural food, but spinach was not. Spinach is a pretty rich vegetable and experience (the two-tank experiment) indicated that snails could persist on it. Food in the plexiglass containers was provided *ad libitum*. It was changed twice per week and at each change containers were cleaned. Each snail was weighed once per week. Water in the tanks was aerated, filtered, and completely changed once (1 September) during the experiment. And so the arrow was in flight. Would it land in the circle predicted? We monitored the experiment until 26 October 1978 (187 days). I apologize for the detail of this explanation of experimental protocol, but the reader should be aware of the care, attention to detail, and a little of the tedium involved in an experiment's execution.

Statistics are not evil, but I will spare the reader a trip through the statistical analyses of these data. Suffice it to say that there are statistical justifications for the conclusions I mention below. Results from this experiment were quite striking. Starved snails clearly lost the most weight of the four sets of snails. It should be no surprise that snails need food to survive and grow, but it's good to know that this applies in the time-frame of the experiment. Fed snails all lived for at least 67 days. During this time, spinach-fed and shrimp-fed snails maintained their weight just below where they started. Snails on a mixed diet, on the other hand, were able to gain a pronounced amount of weight over the same time. In terms of survival from the very start of the experiment on 22 April 1978, all of the starvation control snails were dead after about 120 days, shrimp treatment snails were all dead after about 145 days, some spinach-fed snails lived about 180 days, and the last mixed-diet snail died at 187 days.

So what did these experiments teach us? Because the snail is a neogastropod and it had often been observed in the field being attracted to and feeding upon carrion sources (dead fish, crabs, etc) it had been thought for a long time to be carnivorous. But in the 1950's Charles Jenner recognized that there was not enough carrion around to support the large number of *Ilyanassa* that existed and noted the presence of a crystalline style. Later, in the 1970's, another researcher (R. Wetzel) found that most of the snail's nutrition came from microscopic benthic algae (diatoms living among bottom sediment particles). Clearly this snail digests both plant and animal materials. The surprise was that it could not get along on either one alone. It needed both. My research partner and I tagged the snail as an "obligate omnivore" to encapsulate this idea. We liked the alliteration and thought it had a ring to it.

L^2, as were routinely called at the time, did a follow up piece of work on the diet of this snail (Curtis and Hurd, 1981a). During the summers of 1978 and 1979, in association with field work on the crystalline style to be described later, we had occasion to collect and dissect many snails just taken from the field. They came from the Cape Henlopen sandflat and also from a nearby saltmarsh. We examined snails for the contents of their gut. Our interest was to find out what our obligate omnivore was feeding on in the field. (In doing this we became aware that *Ilyanassa* is frequently infected with larval trematode parasites. There will be more on this later, but this was a seed that would later germinate and come into full bloom.) In all 3,571 snails were examined. Empty guts were found in 26% of them and 57% had only sediment grains in the gut. Undoubtedly these grains had diatoms and other organisms on and among them that would serve as nutrition sources. The list of other food categories found includes: 2% had only macroalgae (pieces of macroscopic algae, in the laboratory snails can not live on these); 9% had a mixture of macroalgae and sediments; 12% of snails from the saltmarsh habitat had sediments and fragments of *Spartina* (a grass common in saltmarshes); less than 1% had a polychaete worm in the gut, sometimes still alive; about 1% of snails from the sandflat had eaten horseshoe crab eggs; crustacean pieces were found in less than 1% of snails, as was unidentified meat. It is clear that *Ilyanassa* eats an array of things in the field. Sediment was the most frequent item found in the gut, which is appropriate for a deposit-feeder. From what was in the sediment (diatoms and other microscopic algae) and other gut contents the snail did indeed get a mixed diet in the field. Basically, whatever the snail could swallow it consumed and we christened the snail a "biological vacuum cleaner," which seemed, still seems, to fit the bill.

Because the style is needed to digest plant material, one might think that it would be present in the stomach when plant food had been eaten. However, the data say no. By looking to see if a crystalline style was present or absent when food of any type was present, we found that more often than not when food was present a style was also present. When we looked for style presence or absence in association specifically with plant or animal food presence, there was no association. The style tended to be present with both food types. Based on the literature we were expecting its presence to be associated with plant material in the gut, but no. Apparently the snail does not digest on the basis of what it ingests. Some other strategy is involved.

The author collects data at the Canary Creek saltmarsh footbridge in the late 1970's.

Chapter 3

A STYLISH AFFAIR

Like all enzymes, amylase is made mostly of protein. In the snail this enzyme is released from the crystalline style rod as it breaks down when spinning against the gastric shield, which is located on the inside of the stomach wall. How can the style, itself of protein makeup, and the amylase it releases coexist with the protein-digesting enzymes used to digest animal material? Some provision for this must be in play. After all, *Ilyanassa* needs to be able to digest both plant and animal food and the style must somehow exist together, at least some of the time, with the ability to digest animal material. It was early July in 1978. The diet experiment described in Chapter 2 was under way in the laboratory. One night I was thinking about this mystery of coexistence. Pretty much on a whim I decided to get up the next morning, collect some snails from the field, and see whether they had a style present. Let me track events, sort of hour by hour, to give the reader a sense of what can be involved in pursuits of knowledge. At the time, Cape Henlopen was kind of a lonesome place, an out of the way state park that not too many people visited. There was a park superintendent and some regulation, but the place was minimally policed and there was great freedom as to where you might go and what you could do. You didn't even have to pay to get in! That next morning there was a low tide at 7 AM. I used a bicycle to get out there by 6:30 and selected a site where snails could be collected. An isolated place along the beach, about half a mile from the laboratory, was chosen and the collection site was set up below the average low tide line at the base of the beach where there were lots of snails. To be able to come back to the site again, even at high-water, I put a weight there with a buoyed line attached. On that visit, and on each thereafter, I measured air and water temperature and water salinity. At 7 AM 10 snails were collected and rushed back to the laboratory. I wanted to see how many had styles present. After measuring shell height the shell of each had to be cracked and the snail body removed. While looking through a dissecting microscope, the stomach was cut open with a small pair of scissors to look for a style. While doing a dissection, in addition to looking for a style, I also recorded snail sex and gut contents. In this batch six snails had no style, three had partially formed styles, which were either in the process of forming (waxing) or dissolving (waning), and one had a fully formed style. It took some 30-45 minutes to get the laboratory work done and by 8 AM I had biked my way back out to the site and was ready to collect 10 more snails. The tide was starting to come back in now. Among these 10 snails, seven had no style, two had partial styles, and one had a fully formed style. I didn't know what to expect, but at least that was consistent with the previous sample. At 9 AM the water was more than ankle deep and all 10 snails lacked any trace of a style. Hmmm. Among the 10 AM batch six snails lacked styles, but four had partial styles. Given the previous sample I figured these styles were waxing. At 11 AM six snails lacked a style, one had a partial, and three had fully formed styles. By this time the tide at the collection site had risen to about chest deep and on each trip I had to go under water and feel around for snails. It was a nice warm day and that was fine, for now. Not only the water was rising. So was my interest and curiosity. I began to suspect that I was in for a long haul. At 12 PM the tide was high and I had been on the job for about six hours. I submerged and got my 10 snails. Three lacked styles and seven had fully formed styles. At 1 PM it was four without and six with full styles. In each of the 2 and 3 PM samples only one snail lacked a style and the rest had fully formed styles. By this time the tide was about half way down. Nine hours now, I hadn't figured on this thing going this long. I had brought no provisions for the day and working alone, if I wanted to stay on schedule, I couldn't go get any. There were some crackers, cookies, sodas and such around the laboratory and I was subsisting on those, still going pretty strong. At 4 PM two snails were without and eight had fully formed styles, but at 5 PM one had no style, five had partials (probably waning), and four had full styles. Maybe the style is about to be lost again? At 6 PM four snails lacked a style, one snail had a partial, and five had full styles. At 7 PM, about low tide again, two lacked styles, three had partials, and five had full styles. The 8 PM batch consisted of eight snails without styles and two with full styles. Here comes the tide again. It was dark for the 9 PM sample, but the water was still about knee deep. Seven snails lacked styles, one had a partial, and two a full. At 10 PM six lacked styles, two had partials, and two had fulls. Some sixteen hours into it now, I was getting very tired, running on crap for food, constantly wet, and making field trips in the dark. Only my cursed curiosity was keeping me going. Two snails lacked styles in the 11 PM sample, five had partials, and 3 had fulls. I persisted until 12 midnight. By this time the tide was almost high again. I had to wade down the beach face to get my sample. I'm about six and a half feet (two meters) tall, the water at the site was almost up to my neck, and I will readily admit that in the dark in this lonely spot it was damned spooky going under water, feeling around for snails, and getting the sample. Anyway, two snails lacked a style while eight had fulls. Whew! I could reasonably be satisfied that the style had come back and that I had reached a sort of logical stopping point. And stop I did. In the laboratory I put things in a little order and on my wet, now chafed, ass I made a middle of the night ride home. It was two miles or so. I stayed awake long enough to take some insulin and eat, and then indulged in the long sleep of the exhausted.

In the event that tiredness may have set in following the last paragraph, allow me to summarize what happened. On the day samples were taken a low tide occurred about an hour after dawn. In this period few or no snails possessed a style. As the middle of the day and high tide approached snails seemed to regenerate styles, but as the afternoon wore on and low tide approached again they lost them again. Then after dark they regained their styles. By midnight the majority of snails had a style. Apparently, the style waned in association with low tides and dawn and dusk and waxed in association with high tide in the middle of the day and night. The lesson is: within a single day's time a snail's style waxes and wanes, cycles between present and absent. That's pretty clearly what happened, but new information always leads to new questions. It never seems to end. Is this a general pattern, or did it just happen that way on the day I happened to take samples? We have a good indication (Chapter 2) that in the field snails tend to have a style regardless of what they had been eating. So that doesn't explain style waxing and waning. Here we have an indication that styles tend to be absent around dawn and dusk in association with low tide, but was it the tide or time of day cuing the rhythm? What happens when dawn and dusk are not associated with a low tide? Does style cycling occur only on Cape Henlopen? Answers to all these questions required further study. If I wanted to know, which I did, there was more work to do.

The experiment described above, really just a drawn out observation, gave a strong signal as to the waxing and waning of the crystalline style in *Ilyanassa* in the field. Furthermore, it led to several interesting questions. Naturally I couldn't keep my mouth shut about it and before long there were others willing to aid in pursuing the issue. Larry Hurd, the other part of L^2 spoken of in Chapter 2, and another faculty member from Biological Sciences, Howard "Buck" Cornell, were intrigued, as were two quality graduate students, Linda Collins and Kathy Corbett. All these people helped to perform further marathon studies during summer 1978. In doing these studies, for one important thing, we knew what to expect and provisioned ourselves with supplies for the grueling procedure of hourly collecting 10 snails and getting them processed before it was necessary to go out and get the next 10. I will spare the reader portraying these studies on an hourly basis and treat them more in summary. To address the first question, were the results peculiar to a certain day, I waited two weeks and initiated another study. This amount of time allowed the tides to come around and be about as they were in the first study, low tides around dawn and dusk. The first 10 snails were again collected at 7 AM, one hour before low tide. Snails were collected and processed each hour until 9 AM the next day. Larry Hurd was my partner in this study and we stayed awake for about 30 hours getting it done. It was quite an exercise, but entertaining and rewarding. There were of course some minor differences from the former study, but in essence results were the same. Snails lost their styles in association with dawn and low tide, regained them with midday and high tide, lost them again at dusk and low tide, regained them during the night, and lost them again at dawn and low tide the next morning. So, the cycle was not peculiar to a day. Waxing and waning of the crystalline style in *Ilyanassa* stomachs seems to go on every day, at least on Cape Henlopen.

The next study was done a week later. I wanted to see what happened when dawn and dusk were associated with high tide rather than low tide. Waiting a week accommodated this. The first sample was taken at 7 AM, one hour before high tide. Hourly samples were taken until 6 AM the next morning. In essence, snails were mostly without styles in the hours after dawn and also around dusk. At other times of the day they mostly had styles present. So, it seems it was the time of day (particularly dawn, but also dusk) that was of key importance in style waning rather than the occurrence of low tide. That surprised me.

The remaining two studies were started at other times of the day than the previous three. About six hours pass between low and high tides and lows and highs occur roughly an hour later each day. The next study was done two days after the last when dawn and dusk were not associated with high or low tide. The first 10 snails were collected at 3 AM, about an hour before low tide, and samples were taken until 9 PM that night. I have to confess that I dosed off after the 10 AM sample and missed the 11 AM trip. Oh, well. Snails mostly had a style present before the sun came up and during low tide, but lost it after dawn while the tide was coming in. By noon they had it back, only to lose it again in late afternoon before dusk. In the two samples after dusk most snails again had a fully formed style. This was further confirmation that the dissolution of styles was associated with time of day (dawn and dusk), rather than low tides.

The last study in 1978 was done 10 days later. In this one the first snails were collected at 5 PM about two hours before high tide. Samples were taken hourly until 3 PM the next day. It was a moonlit night. The movie "Jaws" had recently been released. I can recall Buck Cornell standing on the beach around 4 AM making the heartbeat-like sounds of the movie's theme music while I was in chest deep water intermittently submerging and scratching for snails. Had he the foresight, he probably would have preferred to rig a tape player and waft the music at me when I thought I was alone out there. Always interested to see science advanced, he liked to help out with comforting contributions. In this study there was little indication of style loss around dusk just after we started taking samples. They mostly had styles throughout the night. High tide was at dawn the next morning and snails definitely lost styles shortly thereafter. They developed new styles before we stopped sampling in the afternoon.

I concluded from these studies that crystalline styles were regularly lost at dawn and often at dusk, particularly if these coincided with a low tide. There appeared to be a synergistic effect of these two factors. If low tide and dawn or dusk occurred together, style loss would surely occur on those occasions, but especially dawn and sometimes dusk would do the trick regardless of tidal stage. It's hypothesis time. Why would such a rhythm be of value to *Ilyanassa*? How would it help them to survive and reproduce offspring? Here's the way I put it together. The snail is a biological vacuum cleaner. It crawls along extending its proboscis with the mouth and radula (a rasping "tongue") at the end taking in a more or less steady stream of food, mostly sediments. Associated with the sediment, between the particles, are microscopic algae, bacteria, and small or sometimes large (relative to snail size) animals, in other words, a mixed diet. The problem is that to digest plant material, which requires a style and its amylase, the snail cannot at the same time be digesting animal material because that would require protein-digesting enzymes (proteases), which would digest the style and destroy the amylase. What to do? I suspect (and that's all I can do with existing knowledge) that protease production is under rhythmic control. Around dawn or dusk the snail begins producing proteases. These cause the style to be dissolved and animal food to be digested. The value of this is that all a snail must do is keep a stream of sediment, and anything else it can consume, coming in. A snail does not have to specifically seek out vegetable or animal material to feed on because, according to the rhythm, when proteases are not being produced the style can be present and plant material can be digested, when proteases are present animal material is digested. The snail digests a mixed diet and is healthier, which means it is more likely to survive, grow and have babies, making it a successful organism. Later, Larry Hurd published an independent study showing that female snails fed a mixed diet produced more packages of eggs (fertilized eggs in capsules) than females fed just animal or plant material. That, I conclude, is the value of the waxing and waning of the crystalline style and so I said in the paper (Curtis, 1980) I ultimately published on this work.

The summer of 1978 was about over. It was late August and I had to think seriously about getting my courses ready for the fall semester. Students, lectures, laboratory exercises, reports, quizzes, and tests take a lot of time. Teaching can't be done well if you don't put in the time. It had been an enlightening summer of research and I felt as though I had accomplished something. I found out some interesting new stuff. As part of a university, administrators of my academic unit had to tolerate faculty research activity, but did not much value it. It was necessary to play down my research and get on with teaching. Well, that too is a rewarding enterprise.

In the next summer we (L^2) did two things relative to the crystalline style. We tested the effect of different salinities on style occurrence (Curtis and Hurd, 1980) and did further studies on style waxing and waning in the field (Curtis and Hurd, 1981b). Let's drop the latter for a moment and consider the effect of salinity on whether or not snails have a style present. It had been known for a long time that *Ilyanassa* occupies the parts of an estuary where salinity is higher. The biologist Abigail Dimon made this point, among many others, in a 1905 monograph.

First, here's a bit about why salinity matters. An estuary is a place where freshwater (river flow) meets and measurably dilutes the salt content of seawater. Recall that full strength Atlantic Ocean water is 35 ppt salt. Cape Henlopen juts into the ocean right at the mouth of Delaware Bay. On the Bay side, where the snails are, when the tide is in salinity will be perhaps 31-32 ppt. When the tide is out salinity is often cut about in half. If it is raining, salinity may fall to 10 ppt. Farther into the Bay, in the adjacent saltmarshes, it goes even lower. Therefore, in contrast to oceanic organisms, estuarine organisms have to deal with variable salinity. So what? Why would salinity changes present a physiological challenge to an organism? Basically, it's about water balance, specifically the concentration of water in the organism's cells and tissues. If the concentration of water in an organism gets too high or low, its physiology is upset and it goes into stress, death in the extreme. Life's chemistry is, after all, water-based. When salinity goes up, for example, there is more salt and less water in the water outside around an organism. At the same time an organism's cells and tissues have relatively less salt and more water in them, i.e., concentration-wise there's more water inside the organism than there is outside. Across the cell membranes on the surface of the organism water moves (diffuses) from where it is more concentrated to where it is less concentrated. In short, in this circumstance the organism will lose water. The opposite happens when the external salinity is low; the organism gains water. If you have an organism like *Ilyanassa*, that cannot regulate the uptake or output of water across its body surface, as salinity changes the water content inside the organism changes. If external salinity change is significant, the organism's water balance will be off and it will be in physiological stress. I have spared you my whole lecture on osmosis, but that's the nut of it and why salinity matters to an estuarine organism like *Ilyanassa*.

So, this snail does not occur where salinity is too low, but exactly what does low salinity do to snails to bring this about? The style is an important adaptation possessed by *Ilyanassa* individuals. An adaptation is any feature of an organism that increases its probability of survival and reproduction. This one, the style, allows the snail to use abundant plant material as food. Could low salinity affect the crystalline style? We didn't know if it did and we were not testing a particular hypothesis. A suspicion was held by L^2, but there was no prediction, just the question. In essence our plan was to expose snails to different salinities in the laboratory and see whether they had a style, but not so simple. What was going on in the field at the time of the experiment? Was a style loss occurring there at the time? If so, we might observe style absence in a salinity treatment in the laboratory, but it might have nothing to do with salinity. Somehow we needed to control for this possibility. We also wanted to know whether any observed effect of salinity on styles was caused by a direct action on the style itself or was mediated somehow through the snail's physiology.

At 2 PM on 19 July 1979 at about mid-flood tide we haphazardly collected 40 snails from the Cape Henlopen sandflat. Salinity at the time was 28 ppt. In the laboratory we had four containers filled with bay water of four salinities, 28, 20, 15, and 10 ppt. We divided the 40 snails (randomly) into four groups of 10 snails each and put a group into each salinity. At 2:10 PM groups were submerged. In the lower salinities snails tried to escape and had to be gently pushed back in. All groups were kept in their salinity treatments for one hour. Snails were removed from all treatments simultaneously and all were dissected within 50 minutes to check whether styles were present. In the 28 ppt treatment eight snails had styles (five fully formed). In 20 ppt seven did (four fully formed). In 15 ppt eight snails had no styles and two had partial styles. In 10 ppt it was seven without and three with partials. Snails in the lower salinity treatments (15 and 10 ppt) had a style less often. But, what was going on out in the field? We collected and dissected 10 additional snails from the field on three occasions, once when the experimental snails were collected, once 50 minutes later, and once 135 minutes later. All 30 snails in these control samples had styles. Therefore the experiment was not done while a style loss event was occurring in the field and differences in style occurrences among experimental groups of snails were indeed caused by different laboratory salinities. Were differences in presence/absence/state of styles at different salinities brought about by a direct effect on the styles themselves or somehow mediated through the snail's physiology? We removed fully formed styles from snails, exposed them to salinities of 28, 20, and 10 ppt, and watched through a microscope to see how quickly they dissolved. Ten styles were observed at each salinity. Removed styles dissolved at about the same rate in all salinities, indicating that salinities had no differential effect on the styles themselves. Style loss was somehow mediated through the physiology of the snail.

Ilyanassa can withstand episodes of lower salinity. They can handle even fresh water for a time without permanent damage. They just temporarily shut down. In the field they occur where salinity is generally 20 ppt or above. Results of the salinity study suggest that the reason for this has to do with that important adaptation, the crystalline style. Wherever the salinity dips below about 20 ppt too often snails would not be able to maintain styles, would not therefore be able to digest plant material (their main source of nutrition), and would become unhealthy. They move away from such areas and thus are found where salinity is higher. Admittedly this is laced with speculation, but it is now an informed hypothesis that could in principle be tested. I can envision keeping snails in the field (perhaps in cages) for long periods where salinity is marginal for style maintenance and looking for evidence that styles are indeed seldom present and the snails unhealthy. I never pursued this angle. A person more agile than I am now would be required to chase down such a field experiment.

The studies done in summer 1978 on the waxing and waning of the crystalline style gave a good idea of what goes on in the field. But, questions remained. Would style cycling occur in another year; does it occur at more locations than just on Cape Henlopen? A particularly vexing concern was that no criterion existed for deciding whether a style loss event in the field had actually occurred, as opposed to a random sampling event. In science you want to be able to make such decisions based on specific (statistical) criteria. It simply won't do to simply wave one's arms about emphatically and say assertively "It sure looks like it did (or didn't) happened." During the 1979 studies we came up with a specific criterion for style loss in the snail population and this aided in the interpretation of data.

Let me deal first with the criterion for judgement. I will not take you through every detail, but I want to provide enough for a basic understanding. Suppose you were to do an experiment many times that consisted of tossing a coin 10 times, 10 tosses of the coin is one experiment. If the coin is balanced (your hypothesis), overall you would expect 50% heads and 50% tails, but in any particular 10 tosses of the coin you might get 8/10 heads, or 5/10 heads, or 2/10 heads, or some other number of heads out of 10. The question is: assuming the coin is balanced, and therefore expected overall to give heads 50% of the time and tails 50% of the time, in what proportion of experiments would you expect to get 8, or 5, or 2, or some other number of heads out of 10. There is something called the binomial expansion that allows you to calculate how often you would expect to see, for example, 8/10 heads even if the coin is actually balanced. The expression to be expanded is $(0.50 + 0.50)^{10}$. There are only two possibilities and the two terms (the bi "nomials") are the overall hypothesized probabilities of heads and tails; the term in parentheses is raised to the tenth power because your experiment consisted of tossing the coin 10 times. Should you be desirous of doing so, you can look in a statistics book to reveal the mechanics of doing the calculation. In the end you have a prediction, again if the coin is balanced, about how likely it is to get 10, 9, 8, 7, 6, 5, 4, 3, 2, 1, or 0 heads out of 10 tosses of the coin. It doesn't actually, but suppose your calculation tells you that on your hypothesis you expect the 10/10 heads result only once in a 1,000 runs of the experiment. If when you actually do the 1000 experiments you see 10/10 heads, say 100 times. You then have good reason to reject the hypothesis, the coin is likely out of balance in favor of heads.

It turned out that 62% of the 1,650 snails taken from the field in Cape Henlopen studies done in 1978 and 1979 had partially or fully formed styles present in their stomachs. It follows that styles were absent in the other 38% of snails. Moreover, this split between the haves and have nots was quite consistent among studies. Therefore, in percentages we have a 62:38 split in terms of the probability of having or not having a style in the Cape Henlopen population. This is comparable to the 50:50 split in the coin example. We took samples of 10 snails each so the binomial expression to be expanded is $(0.62 + 0.38)^{10}$. When worked out this told us how likely it was to collect *at random* from the Cape Henlopen snail population a sample of 10 snails containing any number possessing a style. That is, 10/10 snails possessed a style, or 9/10, or 8/10, and so on down to 0/10. These expected values could then be compared with the numbers of snails out of 10 possessing styles that we actually observed in samples. It turns out that there is a very small chance (under 1 %) of collecting 10/10 snails with a style and less than 1 chance in 20 (5%) of collecting 10 snails at random among which three or fewer possess styles. Because it was so unlikely at random, this became our criterion for a style loss event; if a sample had 3 or fewer snails with styles, that was a style loss event. We collected 30 samples like this, but would only expect to collect 7 such samples. In other words, we collected many more samples of 10 snails with 3 or fewer having styles present than we would expect by chance alone. Style loss is not occurring randomly. Something else is going on. A reader following the argument will suspect that many of these samples were taken around dawn or maybe dusk.

So what did we do in summer 1979 with respect to style waxing and waning? Six more studies, of 27-28 hours duration, were accomplished. These were similar in design to those done in 1978, 10 snails an hour. Larry Hurd and I were involved in all studies and graduate students Linda Collins and Kathy Corbett also put in long hours collecting and processing snails. Thank you ladies for your expert efforts. A question was: did a similar waxing and waning of the style occur in another habitat besides Cape Henlopen. On 15-16 June we did simultaneous studies on Cape Henlopen and in a nearby saltmarsh [about 5 kilometers (3.1 miles) distant]. We had a team of two at each site, Linda and Kathy at the more hospitable Cape Henlopen, and communicated by phone. The Cape Henlopen site was similar to those used in 1978. Canary Creek saltmarsh is quite large, some 200 hectares in area. A hectare is 10,000 square meters and 200 of them comes to about 494 acres. There is a main channel (Canary Creek) with mosquito ditches emanating from it. I understand that the mosquito ditches were dug in the 1930's to drain marsh water and cut down on the mosquito population. It does not seem to have worked. At certain times mosquitos would kidnap and devour you. Further, the exposed banks of the ditches were great places for biting flies to breed. Greenheads could be utterly savage. The marsh site we chose was about 300 meters from the main creek up the first ditch on the right as you enter the saltmarsh via Canary Creek. At the time there was a narrow footbridge (a plank) spanning the ditch there, which was about 2 meters wide and a bit more than a meter deep. (See the frontispiece accompanying this chapter.) When the tide was in there would be about a meter of water in the ditch. Snails could be collected with a long- handled scoop off the footbridge, a very nice feature especially in the dark. The meter or more depth of the water around high tide on the other hand presented a problem. The snails were on the bottom and that is where we needed to get a water sample to measure temperature and salinity on each visit. Even in a meter or so of water there could be stratification, water of different characteristics at different depths. Colder water is heavier and likely to be near the bottom, as is saltier water. Getting down into the ditch on foot would likely destroy any stratification present. There were several pieces of equipment in existence to trap water at certain depths without getting into the water, e.g. Nansen bottles, but they were typically too big for this job and anyway they were out of our price range, which was close to zero. Our solution is typical of how such problems get solved in these situations. This one involved the sacrificing of a broom. You use that which is available. This did not win us any points with the maintenance guy, those pesky scientists acting up again, but we took the broom handle and affixed to one end a water bottle holder that had been used to clamp drinking water on a bicycle frame. In the laboratory there were a number of what were called salinity bottles around. They fit nicely into the bottle holder, which could clamp the bottle tightly, and they had glass stoppers. A string was attached to the glass stopper on the bottle and run to the other end of the broom handle. At the undisturbed site we could then stopper the bottle, put it down to the bottom of the ditch off the foot bridge, then yank the string pulling the stopper, and the bottle would fill with bottom water. The temperature and salinity of this water could be measured when the bottle on the end of the broom stick was retrieved to the surface. Such was the highly technical, but cost free way we solved that particular problem.

The 15-16 June simultaneous study was done when low tide occurred in association with dawn and dusk, the first sample was taken at about 7 AM, the last sample at about 10 AM the next day. A week later (21-22 June), when high tide was associated with dawn and dusk, we executed another simultaneous study involving the same two sites with first and last snail samples collected at about the same times. Two more studies were done singly at another site in the marsh, at the confluence of the ditch with Canary Creek. These were done on 7-8 July, when high tide occurred at about dawn, and on 27-28 July, when low tide was around dawn.

In the first simultaneous study styles on Cape Henlopen behaved as they did in 1978. There was a style loss event (three or fewer snails had styles) just after dawn on both days and also around dusk. During the middle of the day and night styles were regained. At the marsh site, however, there was no such pattern. There were style loss events, but they were not well-associated with times of day, or tides. Low salinity was involved in some of these. In the second simultaneous study again Cape Henlopen styles behaved pretty much as they had before. There was a style loss event after dawn (while the tide was high) on the first day. Styles came back after that. There was also a drop in style occurrence in late afternoon, then a regaining, and finally a definite loss event in the hours after dawn on the second day when the tide was high. At the marsh site at this time snails had styles pretty much throughout the study. They did, however, lose the style around dawn on the second day. Well, here's a chink in the armor, apparently waxing and waning of the crystalline style is not everywhere the same. We were curious to see if this was specific to a particular marsh site. That site was after all way up within the confines of a ditch where conditions (e.g. salinity, light, water current strength) could be very different. So for the last two studies we moved 300 meters down the ditch to its mouth on Canary Creek, where conditions wouldn't be quite so different from those on Cape Henlopen. In the first mouth-of-ditch study high tide was just after dawn and there was a style loss event at that time. The style was then regained and lost after low tide in mid-afternoon. By dusk styles were back only to be lost again the next morning at 9 and 10 AM. This was more like what happens on Cape Henlopen. The second mouth-of-ditch study was done when low tide happened at dawn and before dusk. Results of this study were just like what usually happens on Cape Henlopen. There was a definite style loss just after dawn (and low tide) on the first day. Styles were regained and then lost again just after low tide before dusk. At night they were regained again and there was another loss following dawn and low tide the next morning. So it happened in two separate years. As to pattern as relates to location, one marsh location produced essentially the same pattern as that seen on Cape Henlopen. The other clearly did not. Why is a matter of speculation.

I will summarize style waxing and waning on Cape Henlopen in 1978 and 1979, using the criterion of 3 or fewer snails out of 10 having a style for a style loss event. There was a total of 32 style loss events. When dawn occurred with low tide, style loss occurred in 5/5 times. When dawn occurred with high tide, style loss occurred 3/4 times and when dawn occurred with some other tidal stage, style loss occurred 1/1 times. Taken together this means style loss events occurred on 90% of dawn or just post-dawn periods observed. When low tide and dusk co-occurred a style loss occurred 3/3 times, but when other tidal stages co-occurred with dusk a loss event occurred only 1/4 of times. This means that a style loss event occurred with dusk at 4/7 (57%) of the time. At times other than dawn or dusk style loss events occurred even less often. Style loss is most closely associated with dawn, with dusk next, especially if low tide occurs at the same time of day. Of course we need to reckon with the possible role of lower salinity at low tide. On occasion, style losses in the marsh in 1979 and on the sandflat in 1978 could be attributed to low salinity because salinity dropped to the 16-10 ppt range. Sometimes this was confounded with time of day so the loss could be attributed to either factor. We must thus consider not only dawn (and dusk), but also low salinity when trying to predict style loss events.

We should take a moment now to review what has been found out about what is going on with *Ilyanassa* from the extensive work done so far. It had been known for some time that this snail has the rare ability, among the neogastropods, to digest both plant and animal material as food. Based on evidence from the laboratory and field the point was made by this work that snails need and in fact get a diet with both components in it. Why do they need such a diet? Well, it is because only on such a diet do they survive, grow, and reproduce and this makes the crystalline style an important adaptation. Only by means of the style is it possible to digest plant material, the snail's main source of nutrition. The problem of how herbivorous and carnivorous habits can coexist in a snail species is addressed by daily style waxing and waning. True, the style goes away at salinities below about 20 ppt, but at these low salinities the snail is pretty much disabled and cannot feed. So this can't explain how style waxing and waning aids in digesting a mixed diet. Other times of day may be involved, but the style most regularly goes away (dissolves) primarily at or just after dawn and is regenerated after that. Why does it go away then? Well, I must admit I don't actually know this based on evidence, but it is logical to strongly suspect that at this time protein-digesting enzymes, associated with carnivory, begin to be produced. These enzymes would digest the protein-containing style, as well as the plant-digesting enzyme(s) made by it. Thus the style disappears and the ability to digest plant sources of nutrition with it. When a style is there, most of the time, plant material is digested. When not, animal material is digested. Because a more or less steady stream of sediments and their contained plant and animal foods is taken in, cycling of style presence and absence makes it possible for the snail to obtain (digest) a mixed diet. This understanding (hypothesis) is subject to improvement (further testing), but as it is it helps us to grasp how and why this important species of snail is so successful.

The stylish affair comes to an end. At 9 or 10 AM when L^2 finished studies during 1979 the unit had been up and working steadily for some 30 hours each time. Needless to say, we were both exhausted and very ready to eat and crash. Down on the beach in Lewes at the time there was the De Braak Inn. It was kind of a classy old local place named after a ship that had long ago gone down off Cape Henlopen. It had low ceilings, live entertainment sometimes, a kind of darkness to it, a bar, and good breakfasts were served. We used to go there to take care of the eating part of our needs at the finish of studies. On most occasions we were well behaved. There's no denying that L^2 used to knock back a few beers while discussing all sorts of issues and we were bad boys after at least one style study. We decided to have pitchers of beer with breakfast. I guess the statute of limitations has run out on this so I can tell the tale including any illegal activity. We had breakfasts of eggs, sausage, home fries, rye toast, orange juice and coffee (a mixed diet). So far, there's nothing illegal there, but we were two youngish, indestructible guys (so we thought) and we then started in on beer and marveling at our recent great contribution to science. Look out! Along about the third pitcher we were getting pretty giddy, making a lot of noise, and causing a disturbance in the place. The lady that owned and ran the Inn was polite and we did not offer any argument, but she summarily kicked us out and we deserved it. It was about 11 AM and we were both severely messed up. We managed to get outside along the side of my 1976 Ford pickup trunk. I remember it was blue and we were both clinging to its bed gunnels somewhat desperately. I'll not say who it was, but after a while, to put a nice cast on it, one of us fell asleep right there next to the trunk. It was no small feat, but somehow one member of this crack research team managed to roll the body of the other member into the bed of the pickup and without incident drove the half mile or so back to my place. Probably some illegality involved there, but it was almost 30 years ago. We attempted a game of backgammon, but sleep quickly claimed us both. One of us landed on my sagged couch, the other the floor. Such is sometimes the illustrious way great research is concluded.

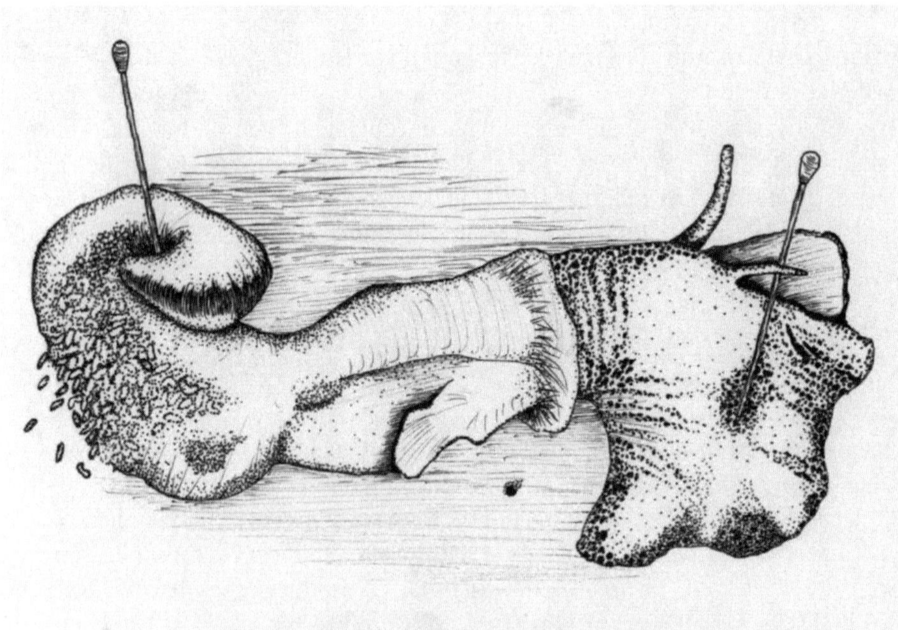

Ilyanassa obsoleta, shell removed, infected with the trematode *Himasthla quissetensis* (parasites spilling out of body at left, snail head and foot at right)

Chapter 4

A DELVING INTO PARASITES

It was 1980 and summer was coming on. It was at about this time that I had the first of my diabetic foot problems. A persistent open sore developed. This one eventually went away, but it was a harbinger of things to come. Be that as it may, teaching was about to be over for a while and L^2, as a unit, was thinking about what to investigate next. In the process of doing crystalline style studies we had become aware of the many trematode-infected *Ilyanassa*, especially on Cape Henlopen. In some samples these parasites infected most, if not all, snails and they stood to have consequences. That seed I spoke of in Chapter 2 was about to germinate.

By this time we had acquired the ability to identify the species of most trematode species infecting the snail. We were not perfect at it, but we had a good grasp on which species were which and, importantly, were willing to admit it when we weren't sure. The publications of previous workers, dating back as far as the 1920's and 1930's were a source of very helpful drawings and other information about these species. Let me mention just a couple. This is a tale of the pursuit of knowledge and knowledge always builds on previous knowledge, somebody else's work. In 1926 H. M. Miller and F. E. Northup published the earliest study I'm aware of on the trematodes found in *Ilyanassa* (then *Nassa obsoleta*). They worked around Woods Hole in Massachusetts and named some of the species they found. Just to show you there is rhyme and reason to scientific names, a species of *Himasthla* was found in snails in Quissett Harbor and was named *Himasthla quissetensis*. A snail infected with this parasite is illustrated in the frontispiece for this chapter. At Woods Hole (and elsewhere) this work was the foundation for future work. A name associated with much of that work on trematodes found in *Ilyanassa* is H. W. Stunkard. I never met him, but he published much and had a long, distinguished career in parasitology. There were of course others, notably J. J. McDermott working in New Jersey. Collectively, through their archived publications, they made it possible for us to figure out how to identify trematodes infecting *Ilyanassa*. That's the way it often works when you are doing research. You get help from the literature, but nobody sits down and personally instructs you. There was no one available at the University of Delaware or close by with the expertise, time, and willingness to do that. You must use your education to figure things out on your own. Mine came from the University of New Hampshire where, in getting my Masters Degree in Zoology in the 1960's, I first really learned about a few types of parasites and a variety of invertebrate marine animals.

There are no common names for the species of parasites with which we must deal here. It is not important that a reader keep in mind, be able to pronounce, or remember parasite generic and species names. It is only important that they be recognized as different parasite species. To save a bit of space and cut down on scientific name usage I shall employ a method throughout. In a chapter when a parasite is first mentioned its name will be written out in full with its abbreviation in parentheses after it. After that in the chapter I will use only the abbreviation made up of the genus and species initials. So, for example, *Himasthla quissetensis* will be abbreviated *Hq*. Other parasite species will follow suit. So they all can be found in one place, in my closing pages I have placed a table listing the names of all the trematode species encountered in Delaware *Ilyanassa* together with their abbreviations and some other information.

For the sake of completeness let me list the infecting species we were aware of at this time and encountered in this work, indicating which were frequently found in snails and which were more rare. It was tedious tracking these identifications down and I love to throw the names around, but as suggested above I shall forebear. Species we found fairly regularly and could identify with confidence were *Hq, Lepocreadium setiferoides (Ls),* and *Austrobilharzia variglandis (Av)*. Another infection was common, but we were unsure of its identity at the time. It turned out to be *Zoogonus rubellus (Zr)*. Rarely, we also found *Stephanostomum tenue (St), Stephanostomum dentatum (Sd)*, and *Diplostomum nassa (Dn)*.

Larry Hurd and I decided we would like to characterize the Cape Henlopen *Ilyanassa* population with respect to the distribution in it of differently aged, sexed, and parasitized snails. Why bother? We had noted in the literature that when people assessed such things with regard to this snail, and others, studies were usually based on only a few more or less haphazardly taken samples. The samples would be analyzed, generalizations would be drawn from the results, and applied to the whole population without really knowing whether the samples were representative of the whole population. We thought that if snail age, sex, or parasitism was important in this snail population, we had to know how these snail features were distributed in it. At the time there was a term afoot for what we needed. We needed an "intensive" study to find out what we wanted to know. And so, we started on this journey. They all begin with a single step.

Adult snails on Cape Henlopen do not exchange snails with other populations of *Ilyanassa* in the area. They are isolated, discontiguous. However, they are not genetically isolated because there is a planktonic larva, the veliger larva, in the life cycle of this snail. (Larva is the singular, larvae is plural.) Recall what planktonic means. The larvae are free in the plankton for a time (a week or two, possibly longer) drifting about at the mercy of prevailing currents. With luck, eventually the larvae become mature, drift over a place to settle (e.g. Cape Henlopen sandflat), settle, and metamorphose into bottom-living juvenile snails. Veligers have a genetic makeup and larvae produced from adult snails on Cape Henlopen could settle somewhere else and veligers produced elsewhere could settle on Cape Henlopen. Planktonic larvae keep genetic contact among and unify the separated populations of this snail.

Adult snails on Cape Henlopen occur in a band along shore about 2,000 meters long. They do not extend offshore below the line of the very lowest low tides. There is some kind of barrier there. What is it? I have some ideas, but it is a wide open research problem. Is anyone interested? The sandflat runs mostly southwest to northeast and the band of snails is less wide to the southwest (at most about 50 meters out from shore) and wider to the northeast (up to about 200 meters or more). Snails eventually peter out as you approached the tip of the Cape at the northeast end of the sandflat near the ocean. This is a lot of area and a big population, at least millions of snails, to characterize. My partner in L^2 worked mostly on the ecology of terrestrial habitats and had a supply of very neat and tall metal marker stakes. Early that summer we set about using a tape measure and the stakes to mark off mostly 50 meter intervals along the beach. We placed stakes along about 1,200 meters (about 3/4 mile) of shoreline, putting the markers up in the dune grass so they wouldn't be vandalized by passers by. I use the term with greatest respect to the Vandal "barbarian" people. Sheesh! I sound like a politician. Anyway, with these markers in place we spent the rest of summer sampling snails. At some positions, if the band of snails was narrow, we would take just one sample on the sandflat close to shore. If the band was wider we might take more samples out from shore to the limit of snail distribution, sometimes on the sandflat along the edges of sandbars, sometimes no sandbar was followed and samples were just taken from the sandflat. We measured our way to all sample positions so we knew where samples were collected and how they were spatially related to each other. There is a fine diagram of the disposition of samples in the paper we finally published on this work (Curtis and Hurd, 1983).

When we would get to a sample point, we would usually collect 40 snails. They all had to be dissected and that was as many as we could effectively handle. It's always easier to collect samples than to analyze them. We would pick up snails at the sample point first and move radially outward if necessary (as far as 3 meters sometimes) to finish the collection. In all, in this specific study we collected 56 samples from all over the sandflat and took data from 2,235 snails. For some purposes we also used snails collected during the crystalline style studies, as well as samples collected since 1977 to assess size (age) and growth of snails. In the end we had a lot of data from all over the sandflat that could (we erroneously thought) be used to make our characterization of the snail population. This was before the days of desktop computers and it was a challenge to keep track of all the data and ultimately get it all analyzed. It was all written down by hand, filed, and analyzed with a small, but capable calculator. Looking back on it all now from this computerized age I can't imagine doing that, but that was the way things had to be done then. It was good practice for projects to come, which were even worse in this regard. It was not until a couple of years later, sometime in 1982 or 1983, that I got my first computer, an Apple IIe, and started learning what could be done with it by way of data handling.

To understand size and age relationships you can do something called a size-frequency analysis. Previously, from October 1977 through November 1979, on nine different occasions snails were collected from Cape Henlopen and measured for shell height. Snails totaled 5,541. On each of those nine occasions each snail was placed in a size-class based on its shell height. That's simple enough to do. For two examples, a snail from 5.5 to 6.4 millimeters went into the 6 mm size-class and one between 14.5 and 15.4 mm went into the 15 mm size-class. The value of doing this is that it makes data more manageable. The purpose of these nine sampling occasions was to reveal all size-classes of snails present on the sandflat at the time. Trust me, without going into it, great effort was expended to see that this was accomplished. What did we learn? First, on any one of these occasions there were typically three groups of snails present on the sandflat. The groups are alternatively called cohorts because all the snails in a group originated, settled as juveniles, on the sandflat at about the same time. In October 1977, for example, there was a cohort measuring from about 4 to 6 mm in shell height, then a break (no snails), then a cohort from about 10 to 14 mm, then another break, and finally a group from about 16 to 22 mm. You can get an idea of growth over time by watching cohorts change in the size-classes that make them up. Let's consider the 4 to 6 mm group observed in October 1977. That cohort of snails was still present, and still 4 to 6 mm, in January of 1978 and again in April of 1978. They had not grown over the winter. But, by June of 1978 snails in this cohort were bigger. Then, they were between 5 and 8 mm and had grown some since April. By September of 1978 the group had grown to 8 to 13 mm. In May of 1979 this cohort was still there, but had grown a little more (now 9 to 14 mm). Initially, in October 1977, these snails made up the small group of snails. Now (May 1979) this cohort made up the middle size group because new juveniles had settled on the sandflat to form a new small cohort. By November of 1979, and the last sample, some in the October 1977 cohort we have been following had grown to about 18 mm. Now they were part of the largest of the three size-groups on the sandflat. It had taken them three years to get there (1977 their settlement summer, 1978 their second summer, and 1979 their third summer). So it takes about three years to grow into the largest size group. Some snails in this largest group were larger than 18 mm. They must have been older than three years and still growing. That largest size group must then be considered a composite made up of snails three or more years old. Put another way, in their first summer snails settle out of the plankton at about 1 mm and grow to 5 or 6 mm. They appear not to grow during winter. By the end of second summer they reach about 10-13 mm and by the end of the third they reach about 18 mm. There are consistently three cohorts of snails on the sandflat because each year new juvenile snails are recruited from the plankton. The population is yearly rejuvenated. In the early 1980's, at the time of this work, the snail was thought to live from 5 to 7 years. That idea will get corrected upwards later on in this tale. We were not the first to understand this much about the relation of snail size to age. Back in the 1960's a Woods Hole scientist, R. S. Scheltema, had gotten similar results for the growth dynamics of *Ilyanassa* located there. It was good to know that we could analyze Cape Henlopen snails and come up with about the same result. Love that beautiful consistency.

We dealt with something else in this study, parasitism. It has its beauty, but is inconsistent and wild. In all we had examined 3,638 snails for trematode parasites and also measured the shell heights. We didn't know it at the time, but in the lingo of parasitology the term used for the percentage of snails infected, out of all those examined, is prevalence. It can range from 0 to 100 %. With parasitism and size data a size-prevalence curve, or relationship, can be drawn. What you do is figure the percentage of each size-class that is infected. When you do this with these data what you find is that snails in or under the 15 mm size-class (snails 1 to 2 years old) are seldom infected. Less than 5% of them carry trematodes. Size-classes from 16 to 18 mm, i.e. snails in their third summer, are about 10 % infected. When you get to 19 mm snails they are a little over 20 % infected. Percent infected goes up rapidly from there: 20 mm snails are nearly 30 %; 21 mm snails are nearly 50%; 22 mm snails are pushing 70 %; and 23 mm snails are approaching 90 % infected. When you get to the largest snails, 25 mm or more, almost 100 % are infected. Briefly, percent infected increases exponentially with snail size to nearly 100 % among the largest snails. This relationship between size (age) and percent parasitized suggests strongly that once a snail is infected it stays that way until death. There is no decrease in percent infected among larger snails. Why would it be that larger snails are more likely to be infected? The simplest hypothesis is that they have been around longer (they're older), have been exposed to infection by trematodes longer, and as a result more of them have picked up infections. There are other possible explanations and we entertained them, but this is the simplest and most likely. Consequences for the way the Cape Henlopen *Ilyanassa* population must function are severe. Trematode-infected snails are virtually always sterilized because the gonad (ovary or testis) is destroyed by the parasites in their tissues. An uninfected snail becomes sexually mature, i.e. you first start observing eggs or sperm, when they hit about 14 mm at the beginning of their third summer. They cannot reproduce before that and their future reproduction is jeopardized by trematodes. On Cape Henlopen they likely have that third year to reproduce and probably the fourth, but beyond that their probability of becoming infected and sterilized goes up rapidly. We used a Latin phrase, *fenestra temporalis*, to describe this temporary window for reproduction. Somebody stop us.

We looked into this reproductive window a bit further. On the southwestern end of the sandflat less than 50% of the snails in samples were infected, often much less. On the northeastern end more than 50 % of snails in samples were infected, often much more. The dividing line was about half way along the 1,200 meter length of our of sampling. When you look at Cape Henlopen sandflat during low tide it is just a monotonous expanse of snail-filled sand with nothing much to break it up except low (at most about half a meter high) sandbars here and there. There was no apparent physical barrier to create such a divide in frequency of infections. This is one of many instances of unexplained spatial heterogeneity (things being different in different areas) in this seemingly monotonous habitat. We examined the prevalence of trematodes in each size-class of snails from the southwest and northeast ends of the sandflat and used a statistical test to decide whether a size-class exhibited a different prevalence on one end or the other. Snails of 24 mm or more shell height were about 95 % infected in both zones. So, there was no difference for snails of that size; prevalence was very high regardless of from which end of the sandflat area they came. However, snails in size-classes from 23 to 19 mm showed differences. Snails in these five size-classes were more often infected in the northeast zone. And, not just a little bit more, it was quite a bit more. In the language of statistics, significantly more snails in these size-classes were infected on the northeast end of the sandflat. As for snails smaller than these five classes, 18 mm and down, they were seldom infected in both zones. This means that the reproductive window is closed more quickly for snails living in the northeast section of the sandflat because trematode infections come in sooner in a snail's life. As a result, the reproductive output for the Cape Henlopen snail population is weighted toward the southwestern section. We surely did not expect that.

Why would parasites be more frequent in one section of a sandflat than another? This an example in the small of a general challenge to ecological parasitology. Usually at larger spatial scales, the challenge is to explain why parasites are more prevalent in hosts in one place than in another. Snail infections come from other hosts in the life cycles of the trematodes. In this case, those other hosts are mainly shorebirds (e.g., gulls) and fish. To answer this question their behavior would have to be considered. Maybe birds and fish hang out more in the northeast section. Maybe, also, infected snails move into that section. How much do snails move around anyway? (We shall get into that later.) Or, maybe infections are for some reason more successful there. Maybe, maybe, maybe, or possibly there is some combination of these factors at play that produce this difference in trematode prevalence in snails on two ends of a sandflat. We can speculate, but didn't know then, or even now, what brought this about. This is a complicated problem that would require a variety of additional data to solve and most of those data would certainly be beyond our resources to collect. Do I think the problem should be solved? Yes, I do. It would be an important ecological phenomenon to understand, but at the same time I recognize that not that many people or agencies, with resources to give, worry about this particular problem and money, personnel, time, and equipment are not likely forthcoming. Well, at least we have shown that the situation exists and recognized some of its consequences.

With respect to parasitism there was even spatial heterogeneity in the snail population within each of these two zones. In both zones there were areas where snails showed a higher prevalence and other places where snails were more often uninfected. In the northeast section, for example, there were places where patches of mostly uninfected snails existed in the midst of a sea of mostly infected snails. In some cases snails infected with particular species showed a tendency to be located in different parts of the habitat. Snails infected with any species might be found anywhere on the sandflat, but snails infected with *Hq* tended to occur close to shore, while snails bearing *Zr* tended to occur well out from shore. As I said above, such demonstrable heterogeneity between and within these zones was mysterious. The reason(s) for it must be biological, as opposed to physical, because there were no obvious physical barriers to explain what was causing it. This was to be further investigated later, but never completely solved.

There were 4,664 snails collected from Cape Henlopen between 1977 and this work in 1980 that had been examined for sex. The sex ratio in this large sample was 64 females to 36 males. This gave us an expected sex ratio for the sandflat in general. How did sex ratios in specific samples compare with this expected ratio? There was much spatial heterogeneity with regard to sex too. Among the 56 samples collected in 1980, 24 had sex ratios statistically at variance with the 64:36 split expected. In 14 samples there were too many males, while in the other 10 too many females. Thus, if we had characterized the Cape Henlopen *Ilyanassa* sex ratio based on one of our 1980 samples, we would have been wrong nearly half the time. Even a few samples (which ones to choose?) as a basis wouldn't have done much better. Sex ratio spatial heterogeneity makes prediction as to the general situation very difficult. Some of this heterogeneity in the distributions of the sexes appeared to depend on differential responses to sources of food by females and males. Trematode infections might also play a role. The roles of these factors will be developed further in Chapter 5.

Another issue raised by the spatial heterogeneity inherent in the Cape Henlopen snail population, and probably by extension in other populations as well, is the question of the snails collected and used in laboratory (or field) experiments. In our experiment on diet from Chapter 2, for example, the snails experimented upon were small (young) and unlikely to be parasitized, but was there some unaccounted sex bias? Were experimented upon snails mostly male or female, could that have affected the experiment's outcome? When we experimented with the effect of salinity on crystalline style occurrence (Chapter 3), we had to dissect snails and would know if they were mostly of one sex or mostly parasitized. In that case, they weren't. However, in many laboratory and field experiments described in the literature a limited number of snails from one place in a habitat are used and these are unexamined for sex, parasitism, and sometimes even size. An experiment might easily be run on mostly one sex or the other, or mostly infected snails, or, or, and the effects on results would be unknown. In the language of experimentation these would be unaccounted for sources of variation possibly affecting results. Particularly with respect to parasitism, it is known that behavior and physiological functioning of host snails may be affected. In our paper on this work, rather clumsily in retrospect, in comparison with our shining "obligate omnivore" and "biological vacuum cleaner" strokes of genius, we referred to uninfected snails as "functional" snails and infected snails as "apparent" snails to address this concern. Clumsiness aside, our concern was that researchers should look to see whether they were experimenting upon mostly one sex or the other, or a biased group of infected snails. Given the extent of spatial heterogeneity in *Ilyanassa* populations, a biased set of experimental snails is at least quite possible.

Species that exert an important or controlling effect on the community of species around them, by virtue of competition with or predation upon other organisms, are referred to as "keystone" species. If a keystone species is removed, the effect on the species composition and very nature of the community is dramatic. In many *Ilyanassa* populations, certainly as exemplified by the one on Cape Henlopen, trematodes are very prevalent. In the sandflat case seven trematode species, that we knew of at the time, cycle through *Ilyanassa* individuals, some at considerable prevalence. We felt that *Ilyanassa* could be a keystone species in the estuarine communities of which it forms a part by virtue of its role as a host for trematode species, rather than because it was a competitor or predator. This is an idea that we may safely say has not taken hold in the minds of other researchers, but maybe its time will come. The snail is the single host necessary in the life cycles of all these parasites. From the snail parasites fan out to all the other hosts in their life cycles. The variety of other hosts used by these parasites is collectively broad. Beyond the snail, we were able to count a total of 35 taxonomic groups (species) involved in all of the life cycles involved. Currently, there would be more. Anyway, there are at least 35 other populations, other species, in estuaries that carry parasites that must cycle through *Ilyanassa*. This could constitute a subtle, extensive, and complicated network of interactions. Effects on these other host populations are presently mostly unknown, but were the snail removed, driven to extinction, we might expect major repercussions. It is perhaps unlikely that extinction could happen to an abundant, widespread snail such as *Ilyanassa*. But, it's possible. The passenger pigeon once literally clouded the skies of North America and that species is now extinct, gone forever. Estuaries are under great pressure from the effects of mankind's population and activities. Who's to say what can happen?

Larry Hurd taught me many things in the process of doing this and other projects. One thing in particular comes to mind here and that was to shoot high when you submit work for publication. There is bound to be a ranking of journals. Some are more general, widely circulated, and highly prestigious. Some are more specialized as to content and less so. Don't get me wrong. Scientific journals of all sorts are worthy enterprises and to be respected. I have some experience with this and the editors and peer reviewers of submitted articles take their responsibilities seriously, perform a lot of work, and try their best to produce quality output. They are, after all, collectively the keepers of the gateway to scientific knowledge.

A carrion response experiment set up on Cape Henlopen in 1983

Were it not for the efforts of an author, an editor, referees, and an up and running journal, some paper we needed as a reference to do this work might not have been there and available to us. Up to this point, I thought that much of my (our) work on this snail was of no or marginal interest to others. I did it mostly because I thought it a valuable thing to do and because I was curious. If others found it useful that was good, but there is no escaping that I was mostly doing it for myself. As a result I tended to submit manuscripts to more specialized journals. Whether to a specialized journal or not, submitting articles for peer review is a grueling process. Some were rejected and that's always painful. Even if there is reasonable cause for rejection, much effort and heart and soul goes into the preparation of even a brief manuscript. Fortunately, others were accepted, usually after much revision and wrangling. Larry Hurd made the suggestion that we submit the manuscript we had prepared on this work to the journal *Ecology*. I didn't think there was a chance it would be accepted for publication there and I don't think Hurd did either, but lo and behold it was. In our subject area back then, and still, getting a paper published in *Ecology* was a pretty high profile thing to accomplish. The referees and editors are demanding and you really need to have something new and important to say. The paper we published is not the final word on anything, but it has influenced the thinking of other workers. That is a large part of what scientists try to do. Marine ecological parasitology has seen an insurgence of interest in the last 20 or 30 years. This paper is an early one in that insurgence. It is perhaps not too presumptuous a characterization to call it foundational. Our success is evidenced by the number of times it has been cited by other workers since it came out. I can't thank my partner enough for leading me to that experience.

Chapter 5

THE FLESH-EATING SNAIL

In 1982 and 1983 I did some field experiments on the factors controlling response to carrion in *Ilyanassa*. The official report of results and conclusions was published in Curtis (1985). For me the earliest, high profile recognition of the ability of parasites to alter the behavior of their hosts was a 1962 article in the weekly journal *Nature* by the late Miriam Rothschild, a famous British biologist. She was full of wisdom and very dedicated. I remember reading in one of her papers that she had a field experiment on snails destroyed by enemy action in WWII. That was unlucky for sure, but to me inspiring. She went on. In this work I was lucky enough to discover an example of parasites altering the behavior of *Ilyanassa* about 20 years after her article. My work has been cited as an example of parasite-altered host behavior by a number of other workers over the years. That is rewarding, but in my opinion not absolutely necessary. You never know under what circumstances, when, or if your work will figure into someone else's research. Other researchers may be argumentative, or laudatory, or need support, or they may simply ignore you. You just never know what will happen, but you can't let the attention of others be your driving force. That should be finding out new stuff. My friend Buck Cornell used to say that the busyness of scientists is a necessary characteristic in the development science. In the past I had made complicated observations in the field, as with the crystalline style, but had never done an actual field experiment, with controls and all the trappings. This one turned out to be a burning mass of difficulty in its design, execution, and analysis. The paper that came from it is a complicated one. The issue was the snail's attraction to dead flesh (carrion), one of *Ilyanassa's* most notable and striking behaviors. That is, if in point of fact anything a snail does can be considered "striking."

In Chapter 4 I noted that on Cape Henlopen, based on 1,000's of examined snails, the overall sex ratio is 64 % females to 34 % males. Once, I collected all the snails piled up on two crab carcasses. In all there were 118 snails and upon dissection the sex ratio was 92.4 % females to 7.6 % males, way off from the 64:36 split that might be expected. Another gathering of 25 snails on a dead razor clam was composed entirely of females. Why would there be more females than males on carrion? Was this a general phenomenon or just a local happenstance? In either case, this observation got me going. What if I placed carrion in the field? Would more females than males be drawn to it and begin to feed? Well, maybe so, but that wouldn't really answer the question. Given the spatial heterogeneity in sex ratios discussed in Chapter 4, quite possibly the available population from which snails were drawn to the carrion was mostly female in the first place. In any experiment a control for that would have to be instituted. That is not a simple control, as even small areas of the sandflat can harbor tens of thousands of snails. How might they be assessed? And, what snail characteristics should I look at? Sex obviously, but what about the sexual condition of snails. They might be reproductive or non-reproductive. Larry Hurd and I had come up with a system for categorizing snail sex and sexual condition. Snails copulate and the male introduces sperm into the female's reproductive tract. On the right side toward the head end of a female snail there is a bright red structure (the capsule gland) that serves to package fertilized eggs into egg cases (capsules). Egg capsules are later emitted and attached to any available firm substratum (shells of dead clams, or crabs, or pieces of wood, or cement blocks that might be scattered around). If a snail was a female with a capsule gland present and an egg case visible inside it, that female was designated "fec." Through a microscope you could see the egg case by pressing on the capsule gland with a probe. If there was a capsule gland, but no egg case visible, the female was assigned the abbreviation "fg." Non-reproductive females with no capsule gland or egg case were assigned "f." Such females were often parasitized and had no eggs in the ovary. In the 1950's Charles Jenner found that males reabsorb their penis after the reproductive season sometime in late summer. The penis is a large (relative to the size of the snail), hatchet-shaped affair located behind the right tentacle on the head. Females store sperm over the winter for use in fertilizing eggs before packaging them into capsules and attaching them to substrata in the spring. A male with a penis present was designated as "mp." These were usually reproductive. Non-reproductive males without a penis were labeled "m." Often such males were parasitized and the testis was destroyed. So there are 5 sexual categories, fec, fg, f, mp, and m. Also, larval trematode infections might have a role in carrion response and so might snail size, water temperature, and salinity. Lots of factors might be important, but which ones were? That was the problem.

Here's the way I ended up designing my field experiments. There is a depiction of one experiment in the frontispiece for this chapter. Between April and November, when snails were active, I would go out on the sandflat and identify an area with a moderate number of snails present. I didn't want to be overwhelmed by hordes of snails. On an ebbing tide an area had to have a flow of water in one direction and water had to be no more than 10 cm (about 4 inches) deep. In a situation like this I would have about two hours to do an experiment (one on either side of low water). After that the water quickly got to deep to work and I had to hustle to get things accomplished. Having selected a site, I set up an enclosure one meter wide and five meters long fenced off using "hardware cloth." The water current ran through the long way. The fencing material had a mesh size that would retain snails within the enclosure and, more importantly, would keep external snails out. It gave me a smaller, defined population of snails with which to deal. What were the characteristics of the enclosed population? To find out I used a 0.09 square meter sampling frame (about one foot on a side) to collect control snails before carrion was placed in the enclosure. I would have to feel around in the sediment within the frame to locate most snails. I took the first 10 snails I ran into from each frame position and counted the rest, disturbing them as little as possible. After collecting snails from five uniformly placed positions along each long side of the enclosure I had a sample (actually 10 samples) of 100 control snails. This would give me a measure of how many and what sorts of snails were available to respond to the carrion when I placed it in the enclosure. At time zero of the experiment carrion (fresh shrimp or clam meat) was pinned with a small wooden stake to the sediment inside the upstream end of the enclosure. The water would run over the carrion, stimulatory substances would be released from it, and those substances would travel to the snails within the fence. Snails outside the enclosure would be stimulated too and they often piled up outside the fence, but they couldn't get in to screw up my experiment. Snails within the enclosure would respond to the carrion by crawling to it, climbing on, and beginning to feed. Responding snails were collected as they mounted the carrion and kept in the order they came in in a compartmented tray. Frequencies of types of responding snails could then compared with frequencies among the control sample, which was collected before the carrion was placed. Because I had counted all snails in the 10 control samples, I had a good estimate of how many snails were available to respond inside the enclosure and it was possible to calculate percent responding. Of course, the weight of the carrion used, air temperature, and water temperature and salinity were all measured. Experiments went until there were no more snails responding (at most about one hour).

So that was my design. The people that help out on research projects should be recognized and thanked. A student at the time, Ann Barse, helped out with some of these experiments. She and I did a later study, in which she was more involved. She became my first undergraduate co-author. There will be more about her later. On the Fourth of July in 1982 I did the first experiment. From then until mid-November three more were done. In 1983, from the end of April to early November eleven more experiments were executed, one of them in Canary Creek saltmarsh in the ditch that had been used for crystalline style studies. Many of these studies went more or less without a hitch. Others were troublesome in one way or another. For example, I had to learn, through toilsome experience, that trying to do an experiment when salinity was below 20 ppt was a waste of time. Snails were unable to respond. In the marsh, mosquitos and greenhead flies fed heartily on me. Another issue, in one experiment in particular, was the "swimmer's itch" with which I came down. This malady is also known as "clam-diggers itch" and Cape Henlopen is a source of it from time to time.

This affords a chance to elaborate more on the trematode parasites in my tale. Trematodes are parasitic worms. The trematode group I want to mention here belongs to the family known as schistosomes. In many parts of the world humans serve as the definitive hosts for these parasites. A definitive host is the one in which adult worms live and where they reproduce sexually. The adult worms of this group are also known as blood flukes because they live in the blood vessels of the definitive host. In humans these worms, which are gotten from parasite larvae produced in freshwater snails, cause a debilitating disease known as schistosomiasis. There are some 200 million cases of this disease worldwide and it is a major health problem. The schistosome involved with our snail [*Austrobilharzia variglandis (Av)* from Chapter 4] is not a human parasite. It uses birds (e.g., ducks, seagulls) as definitive hosts. The adult worms lay fertilized eggs within the bird host and when the bird eliminates wastes (defecates) the eggs of the parasite are released into the external habitat (e.g., the Cape Henlopen sandflat). An egg develop into a microscopic, swimming larval stage (a miracidium) that finds and infects an *Ilyanassa*. The snail is termed an intermediate host in the parasite's life cycle, a host wherein the parasite undergoes reproduction, but sex is not involved. Such reproduction is termed asexual, with sperm, egg, and fertilization not involved. The parasite sets up an infection in the snail, stemming from the infecting larva, which consists of a stages called sporocysts. Inside the snail, inside the parasite sporocysts, another larval stage is produced, the cercaria. ("Cercaria" is singular, "cercariae" is the plural.) This is a swimming larva that breaks out of the sporocyst and leaves the snail to infect a bird host, thereby completing the life cycle. It's those free, swimming cercariae that are the problem. They don't always try to infect birds. Most any warm-blooded organism will do for a try, including one running a biological experiment. When you do one of these experiments you are on your hands and knees in shallow water on the sandflat, in one place for an hour or more putting up enclosure fences, taking control samples, placing carrion, and collecting responder snails. If there happens to be even one snail in the vicinity carrying a mature infection with the schistosome (fortunately they are fairly rare), you make a wonderful target for cercariae. They swim to you and penetrate your skin. You are the wrong host, but that doesn't stop penetration. For a penetrating cercaria it's curtains because your human immune system attacks and destroys it. Bad mistake for this cercaria, but there are many, many cercariae and the parasite can afford to spend them. Enough will luck into a bird host and the life cycle will be carried on. Wherever a larva penetrates your (human) skin a red welt is raised that itches like hell for about two weeks. That's swimmer's itch. In one unfortunate experiment there was an infected snail somewhere close by and I got hit over and over again. The lower two thirds of my legs, except where I had on wading shoes, were covered with red welts, as were my hands and half way up my forearms. No ultimate, long-term harm was done, but there were snails to dissect while dealing with my miserable, itching two weeks. I felt stupid, but learned to be more careful about how long I knelt in the water. I've gotten swimmer's itch occasionally since, but never so severe. All in all, however, these experiments were a blast to do. The sandflat is generally a pleasant place to work. The wind keeps most biting insects at bay and once you understand some of the wonders around you there's a lot to keep you entertained.

In these field experiments, what factors were important to consider? One potential problem was how snails of certain types (females, males, snails carrying certain trematode infections) were distributed in space within enclosures. Patterns of how organisms are distributed in space are referred to as dispersions; there are basically three patterns, uniform, random, and aggregated (or clumped). Why worry? Well, for example, suppose that there was a clump (an aggregate) of females close to where the carrion happened to be placed in an enclosure. If females showed up on the carrion in greater numbers than males, the reason for that would probably be misunderstood. It would be attributed to the femaleness of snails rather than to the position of that clump of females in the enclosure. The 10 control samples taken in each experiment allowed me to test for such a problem by calculating something called the coefficient of dispersion. It turned out that females, males, and snails infected with certain trematodes were dispersed pretty much randomly within enclosures. That is, there were no concentrations of certain types of snails anywhere within the 5 square meters of the enclosures. So this could not be used to explain enhanced responses by any group of snails. With regard to size, most snails in experiments were in excess of 20 mm shell height. Overall, responding snails averaged 21.8 mm and control snails averaged 22.0 mm. Not much difference there so snail size was not an important factor. As mentioned above, salinity was important. Experiments had to be run where and when it was 20 ppt or more.

With these concerns settled or in mind, what was learned from the joys, trials, and tribulations of these experiments? First, there were estimated to be anywhere from 4,295 to 140 snails available to respond to carrion in the various experimental enclosures and response ranged from 0 % (where salinity was low) to 47 %. The message of this result is important: not every snail responds to available carrion. Many are uninterested or inhibited for various reasons. Luckily, the 47 % response occurred when there were only 194 snails in the enclosure. Had that happened when there were 4,295 in there I would have been swamped dissecting snails. As it happened only 3 % of the 4,295 snails responded, but that was quite enough. That made 129 responding snails to dissect on top of 100 control snails. To help the reader better understand what went into processing snails it's worth covering what was done during dissections. Each dissection involved measuring snail size with calipers, carefully cracking the shell with a hammer on an anvil, separating the snail's body from the broken shell pieces, moving to the dissecting microscope and determining sex and sexual condition, and not least searching for trematode larvae within the snail's tissue. It was also essential to keep track of the data obtained for later analysis. By the time I was doing these experiments I had developed the ability to confidently identify trematode larvae infecting *Ilyanassa*. It was always a challenge, but I could almost always do it. If found, sure identification of larvae usually required pipetting a few onto a glass slide in a bit of seawater and moving to the compound microscope, which had greater powers of magnification. Fairly often there was more than one trematode species infecting a snail and this complicated matters. Dissection of some snails went quickly (maybe 5 minutes), but others required a half hour or more. This all took much longer than the field experiment itself did, but that was no surprise. The reader should understand that extracting data from an experiment usually takes more time than doing the actual experiment.

Females responded to carrion more frequently than expected. This is one of the points to be made, but on what basis do I make this and other claims. Well, it's expected based on the control sample, which recall consisted of 100 snails collected before the carrion was placed. Forgetting about sexual conditions for the moment, suppose that 100 snails responded to the carrion in an experiment and 70 of these were female. Is this more than expected, fewer than expected, or about as many as expected (a positive, negative, or neutral response)? If only 35 control snails were female, we would have good reason to think there was a positive response on the part of females. The response was 70% females when we would only expect only 35%. Put another way, twice the females responded compared to what we would expect to respond. In fact, there is a statistical procedure called a goodness-of-fit test that tests the fit of observed results to expected results. This test helps decide whether more, or fewer, or about as many females as expected responded. I'll not take the reader through all of the details. There's no particular point. In statistics books this test will be discussed as the chi-square goodness-of-fit test. With it we can judge whether the number of fec, fg, f, mp, or m snails responding to carrion was as expected or not based on the control sample. Snails in these sexual condition categories might each show a neutral response to carrion if they responded in the numbers expected, or they might show a positive, or negative response.

Experiments were grouped for analysis into breeding season [females with egg cases (fec) present] and non-breeding season (no fec present). Let's consider only uninfected (trematode-free) snails for the moment. During the breeding season fec females showed a positive response to carrion while snails in all other sexual condition categories were neutral. However, when out of the breeding season, fg females showed a positive response, f females were neutral, and both categories of males had negative responses. Now let's consider parasitized snails. During the breeding season infected f females showed a positive response to carrion. Infected snails in other sexual conditions were all neutral. In summary so far then, uninfected fec females, and even infected f females, showed a positive response during the breeding season and fg females were positive out of the breeding season. Males were either neutral or negative in their response to carrion both in and out of the breeding season and whether or not infected,. The question immediately arises, why were females more interested in carrion than males? Based only on these experiments this question cannot be answered, but if other information is brought in a very probable answer emerges. It has to do with reproduction. Sperm are cheap, but making eggs is metabolically expensive and so is making the packaging in which fertilized eggs are released (the egg capsules). They both require much energy. Particularly important is the fact that a female putting an egg capsule together requires protein to make it. Males are free of making eggs and capsules and females are more interested in meat because they need the extra protein for reproductive processes. Interestingly, during the breeding season parasitized f females, which are virtually all non-reproductive, nevertheless were interested in carrion. In this regard they were behaving like uninfected, reproductive females. Perhaps some were infected by *Lepocreadium setiferoides (Ls)* (see below), but I am still puzzled by this.

In this work nine trematode species were found infecting *Ilyanassa*. There is an important addition to the list given in Chapter 4 and that is *Gynaecotyla adunca (Ga)*. I will have much to say concerning this species in the next chapter. Anyway, parasites were found one species at a time in snails (single infections) and also in combination with other species (multiple infections). In fact, 9 % of 2,111 snails dissected in this work were multiply infected. Among snails in general, this is an unusually high percentage. Not only was the percentage high, there was also quite a variety of multiple infections, 11 double-species and two triple-species combinations. I won't go through the carrion response of infected snails species by species. Let it be enough to say that during the breeding season no parasite infection was associated with either enhanced or negative effects on carrion response of their host snails. Parasitized snails always responded in expected numbers. The important factor then was sexual condition. However, out of the breeding season (i.e., before May and after July-August) certain species did exert influence. Snails infected with most species, or combinations of species, were neutral in their response to carrion, a few were negative. However, there was one glaring exception. Snails bearing infections with one trematode species, *Ls*, always showed a positive response. This was true whether I analyzed both sexes combined, or females alone, or males alone. As an aside, snails doubly-infected with *Ls* and *Zoogonus rubellus (Zr)* also showed a positive response to carrion, even though *Zr* when infecting alone revealed a negative response. One can imagine a conflict between parasites for control of the host snail.

In the middle 1980's when I was considering what this parasite could do to the snails it infected, a famous parasitologist (the late Thomas Cheng) and his colleagues had recently shown that snails infected by *Ls* had a modified, unique protein metabolism. A reasonable hypothesis to explain what was going on is that a *Ls* infection in a snail has a nutritional need for extra protein and satisfies it by altering the behavior of its host with respect to carrion. How a parasite makes a host snail do anything is mostly beyond present knowledge and we certainly don't know exactly how it works in this particular case. The parasite is intimately associated with the snail's tissues and it gains its nutrition from the host. It is likely that an exchange of chemicals between parasite and snail occurs and the snail's physiological state is altered. It could be as simple as: the parasite needs extra protein (for some reason), gets it from its host, and the host snail feels deprived of protein and becomes avid after carrion to make it up. That hypothesis is about as close as I can come to explaining *how* this parasite changes the behavior of its snail host.

There is another issue. Biologists are not satisfied to simply know how organisms are built and how they function. They also seek to understand how they come to be built and function that way. A useful distinction between types of causation comes to mind here that I picked up by reading the work of a well-known biologist, the late Ernst Mayr. If we ask, what is the cause of this organism having this particular structure, physiological feature, or behavior we are asking a question that requires a proximate causation answer. The answer will be something like this: the organism possesses certain genes inherited from its parents and through the action of those genes and the metabolism of the organism the feature in question is produced. Suppose on the other hand we were to ask another question. What is the cause of this structure, physiological feature, or behavior arising in the population of the species in the first place? There are a number of ways to put essentially the same question. For a couple of examples, why did the parent organisms have those particular genes and not others; and why does the observed species' feature have this particular form and not some other? Now we are asking for an ultimate causation answer. By what mechanism did the feature in question arise in the species? Ultimate causation will not deal solely with one organism and the here and now. It has an historical element that traverses the previous generations in the population of the species involved. I have mentioned previously the concept of adaptation as any feature (anatomical, physiological, or behavioral) of an organism that helps it survive and have babies. The existence in a single organism of features that function as adaptations is one thing and is explained by proximate causation. Where they come from in the first place, how they arise in a species, and how they are improved over time is explained by ultimate causation. As the term is used in biology, adaptations have a genetic basis. An organism has an adaptation because it inherited the necessary genes from its parents and the parents got the genes from their parents and so on back down the line. Why would the genes necessary for an adaptation and not some others get passed down though the generations? A successful species is one that (through reproduction) successfully passes genes (and the adaptations associated with them) to the next successive generation of offspring, which then in turn successfully survive and reproduce. The mechanism by which adaptations come to be, i.e., are developed in a species, is through the well-substantiated process of natural selection, something that can only happen across generations in a reproducing population. The process of adaptation is not something that can take place in an individual organism.

If altered host snail avidity for carrion is an adaptation, an issue is: was it developed in the parasite or the host lineage? A snail infected with a trematode is sterile (all indications say permanently so) and a dead end. It cannot pass any genes on to any offspring for any adaptations it possesses. It might have done so before encountering the parasite, but once infected its contribution of genes to future snail generations is shut off. The organism in this pair that can reproduce and pass genes for traits with survival value to offspring is the parasite. If we are dealing with an adaptation (and not some sort of accidental byproduct of parasitism), it must belong to the parasite not the host snail.

Let's trace out how the ability to alter host snail behavior might have developed by natural selection in the parasite population. Probably many generations ago in the *Ls* population, a mutation (a genetic change) occurred that led certain genetic strains of *Ls* to make good use of extra protein. Presumably, this "good use" was to somehow enhance survival and reproduction of the parasite. In the present context the parasite would produce more cercariae. The genetic strains that had the mutation would have reproduced more and passed the genes on to more offspring than strains that lacked the mutation. Through the generations, genes for the ability to make use of extra protein would increase in frequency (become more numerous) in the *Ls* population, i.e., strains that had the genes to make use of an optimal amount of protein would become established in the parasite population. The optimal amount of protein would be enough to help the parasite reproduce more offspring, but not so much that the host's ability to make up the protein by seeking out extra carrion was outstripped. Parasite strains that used too much protein would kill or seriously disable their host vehicle and would themselves be weeded out. The result was that in the early 1980's when I started looking for effects of parasitism on carrion response of *Ilyanassa* on Cape Henlopen, *Ls* was there to observe and with the ability to alter host snail behavior. In short, I am saying (hypothesizing) that this parasite-induced host behavioral modification arose by natural selection as an adaptation in the parasite population. I will have to admit as reasonable the alternative hypothesis that this is not an adaptation at all, but rather an accidental side effect of parasitism. Possibly, but the fact that no other trematode species causes this effect on their host snails argues against this. It is a property solely of *Ls*-infected snails, which suggests that something peculiar occurred in this parasite species' genetic history.

In the final analysis I can't claim to know exactly why this particular alteration of host behavior enhances the survival and reproduction of this parasite, but other cases of parasites altering host behavior are well-known and these are usually presented as parasite adaptations. In most cases the value to the parasite is that by altering host behavior that host's chances of being eaten by the next host in the life cycle are increased. Most studied parasite transfers from host to host in life cycles are accomplished by predation. Therefore, as this line of reasoning goes, if a host is made more vulnerable to predation by a parasite-induced behavioral change then it is more likely that the parasite will make the host-to-host transfer and the life cycle will be completed. In other words, it has survival and reproductive value for the parasite. I agree with this scenario. There is much evidence to support it, but it can't work in the case of *Ilyanassa* trematodes because these parasites are not transferred to their next host by means of predation. As mentioned above, transfer from the snail host to the next host is by means of swimming larvae, the cercariae. Whatever the value of the *Ls*-induced behavior alteration, it probably does not relate directly to transmission to the next host. Rather, it probably has to do directly with the nutritional needs of the parasite. Secondarily this could translate into better transmission to next hosts in the life cycle because more cercariae ("babies") are produced to make the transfer.

In the early 1980's when I was first experimenting with carrion response of *Ilyanassa* I had one of my earliest undergraduate students working with me. She had a classy name, Mimi Le Maire. This was before I thought of undergraduates as potential co-authors. This student taught me how motivated and competent undergraduate students could be. I remember that her interest level was very high and she put in hours in the laboratory. She was able to move the phenomenon of enhanced carrion response by *Ls*-infected snails into the laboratory. Her main question had to do with the effects on carrion response when snails were starved for various periods before running the experiment. Basically, if you starve them for too long, the effect goes away. Snails infected with *Ls* don't show a positive response because their positive response is swamped by the other snails in the experiment. All hungry snails, *Ls*-infected and not, responded. The experiments were complicated and required some real cleverness on her part. Neither of us was thinking about trying to publish a paper on her work. Maybe that would have been successful, maybe not. We'll never know. As it was she did her work for research credits and wrote up an excellent report. That report is now stashed away among my notes. That's the sort of thing that sometimes happens to clever "little" research efforts. Her work habits had a lot to do with changing my attitude about what undergraduate students could be asked to do. I guess and hope she got what she wanted out of that special project. She seemed pleased. I got more than I knew at the time. I began asking more of undergraduates interested in working on snails and parasites. Many were up to it and responded by delivering quality work of benefit to us both.

At least some of the spatial heterogeneity spoken of in Chapter 4 is now explained. Some concentrations of females arise because females have enhanced need for protein and respond differently than males to localized food sources. We also know that some concentrations of snails bearing certain species of trematodes stand to be the doing of the trematodes themselves. Miriam Rothschild would likely be pleased. No way has all been explained, but some has and our knowledge of the natural world has increased. We can add this to what has already been found out. At this point we now know that this snail requires a diet containing both plant and animal material and snails ingest such a diet during their field existence. For a snail, digestion of such a mixed diet is problematic because carnivorous and herbivorous habits normally cannot coexist. It is only possible in the case of *Ilyanassa* because the crystalline style daily waxes and wanes. Further it is likely that, because the snail needs a style and salinity affects its occurrence, snails are restricted to estuarine areas where salinity is usually 20 ppt or above. We know that spatial heterogeneity is a likely characteristic of snail populations. Trematode parasites can be abundant in snail populations, but are unevenly distributed in them, and can have a variety of consequences. Not the least of which is that they can influence host snail behavior. This aspect of parasitism will be developed further in the next chapter. Buckle up.

The author's companion Kai and the 3x5 m plot atop the 1984 Cape Henlopen sandbar

Chapter 6

SNAILS POSSESSED

So, Chapter 5 tells us that a snail's attraction to carrion can be enhanced by a parasite. In this chapter a more elaborate and consequential parasite-altered host behavior is to be documented. This one will probably grab your attention even more than the fascinating foregoing. This part of my tale begins with a romantic interest. It was 1983 and that interest drew me to the vicinity of the south shore of Indian River Bay, an estuary about 15 miles south of Cape Henlopen in Delaware. As is not uncommon in these matters, I often had to kill time while she got organized to go places. To handle that I would go exploring at a nearby 1,700 meter stretch of beach near Gray's Point. The beach had no sandflat associated with it. Now the area has many houses, but then it was quite desolate. When I first went down there I was interested to find that there was a population of *Ilyanassa* in residence, home sweet home. But, the population was not what I was used to on Cape Henlopen. This was the key and happenstance observation that started this whole episode. At low tide the whole beach face was littered with snails. On Cape Henlopen I had never seen such a thing. There, as far as I was concerned at the time, snails occurred on the flat off the beach and did not inhabit the beach face during low-water. In retrospect I should have looked more closely at the Cape Henlopen beach. In any case, this was brand new to me. What was going on? Naturally I collected some of the snails from the beach in Indian River Bay and examined them when I got back to the laboratory. They were all infected with *Gynaecotyla adunca (Ga)*, the trematode I had added to my parasite list during the carrion response studies on Cape Henlopen (Chapter 5). Another time I collected a few more snails from that stretch of beach and they too were all infected with that species. What indeed was going on? That species occurred on Cape Henlopen, but there so far I had seen it only infrequently. Maybe the species was just more abundant among snails along that beach, but what were they doing up on the beach in such numbers? The attentive reader will guess what happened next. I began to investigate. Here we go again.

My first move was to investigate that beach and the snails along it in greater detail. I picked five, more less equally spaced, positions along the 1,700 meters of beach and at each sampled the snails at low tide eight times from July 1983 through November 1984. Each time, at each of the five positions, I would collect 25 snails from the beach a meter or more (surface distance) above the waterline, 25 snails from at the waterline (plus or minus one meter), and 25 snails from one meter or more below the waterline. In total 804 snails were collected and examined. There were dramatic (yes, statistically significant) differences between the collective samples taken at each level with respect to parasitism. [I won't burden you with the details here, but the statistical test used was a chi-square contingency analysis. If a reader wishes to pursue, or better understand the design of the test and the results, they are presented in the paper published on this (Curtis, 1987).] Below the waterline there were more uninfected snails and also more snails than expected infected singly with the trematodes *Lepocreadium setiferoides (Ls)* and *Zoogonus rubellus (Zr)*. However, in this environment (below the waterline) there were fewer than expected snails infected with *Ga*, either singly or in combination with other species. At the waterline uninfected snails and snails with all types of infections were at expected frequencies. The surprise was above the waterline. There, there were fewer than expected snails uninfected or infected singly with the two species mentioned above. In much greater than expected numbers were snails bearing *Ga*, both singly- and multiply infected. Could this have something to do with snail sex? I tested for this and no, sex was not an important factor. These results made it clear that occurring above the waterline was associated with being parasitized with *Ga*. How about that! Snails on the beach being infected with *Ga* is not a matter of that trematode being peculiarly abundant on the south shore of Indian River Bay. It happens *because* the snails are infected with that parasite. In estuarine ecology, once again, parasitism raises its beautiful head. Intertidal animals and plants exhibit what is called zonation. They occur at certain vertical levels between low tide and high tide. The level is set by, among other things, resistance to drying out when the tide is down. *Ilyanassa's* zonation is to hang out at or below about the average low-water level. What happens here is that a snail's preferred zonation is altered by a parasite.

The above samples that I collected in 1983 made me suspect that I had been missing something in the Cape Henlopen snail population. In early summer 1984 I went out and explored the beach face there with greater care than before. Sure enough, if you looked closely there were snails on the beach there too. They were buried in the sand, but left a dimple where they had buried. As much time as I had spent out there, I never noticed them. I picked up a few and brought them back to the laboratory. Yep, they were *Ga*-infected. I did a small study on 12 July to see if there was a pattern on Cape Henlopen too. Along 250 meters of beach I found and collected 25 snails from up on the beach and 25 from down on the adjacent sandflat. On the beach that day there were not many snails and I had to search carefully. Of the 25 snails from the beach 22 were infected with *Ga*, below the beach on the flat none of the 25 collected there were so infected. The results speak for themselves. The same phenomenon occurs on Cape Henlopen. It's just less obvious.

I did one more thing in my summer 1984 explorations. I noted that sandbars out on the Henlopen flats had snails on their peaks. The "tallest" sandbar peaks are about half a meter (about a foot and a half) vertically above the level of the flat. When I examined some of these snails they too were virtually all *Ga*-infected. Hmmm. How long did it take to establish the observed pattern, tides, days, weeks, or what?

This question led to a major effort with regard to this phenomenon. A field experiment was designed that took advantage of the sandbar situation. First, an experimental sandbar was selected and some snails from it were examined to be sure that they carried *Ga*. They did. Just below the highest point on the sandbar I marked out a three by five meter plot. I drove a metal pipe two meters into the middle of it so I could re-establish the plot were the corner markers uprooted. From that pipe I ran eight, equally spaced radial lines, along which every three meters I measured elevation relative to the stake. It was no small deal making these measurements, but they allowed me to make a contour map of the sandbar showing the position of the plot and how elevation changed (mostly decreased) away from the center of the plot. The reader may consult the drawing in the published paper if desired, but it is a topographical representation of the sandbar. The sandbar peak was 4 cm (about 1.5") above the plot center close to the plot. The beach was approximately south of the sandbar. There was a shallow tidal gully between the sandbar and the beach and extensions of that gully on the northeast and southwest sides of the sandbar. These tidal gullies were about 30-35 cm (about a foot) vertically below the sandbar peak. With this sandbar chosen and set up for the experiment I collected 30 samples of 10 snails each around its periphery. These 300 snails were examined for size, sex, and parasites and would be used as an experimental control for comparison with snails collected from the plot near the sandbar peak.

Plot snails were collected in the following way. On Saturday 21 July I went out to the plot during low tide at about 9:00 AM, got down on my hands and knees, used my fingers to feel my way through the sand in the whole 3 X 5 meter plot area, and removed all of the snails within the plot. They were cast onto the sandbar outside the plot at lower elevations. I then graciously allowed the tide to come in and go out. Nice of me. When the tide rose and covered the sandbar it did so over a plot empty of snails. When the tide was low again, about 12 hours later at 9:30 PM, I returned to the plot and felt my way through it again removing all snails. I wanted to see "who" had come back into the plot during the previous high-water. Some plot snails were taken as a sample for examination, the rest were cast out away from the plot onto the periphery of the sandbar. Snails on the sandbar were in limited numbers and I couldn't examine all of the snails collected in the plot because that would defeat the purpose of my experiment by eventually killing off most of the snails visiting the sandbar peak. On that second low tide I again left the plot empty of snails. The sample was taken back to the laboratory and examined as the peripheral control snails had been. I returned on the next low tide, which was the next day about 10:00 AM, and repeated the same procedure. Here we go again. In all I kept doing this for the next 30 low tides. Actually I missed one. That covers about two weeks from 21 July to 5 August. Ann Barse, who also helped out on some of the carrion response studies, assisted on some of these trips. Lucky me, I was in on all of them.

You might think that getting in tune with the tidal rhythm for about two weeks would be relaxing, but quite the contrary. It meant that every 12 hours or so there had to be a trip to the plot with travel, dissecting snails, and keeping track of data in between. Tides occur about one hour later each day. So if there is a low tide is at say 9 PM one night, the next one would be about 10 AM the next day, then there would be one at about 10:00 PM, another at around 11:00 AM the next morning, and the next would by about 11:00 PM that night. And so it went. Luckily, my feet were in pretty good shape during this ordeal. Also, I was watching Ann's dog Kai much of the time here and Kai was great company on those nighttime walks along the beach, except for the time she got into a skunk. It's about three quarters of a mile from the laboratory to the sandbar. As mentioned previously, at the time there wasn't much policing in the Cape Henlopen State Park. You could get away with these trips, and the dog, without getting a ranger all excited. Kai at the 3 x 5 m plot on the 1984 sandbar is shown in this chapter's frontispiece. By the end of the two weeks I was a walking vegetable, ready to become snail nutrition, part of a mixed diet. I had been building a lot of bad karma and that would have been only justice, what with the way I had been treating snails. I imagine a cry going up among them, "Oh no! Here he comes again! Who's time will be up this time?"

When I pulled all the data together they told a pretty complete story about this parasite-altered snail behavior. First, a total of 1,929 snails were recovered from the plot on the 29 low tides and 271 of them were taken as plot snails for examination. The rest were cast back onto the sandbar outside the plot, possibly to come into the plot on the next high tide. Based on the examined 271 snails, 86.3% of snails that came back into the plot during high tides were *Ga*-infected. In the statistical test, a chi-square contingency analysis again, outside the plot (in the peripheral control samples) there were significantly more uninfected and *Zr*-infected snails than expected, in comparison with snails collected in the plot. Importantly, outside the plot on the sandbar periphery there were fewer *Ga*-infected snails than expected. This was true of snails singly infected with *Ga* and also for two double infections involving *Ga*. In contrast, plot snails were composed of the reverse. In comparison with peripheral snails, retrieved in the plot there were too few uninfected snails and *Zr*-infected snails and too many snails infected with *Ga*. Again, this was true of singly *Ga*-infected snails and the same two doubly-infected snails involving *Ga*. The double infections are important because they mean that *Ga* was overriding whatever influence the other parasite species had on the host snail. In brief, snails carrying *Ga* crawled into a plot on the top of a sandbar during single high tides, but uninfected or otherwise infected snails rarely did.

There was another aspect to this parasite-modified host behavior, a diurnal component. Snails infected with *Ga* inhabited the plot on both daytime and nighttime low tides. However, there were 17 daytime low tides and on those there was and average of 30 total snails in the plot (range 4 to 64 snails). Compare this with nighttime low tides. On the 12 such low tides there was an average of 118 total snails (range 66 to 194 snails) in the plot. Clearly, more *Ga*-infected snails came to the sandbar peak and remained for nighttime low tides.

Again we are confronted with how the parasite can cause the host snail to adopt this altered behavior. This is a more difficult case than merely enhancing response to carrion because this time the parasite is causing the snail to do something that is not part of its ordinary behavior. A snail would not crawl up into the higher intertidal zone on the beach or on sandbars if it were uninfected or infected by other trematodes. The best I can do to explain *how* this is caused is the parasite must produce a chemical (maybe one that acts as, or stimulates, a snail hormone) that is the proximate cause for the behavioral change. This is utterly uninvestigated, but should be.

However, why this ability of *Ga* to modify host behavior exists at all can be profitably investigated. First, because this parasite sterilizes snails, as do other trematodes, it can't be an adaptation of the host. In a sterilized host it would be a dead end for genes responsible for any adaptation. They could not be passed on to offspring snails. It must be an adaption that developed by natural selection across generations in the parasite population. What could be the value of this adaptation in terms of survival and reproduction (life cycle completion) for the *Ga*? Upon investigating the life cycle of this trematode further I found, thanks to papers previously published in the 1940's by J. S. Rankin and 1950's by W. S. Hunter and W. B. Vernberg, that the next hosts in the life cycle were species of semiterrestrial crustaceans. Specifically, two species of beach-hopper crustaceans and a species of fiddler crab were used as 2nd intermediate hosts. Fiddler crabs were not present on Cape Henlopen and I eliminated them from consideration, but there were beach-hoppers there. When I would walk along among the dead algae high on the beach around dusk or in the dark with a flashlight they would explode around my feet in the fashion of popcorn popping. Hence they are called beach-hoppers. So, to complete this parasite's life cycle, cercariae somehow had to get from the host snail, which normally lives lower down on the sandflat, to the next host, which lives high up on the beach in a semi-land environment. The cercariae of *Ga* are tiny (only about a tenth of a millimeter long) compared to other species that infect this snail and they do not swim much. When released from their sporocyst and snail they just fall to the bottom of a dish and generally stay there. Swimming would not help much getting to the next host anyway. Beach-hoppers are active, mostly at night, up on the dry beach when it is uncovered by the tides. The snail is the first intermediate host in the life cycle. I don't know why this second intermediate host, the beach-hopper, was originally adopted by the parasite in its life cycle. Perhaps the parasite became tied to using it before it moved to the beach and took up its semiterrestrial existence. Whatever the reason, the parasite uses it now and is quite specific about it. Some parasites use several species as hosts, but with regard to first and second intermediate hosts *Ga* is very specific, it must be *Ilyanassa* and semiterrestrial crustaceans. What I propose is that to be able to get cercariae to the beach-hopper next host, the parasite changes the host snail's behavior such that it crawls to the higher intertidal zone during high tides, releases cercariae during that trip, and cercariae are then exposed to beach-hoppers when they become active at low tide. Thus is the transfer from host to host accomplished and *why* the parasite alters host behavior. This would have great value to the parasite because it would increase the chances of completing the life cycle. Bird definitive hosts feed on beach-hoppers, but absent the snail's modified behavior, i.e., if it stayed down on the sandflat, the crustaceans would seldom be exposed to *Ga* cercariae and therefore seldom infected. The parasite's life cycle wouldn't work.

The only other thing I should mention at this point is the abundance and variety of multiple-species infections I observed during this work with *Ga* involved. This parasite is involved in more multiple infections than any other species. These snails demonstrated the altered behavior even though infected by species besides just *Ga*. Recall that 9% of the snails examined during the carrion response studies were multiply infected. So I had seen this before, but never so much as in this work. From Indian River Bay 28% of snails examined were infected by more than one species, some with two, some with three and 78% of those multiple infections involved *Ga*. In the sandbar study on Cape Henlopen 35% of snails were doubly and triply infected and 89% of those infections involved *Ga*. This parasite can coexist in the same snail with a variety of other trematode species. This observation will play a future role. Further, dominance of *Ga* is suggested. It can take over a snail's behavior despite the presence and effects of other infecting species.

So that's my big discovery. I was pretty excited about it and decided to take Larry Hurd's general advice with this one. I chose to submit the manuscript to the high profile, weekly journal *Science*. Publishing there is sort of what people doing science of all sorts aspire to. There was, and is, no such thing as surely getting published there. As I recall, at the time I submitted this manuscript about 95% of manuscripts were rejected out of hand with no external review. If you made it through that filter, you still had tough reviewers to go through. Anyway, much to my surprise and that of my faculty colleagues it got accepted. It was a single-author paper in *Science*. These days that is almost unheard of, there are few single-authored papers in *Science*. Anyway, when it came out it was a fairly big deal. There was a write-up in the British magazine the *Economist* and also in the *New Scientist*. Nothing ever came of it, but I was contacted by a Canadian and an Australian broadcasting company. I don't know how you would film this not so action-filled phenomenon for a documentary. Ten years later (October 1997), the work was included in an article about parasites and the cool things they do by Jennifer Ackerman in *National Geographic*. Of course the administrators of my academic unit, not scientists, thought it was a yawn and sort of barely nodded their collective heads when confronted with the accomplishment. It was about what I expected from that quarter. I don't blame them too much, as they were largely unaware of how difficult publishing something in *Science* is to do. Besides, they thought of my academic unit as strictly a teaching enterprise. Why was I wasting my (their) time on this sort of thing? That attitude gave at least me a problem. Teaching and research are intimately interrelated. For administrators, responsible for overseeing student academic progress in a university setting, that seemed to me to be an unenlightened and inappropriate point of view. To be consistent, they should have protested or outlawed my use of this work in lectures as an example of science in action. That lecture material intrigued students. It was sort of amazing and interesting. It showed that I was involved and not merely spouting stuff that I had read somewhere in a textbook. Anyway, administrators weren't watching so it was well worth the small risk. It did do me some good professionally. I had already been promoted to Associate Professor by 1987 when this work came out. The main thing this publication did was to contribute to my later promotion to Full Professor when committees higher up in the University reviewed my dossier. But, fame, especially if small, is fleeting and I soon became otherwise involved. Even now though, looking back about 20 years, it is satisfying to have done this.

In the summer of 1985 I pursued this alteration of host behavior with another and more difficult field experiment (Curtis, 1990). A student in one of my courses, Harry Hudson, assisted with this experiment. He did exemplary duty under sometimes very challenging conditions. Also, will wonders never cease? Through the auspices of Biological Sciences, I actually got a few thousand dollars to do this particular work. It meant I could buy some equipment and Harry would rightfully get paid, at least a little. The main problem with the 1984 experiment was that it did not allow the following of individual snails. Individuals are the ones infected with parasites, have a sex, and a size (age). Taking observations at the population level, as in 1983 and 1984, did not let me understand what individual host snails were doing. I needed to be able to identify and keep track of individuals. A faculty member in the College of Agriculture at the University, who contributed to discussions over beer, the late Tom Wood, suggested that I try marking snails with bee tags. These are small plastic disks (2.5 millimeters diameter, less than 0.5 millimeter thick). They come in different colors in batches of 100, a number (1-99 plus a blank) on each one. The try was a success. With clear fingernail polish I glued them onto a cleaned and roughened area of the shell on the side of a snail that would usually be up in the field. If put on with care, these tags were quite durable in the field. I would make extensive use of them. Tom gave me seven batches (700 tags) and off I went. Thanks Tom. The sandbar used in 1984 was still there. It had drifted some as sandbars will and was a bit modified, but similar and I set up my experiment there again. Elevations were re-measured, as in the year before, on the new sandbar and a contour map of it in its new configuration was drawn. This was necessary because I needed to know what the elevations were where marked snails were sighted. The 1985 sandbar was higher (at about half a meter) off the surrounding flat than it had been in 1984. Once more a three by five meter plot was laid out in the peak area of the sandbar.

Preliminary samples were taken to see if there were *Ga*-infected snails near the peak, but not around the periphery. That was indeed the case.

On 15 July 1985 Harry and I collected 250 snails from the sandflat around the sandbar. We took these back to the laboratory and measured (shell height) and marked them individually with bee tags. This meant that each of these snails had a unique number and tag color on it. They were all individually identifiable. After the tags had dried on, snails were stored in an aquarium. Next day early in the morning we collected 250 snails either from high on the sandbar or from the beach face. The beach face rose a vertical meter above the sandbar peak and beach snails were collected at least as high as the plot. These snails were uniquely marked and measured and, after drying, stored in an aquarium. This was all quite a grind for both of us. On the afternoon tide of 16 July we took the two groups of individually marked snails out to the sandbar. The plot was emptied of snails as had been done on each low tide in 1984. Marked snails were divided into 20 groups of 25 snails each, about half sandbar/beach and half sandflat in each group, and on a 10 meter radius circle around the plot center the 20 groups were released. Soon the tide came back in. The plot was visited on the next 16 low tides, i.e., for a little more than a week. Depending on the extent of the tide, one would have up to 3 hours to work on each tide. The plot was searched and all snails within, marked and not, were retrieved and counted. Marked ones found in the plot were identified and recorded. Before snails were cast out from the plot, the whole area of the sandbar and surrounding flat was searched for marked snails. When one was sighted (found), its distance (using a measuring tape) from plot center and degrees clockwise from magnetic north (using a sighting compass) were recorded. One person would stand over the found snail and the other would use the compass (a classy Brunton pocket transit) on a level tripod table over plot center. Miner head lights with belt batteries would have to be used at night. If there was no or little wind, cloud of insects came to the light and you would find yourself at times adversely busy. However, they never drove us off. Later, using the contour map, the elevation down from the plot could be determined for a snail sighting position. In effect it was known in 3-dimensional space where and when marked snails were sighted.

Of the 500 marked snails released, 260 were ultimately recovered and they had been sighted a total of 1017 times during the 16 low tides. A reader who has not had to keep a lot of data organized would not understand how much of a challenge this can be. By this time I was writing Basic programs for the Apple IIe computer to effectively store and analyze data. Computer spreadsheets were not then what they are now and they were not yet part of my repertoire anyway. Many of the 260 recovered snails were sighted several times and positions of individual snails on different low tides had to be accurately recorded, not to mention dissection data on parasitism, sex, and size after recovery. And, of course, many marked snails were sighted during the experiment, but not ultimately recovered and their sightings too had to be kept straight. Above all you don't want errors in recording data, as these would potentially lead you to the wrong conclusions. Night and day sighting data were first written on cards in the field. In the laboratory they were later transferred to a complicated pencil and paper worksheet, and eventually typed into a computer. My Apple IIe programs worked well keeping data organized and available for analysis, but this all took time, attention, and care.

When all this was over a number of things were clear. On low tides 14 through 16 marked snails were recollected as sighted. Some were also recovered a bit later. Let's just consider the recovered snails. These snails were all dissected so I knew what infection was carried by each. First of all, if a snail was from the initial 250 taken from the sandbar or beach, it was much more likely to be recovered (or sighted) than if it was from the 250 initially collected from the sandflat. From the sandbar/beach group 75% were recovered, while among the sandflat group only 31% were. That's because sandbar/beach snails stayed on or close to the sandbar where released and sandflat snails dispersed into a much wider area and were less likely found. We looked often, but marked snails never showed up on the adjacent beach or on another sandbar about 30 meters to the southwest. Such fidelity to a sandbar surprised me. Among 260 recovered snails, in addition to those uninfected, there were six different single species infections. The most frequent among them were snails infected only with *Ga* (106 of them). There were seven different double- and two different triple-species infections. The two most frequent doubles (61 total snails) contained *Ga* and both triple infections (6 snails) also did. In all, 70% of recovered snails were *Ga*-infected. Among the 182 snails that were recovered from the initial 250 sandbar/beach group, 99.5% were infected with *Ga*, some singly, some in combination with other species. The point is, prevalence of infections on this sandbar was high and snails carried a variety of infection types, most of them involving *Ga*.

There were other findings as well. Far and away, as found in the 1984 experiment, most snails found in the plot on low tides were *Ga*-infected and there were more of them there at night than during the day. On nighttime low tides there was an average of 50 snails in the 15 square meter plot area, while during the day there was an average of only 8. This is confirmation, now using individually identifiable snails, of observations made the previous year. Because marked snails were recorded as being at certain distances and directions from the plot center when sighted, it was known (from the contour map) what a snail's elevation was at sighting. From these data it was determined with a statistical test that *Ga*-infected snails (singly or multiply) were found significantly higher on the sandbar than all other snails, including those uninfected and otherwise infected. Not only that, but it could be shown that *Ga*-infected snails occurred higher on the sandbar on nighttime low tides than on daytime low tides.

In the literature trematode infections are generally considered to debilitate their host snail, making them move less. Given this, one would expect uninfected snails to move around the most and infected snails less. I calculated net movement per tide for marked snails sighted on more than one low tide and compared movement by different infection types in a statistical test. This was a Kruskal-Wallis test. I won't explain all the statistical tests I use because I want to keep you as a loyal reader, but in scientific work they can't be done without. This one compared movement by uninfected snails and seven categories of infected snails. What I was comparing was minimal movement per tide. They could not have moved less, but could have moved much farther. Snails infected with *Ga* (singly and multiply) moved around significantly more than uninfected or otherwise infected snails. So they, at least, were not debilitated. That was a bit of a surprise, but I was equally surprised to see that snails infected with species other than *Ga* moved around as much as uninfected snails did. It appears that *Ilyanassa* infected by trematodes are not debilitated in terms of moving around.

The other new major feature of the altered host snail behavior that emerged was the fact that individual *Ga*-infected snails make repeated trips to the top of the sandbar. The same *Ga*-infected snail was observed on the sandbar peak on up to six different low tides during the 16-tide experiment. We wouldn't have seen every marked snail near the sandbar peak on every low tide, as sometimes they would be completely buried and missed, but this is enough to show that snails make regular moves to the same sandbar peak under the influence of this parasite. That was likely why they showed greater movement per tide, they were undertaking repeated migrations to and from the sandbar peak that other snails were not. That certainly got my head up off the desk.

The 1985 sandbar experiment added support to the hypothesis I had come up with based on my 1984 sandbar extravaganza. I came through this one a little less worn out, but not much. This one required more work on each tide. I was fortunate to have Harry's stalwart help. By studying what individual infected snails do, a lot of detail had been added to what was already known, especially when compared to the initial clues from that beach in Indian River Bay. Snails infected with *Ga* coming to a sandbar peak in greater numbers at night had now been demonstrated in two separate years. Now I knew that these snails moved to higher elevations at night than day on the sandbar. They also moved around more than other snails, probably because they were busy making repeated migrations up and down sandbars. All of this further pointed to a parasite adaptation to get cercariae from an aquatic host (the snail) to a nocturnal, semiterrestrial next host (the beach-hopper). I would eventually pursue this further in the 1980's, but there was a health episode to go through first.

Before that, there is an interesting side note. Three snails marked in 1985 that went unrecovered were found in July and August of 1986. By this time the experimental sandbar no longer existed in anything resembling its former form. These snails had obviously survived the intervening winter. I had no information at all of this sort and these rare finds were collected and examined. One was not infected at examination and had grown from 19.9 to 20.0 mm, not much growth for a year. It was seen twice in 1985 around the edge of the sandbar so it had apparently been uninfected then and still was now. The other two had double infections with *Ga* and *Zr* when examined in 1986. The two had grown little, one from 24.7 to 24.9 mm the other from 23.7 to 23.8 mm. These two snails had been seen around the peak of the sandbar in 1985 a total of eight times. This strongly suggests that both were infected with at least *Ga* the year before. They had both survived with their infection(s) through the winter. At the time it was not known for sure whether they could do that. Well, these two did. This observation would lead to much future work.

An attempt was made to do something else in the summer of 1985, something that only marginally worked and never did see publication. Nevertheless, a lot of effort went into it. I wanted to observe the behavior of *Ga*-infected snails on the beach. Why did I feel this was necessary? It's because the beach-hopper next host is a beach-dweller not a sandbar-dweller. On a sandbar there is no place to stay out of the water when the tide is in. The parasite likely can't tell whether the host is crawling up a sandbar versus a beach. In terms of getting cercariae to the next host only the beach would work. However, *Ga*-infected snails often enough climb a beach so members of the parasite population being adapted to alter snail behavior in this way works overall, even if it is a waste of time on a sandbar.

Later in August 1985 experiments were attempted on two occasions. These were done on the beach adjacent to the 1985 sandbar. Harry and I collected 200 snails from the base of the beach and marked them. Snails from the base of the beach were used because they too were very likely to be *Ga*-infected, and indeed they were. They were released in 20 groups placed at the base of the beach along 20 meters of shoreline. The idea was to take hourly observations over 24 hours to see how high up the beach face they migrated and how they behaved. The first time the 24 hours went moderately well. Some *Ga*-bearing snails went a meter or two or three up the beach face with the incoming tide. One got to about six meters. But, it turned out that finding marked snails on the beach was more difficult than on the sandbar, too many confusing pebbles around, and there were few data to report. A number of marked snails could still be found at the base of the beach two weeks later so we tried another 24 hours. This time circumstances were disastrous. The first day went alright, but that night it began to rain. And boy did it. Reports were that it rained nine inches that night. A raincoat did almost no good at all, but I stayed with it through my night shift. I found some plastic tarpaulin along the beach and some poles and managed to put together a crude teepee to keep the rain off a little. It blew over once, but I fixed it and it made things barely bearable. At about 4 AM when Harry came at the appointed hour to relieve me, as I said a stalwart fellow, I left for a while and he stayed trying to take hourly observations. We were both out of our minds, especially me. I should have been smart enough to bag the enterprise. What can I say? The rain stopped the next day and I tried to go on with it, but salinity had dropped so low that all snails were inactivated. This was a bust. A park ranger walked out to our site in the bright sun and bitched about the teepee. So down it came.

The diabetes thing strikes. Things were bad with me when it was time to clean up and get our equipment off the beach. During the activities that summer an open wound had developed on one of my feet and now I was limping around. My pickup truck had four-wheel drive and I had to drive it down on the beach to pick up the gear and clean up. A vehicle on the beach there is correctly prohibited by park rules, but no ranger saw this necessary act of desperation and I got away with it. Still getting around, but badly, I finished that job and the summer and got ready for fall teaching. I made it through the semester, but the wound on my foot didn't heal and got worse. Eventually, during the winter break, I landed in the hospital on intravenous antibiotics for about two weeks. When I got out I had to administer those at home for quite a while. This was in the days before the fairly small pumps and bags they have now that you can carry around with you. I had to hang bags from the ceiling and run antibiotics into my arm through a needle and plastic tube, worrying all the while about embolism-producing bubbles coming in with it. I managed to survive all this and was able to teach the spring semester and do some work on the sandflat in 1986, but very little. I was pretty much disabled. Bummer!

Most of the work that will be described next (Curtis, 1993) was done on a Cape Henlopen sandbar in summer 1987, but some was done on sandbars in the summers of 1986 and 1988. In 1986 I could not do much. That wound on my foot was too fragile. By 1987 I was still limping with an open wound, but it wasn't too bad and I could still get quite a bit of sandbar work done. I just had to avoid getting my foot wet. Most of the time I managed to do that. I had to use a crutch or crutches sometimes and let me say for the record that they work badly on sand. In 1988 I was back in pretty good shape again with no holes in my feet, but I took it easy for the most part. There would be more work on a sandbar after 1988, but this time it would have a different emphasis and be mostly done by a student, one Karen Hubbard, for purposes of an undergraduate thesis. A future chapter will deal with her and the work she did for her thesis and two publications.

My limitations aside, there were still questions about the migrations that *Ga*-infected snails were making. I will ask and deal with them one at a time to avoid them confusing one another. So, the claim is: infected snails make these migrations up the beach and back to promote contact between cercariae and the crustacean next host. If so, when and where cercariae are released during these excursions? They must be released from the host snail somewhere and sometime. There are two basic possibilities, cercariae could be released from host snails during high-water as they move up in the intertidal zone to higher elevations, or cercariae could be released somehow after the receding tide had stranded the host in the high intertidal zone. Because cercariae typically swim to the next host, I suspected it would be the former. But, suspicions deserve a hex. It's always good to test a proposition. There were conflicting indications, prime among which was that fact that the cercariae of *Ga* did not swim very well. The problem was, how could I find out whether cercariae were released from snails during flood tide? What I came up with was to sequester individual snails collected from the sandbar in jars with holes drilled in the screw-on top. The jars were 6.5 cm in diameter and 8.5 cm high (about 2.5" by 3.25") and I shall refer to them as cercarial release chambers. Let me take you through just one snail's experience. After collecting the snail during a low tide on the sandbar, it was placed in a dry jar. The snail was exposed, in its jar, on the sandbar peak, to the flood tide. The water would rise and flow into the jar through the holes in the lid and the snail would then be in water, in the jar, on the sandbar for about six hours until the next low tide. At that time I returned and put a lid, without holes, on the now maybe cercarial release chamber and hauled the jar, water, and snail back to the laboratory for examination.

Of course, I didn't do just one snail at a time. I used a plant press to hold 24 jars at a time. A plant press consists of two, essentially flat, rectangular wooden pieces plus straps that hold them tightly together. The wooden pieces (about 1 by 2 feet) were not one piece of wood. They were openly constructed of slats and would allow water to pass. The bottom piece was big enough to accommodate the arrangement of 24 jars on it in six ranks of four. Before the flooding water reached the sandbar peak, snails would be put in 18 jars. The other six were control jars with no snail. Tops with holes were put on all 24 jars. The other piece of the plant press was put over the top of the jars and strapped firmly into place.

This "package" of jars was then staked to the sandbar peak so it would not get upset during high water. The water would rise and fill all jars and snails could release cercariae into them. The control jars, which contained no snail, were a check on the possibility that cercariae found in chambers could have drifted in rather than being produced by a contained snail. If control chambers had cercariae in them, this would be an important concern. Getting off the sandbar with dry feet was possible only sometimes. I used to use a frame backpack to get dry jars out to the sandbar. Even if using crutches that was pretty easy. Hauling back 24 jars filled with water, plus all the rest of the stuff I had in the pack, was sometimes a challenging affair, especially because the sandbar used for this was farther away along the beach than the one used in earlier years. It did test me, but I don't mean to complain. After all, it was voluntary and in the service of my curiosity. Between the summers of 1987 and 1988 I was able to expose a total of 238 jars for about six hours on high tides, most with contained snails, but some were control jars. The water in each chamber, controls included, was examined for trematode cercariae. If there was a snail enclosed, it was examined by dissection in the usual way. I needed to know if it was infected and with which trematode(s) to try and make sense out of any cercariae found.

As an aside here, in the process of examining these jars after being exposed on the sandbar for one high tide I ran into an old friend from my dissertation research some 15 years before. In the jars some of the planktonic larvae of my worm regularly showed up, and in pretty good numbers. Maybe I had missed the boat. In the old days I would go out with a plankton net, filter great volumes of water, and spend hours searching through formaldehyde-preserved samples to find the planktonic larvae of this worm. Now, here they were at almost no cost. It makes sense. One of the habitats of this worm is the intertidal zone, where they sometimes build sand grain "reefs." Here were the larvae seeking a place to settle and take up adult life. If you want to sample these larvae, better to put some jars with holes in the top on a sandbar than to go through all the business of sampling the plankton with plankton nets, boats, formaldehyde-laden samples under your nose, and all the rest. Not only were these familiar larvae present in these jars, but a whole array of other planktonic larvae as well. Hindsight is bittersweet. This would be a great alternative way to study planktonic larvae, including those of *Ilyanassa*.

In all there were 61 chambers that had *Ga*-infected snails confined during one high tide. Examination of the water in those chambers showed that about 80% of those snails failed to release any cercariae at all. Among the other 20% one did release quite a few, but the rest released very few. At the time of year when these tests were done, a typical snail infected with *Ga* would have 1000's of cercariae in it. The apparent message is that *Ga*-bearing snails do not, or quite rarely, release their cercariae during flood tide.

I was able to do another analysis that backs this contention up. That was what I call the "drift-in" analysis. Cercarial release chambers were exposed essentially flat on the sandbar surface, standing maybe 12 or 13 cm (about 5") high. There were, especially at night, a lot of snails close-by that had a wide variety of infections and would be releasing a variety of cercariae into the water over the sandbar. How often do species of cercariae that the snail in a chamber is not infected with show up in that chamber. When that happens the snail could not have produced the cercariae because it wasn't infected with that species. It (they) must have drifted in from outside the chamber. So the question is, for example, how often did chambers that did not have a *Ga*-infected snail in them nevertheless have *Ga* cercariae? There were 177 chambers, including control chambers, that lacked a snail infected with *Ga*. One of these did have a drift-in *Ga* cercaria, but the rest had no *Ga* cercariae. This means that for *Ga* the drift-in frequency was 0.6%. How do other species compare with this? There were 155 chambers that did not have a *Hq*-infected snail inside, but cercariae of that species showed up in 24% of those chambers. Corresponding percentages for other species were *Ls* 18%, and *Zr* 31%. Apparently there is a cloud of cercariae from most species over the sandbar, but not *Ga*. This is further evidence that *Ga* does not release cercariae during high-water, even from snails free on the sandbar and not in chambers.

A fair objection can be raised. Would snails in contact with the sand on the sandbar, rather than being confined on glass in a jar, release cercariae into the sand? Another set of two experiments pretty well closed the book on *Ga*-infected snails releasing cercariae during high- water. To test this possibility cages were set up to confine snails on the sand in the field on a sandbar. These were circular cages half a meter in diameter and made out of the same hardware cloth used to make carrion response enclosures (Chapter 5). They had a hardware cloth roof, stood about as high as the plant press with the jars in it, and penetrated the sediment deeply enough to prevent snails from going under the side wall. In short, snails could not escape. I had found that surface sand could be collected using a plastic pill bottle. One of these pill bottles was about 4.5 cm in diameter by 8 cm high (1.75" by 3"). By gently scraping the bottle along the sand the top 1-2 cm could be collected for later examination. That examination involved adding some seawater to the pill bottle with the sand sample in it. The bottle was capped and then vigorously shaken to free up any cercariae that might be present. The liquid contents were then quickly poured off into a dish for examination with a microscope. This method took a while to develop and test, but it proved very effective at revealing cercariae associated with the sand.

In 1987 I did two experiments along these lines, one over a daytime and one over a nighttime high tide. They were run this way. On a low-water the cage would be set up on the peak of the sandbar I was using that year. Any enclosed snails would be removed. Before the flood tide submerged the cage, it was thoroughly sampled with pill bottles to cover the possibility that there were already *Ga* cercariae in the sand inside. Eight pill bottles were used. With each I skimmed the sand inside the cage along one of eight equally spaced radial tracks from the center point to the edge of the cage. After this was done the cage, looked at from the top, had the appearance of a circle filled with an asterisk. Once these "before" samples were taken, snails from the sandbar top, likely *Ga*-infected, were gathered and put in the cage. After installing the cage roof, I would command the tide to come in over the cage and then go out. Power is invigorating. About six hours later the cage would re-emerge and at that time the roof of the cage would be taken off, all snails retrieved, and another set of eight surface sand samples (another asterisk) would be taken. A random sample of retrieved snails would be dissected to find out if indeed many were *Ga*-infected. If snails infected with *Ga* had been releasing *Ga* cercariae into the sand during high-water, they should show up in the "after" sand samples.

These two experiments worked wonderfully. The "before" samples in both (all 16 of them) were all empty of *Ga* cercariae. There were 150 snails confined on the daytime low tide and all of them were still there after high-water. I dissected 30 from among these and 90% were *Ga*-infected, all with mature cercariae present. In the eight "after" samples there were a grand total of six *Ga* cercariae. The nighttime experiment, carried out around midnight, went much the same. Confined snails numbered 135 and again I got them all back after the high-water. On dissecting a random 30 of these, 90% were *Ga*-infected, all with mature cercariae present. The eight "after" sand samples produced but nine *Ga* cercariae. Considering that collectively the confined snails on both tides no doubt had at least millions of cercariae that they could have shed, the overall message is clear. Snails infected with *Ga* release effectively no *Ga* cercariae into the water, or into the sand, during high-water. That leaves only the other more problematic option open: if *Ga* is trying to get cercariae to a beach-dwelling next host, they must somehow be releasing cercariae after the host becomes stranded by receding waters. But, where, how does this happen?

I hope I've got you engaged because I sure was. I knew that these cercariae swam only feebly. I also knew they weren't released during high-water into the water or the sand. If my understanding of what is going on (my hypothesis) is correct, *Ga* cercariae must be getting released from snails sometime during low-water. In my time spent around sandbars and especially the beach as the tide was receding, I had noticed that after the water left snails stranded they would still crawl along for a while in the wet sand. Importantly, as they would do this they would leave a trail of mucus behind them. The mucus served to lubricate movement and constituted a snail trail. Eventually the sand would dry out and the snail would bury at the end of its trail. The sandbar, and the beach, ended up with dried snail trails scattered all over it. The snail would remain at the end of its trail for the duration of low-water. It would un-bury when water flooded over it again. Well, maybe *Ga* cercariae were released into the snail trail just as the snail was being stranded in the higher intertidal zone by the falling tide. Let' see.

The technique for collecting a snail trail is similar to collecting "before" and "after" sand samples in the cage experiments described above. On the exposed beach or sandbar you would look for an isolated snail trail, maybe a meter long, with a dimple at the end where the snail had buried. You could skim a pill bottle along the trail and collect the mucus trail and associated sand. You might think of one of these as a snot trail, but they are actually much less gooey than that. The snail at trail's end could be collected as the producer of the trail. If the trail had *Ga* cercariae in it and the snail was *Ga*-infected that would be most interesting. There were two sorts of controls for such a procedure. A stip of sand parallel to the snail trail could be collected to find out if *Ga* cercariae were just everywhere (a no-snail control) and, if the snail at the end of the trail was not infected with *Ga*, that would also be a control making the same point. In all I collected 65 snail trails that had *Ga*-infected snails at their ends. I did not count observed cercariae past 100, but these mucus trails often had hundreds to thousands of *Ga* cercariae in them. There were 28 control trails taken and they almost never had any cercariae in them. I attribute the few *Ga* cercariae found to trail-following. *Ilyanassa* individuals follow the trails of other individuals. When they cross a trail they can even tell which way the trail maker was going as it made the trail. Anyway, if a *Ga*-infected snail follows the trail of a snail not so infected, it might deposit a few *Ga* cercariae in a "control" trail.

How about that! The parasite was releasing cercariae into the host's mucus trail right at the time that the snail was being left stranded in the higher intertidal zone by receding waters. Larvae in a snail's mucus trail was not unheard of in the lore of parasitology, but it had never been observed in a marine situation. It is still the only case of it of which I am aware. I was doing marine ecological parasitology and this was almost as exciting as the basic observation of *Ga* making host snails migrate up and down sandbars and the beach. All this excitement will probably take me down in the end. I should stop, but I'm addicted.

The point had been made with the 1984 and 1985 sandbars that more *Ga*-infected snails made the trip to the high sandbar, and beach, for low tides that occur under dark conditions. I was interested to find out whether this extended to the release of cercariae into the host's mucus trails. That is, do *Ga*-infected snails release more cercariae into trails at emergence on nighttime versus daytime low tides? I divided the already observed *Ga*-infected snail trails and control trails up according to how much of the low tide on which they were produced occurred under dark conditions. There were three categories of low tides, 0-33% dark, 34-66% dark, and 67-100% dark. Among control trails, in a statistical test there was no significant difference. There were no differences in the number of cercariae no matter how much dark there was during low tide. Some control trails produced on 67-100% dark low tides had a few cercariae in them. This was probably due to trail-following. Among trails produced by *Ga*-infected snails, the ones produced on low tides associated with more low tide exposure in the dark had greater numbers of *Ga* cercariae in them. So it did extend down to that level. Somehow, host snails released more cercariae on low tides that occurred in the dark and fewer cercariae on low tides that occurred more in the light. In fact, trails produced on 100% daytime low tides by *Ga*-infected snails never had any cercariae in them.

Another issue was how long *Ga*-infected snails spend making trails on daytime versus nighttime low tides. Five snails that were making trails at emergence in the daytime spent an average of about 5 minutes (range 2 to 9 minutes) trail-making before burial. On a night low tide four were timed and they spent an average of 29 minutes (range 14 to 43 minutes) trail-making. Clearly they spend more time making a trail on nighttime low-waters.

Yet another thing was to find out whether and for how long cercariae, once deposited in the mucus trail, could survive there. There weren't any cercariae in daytime trails so there was nothing to investigate there. In the dark it was a different story. I followed the making of six nighttime trails by *Ga*-infected snails. Each trail was divided into six equal sections. Sections 1, 3, and 5 were collected right away, sections 2, 4, and 6 were collected at the end of low-water about five hours later. Cercariae were present and abundant in all six trails when they were made. In four of the six trails cercariae were still alive after the low-water was over, they had survived in the trail in the exposed intertidal zone for about five hours. In the other two trails the cercariae appeared dead by that time.

All this makes adaptive sense with the activities of the next host, beach-hopper crustaceans, in mind. They are active at night in the high intertidal zone during low tide. It couldn't be otherwise. These crustaceans actually drown if forced to stay in seawater for very long and they are hidden in the seaweed strand line during the day. A parasite needing to get cercariae to that semiterrestrial host from an aquatic snail host would do well to make that snail migrate and get stranded in the high intertidal mostly at night, release cercariae into mucus trails just as they are being left emerged at night, but avoid this when emerged during the day, spend longer making the trail at night, and have cercariae that can survive throughout low-water.

Let's take a day-in-the-life-of-a-*Ga*-infected-snail look at how its behavior matches what would be needed to result in cercariae being exposed to beach-hoppers. Bear in mind that uninfected snails, and snails infected with other than *Ga*, do none of this. I will use anthropomorphic language here, i.e., writing as if the *Ga*-infected snail had human characteristics and "knows" what it is doing. Of course it doesn't "know" anything. It simply responds to stimuli. What those stimuli (environmental cues) might be is an important, unresolved issue. Assume the snail is near a sandbar and we are talking about a day with a low tide around 3 PM. There would have been a high-water around 9 AM. On the flood tide leading to that 9 AM high-water our host-parasite would in all likelihood have "decided" not to make a trip to the sandbar peak. (I use the term "host-parasite" because a *Ga*-infected snail can't be considered as just one or the other. It's a unique unit.) It would likely stay in the lower intertidal zone with all the other snails, but if it did make the trip and was stranded along about 12 PM by receding waters, our host-parasite would have decided to retain its cercariae. It would make a mucus trail at stranding, but not put any cercariae in it. Most likely our host-parasite would have stayed in the lower intertidal during the day and, if it found itself in shallow water or on wet sand, it would not bury and would have busied itself doing what a deposit-feeder might be expected to do, feeding by vacuuming in sediments and their contained nutrients. After all, both the host and the parasite need to be supported. After the low tide at 3 PM the water would again begin to flood. We can imagine the host getting a decisive signal from the parasite. "Get going, we have a job to do." The host-parasite would move with the flooding water to the higher intertidal. By about 6 PM the water would reach the top of the sandbar. There the host-parasite moves about feeding until the water drains off the sandbar peak. It's approaching dark now. The host-parasite will likely be stranded on this ebbing tide. It dumps hundreds or thousands of *Ga* cercariae into the mucus trail that it spends about 30 minutes making as it gets stranded. Most of those cercariae will survive in the trail until the water re-attains the sandbar peak about 3 AM. During those five or six hours the host-parasite is stranded and buried on the sandbar. However, the cercariae in the trail on the sandbar surface would be exposed to beach-hopper next hosts. They would be expected to be active in the exposed intertidal zone on this nighttime low tide. At about 9-10 AM, when the water gets back to the buried host-parasite again, this time it comes out of its burial and decides to migrate back down off the sandbar where it will likely remain all through the next day. When the tide begins to flood for the next day's late afternoon high tide it makes another trip to the sandbar peak, deposits more cercariae in a mucus trail at stranding, and cercariae have another shot at infecting a nocturnal, semiterrestrial next host. With the next receding water it migrates down off the sandbar and spends the day in the lower intertidal zone. And so it goes day after day. The host snail is doing things it would not ordinarily do. It is possessed by the parasite residing within its tissues.

There are still things to be looked into. The step to the crustacean host needs to be investigated. How often is that accomplished? Maybe the mucus trails are attractive to beach-hoppers as a source of food and that's how they get infected. Once in the beach-hopper it could be that this host makes itself more likely to be preyed upon by a definitive host (e.g. a seagull). It could do this by being active during the day. I have seen a few hopping around then, exposed to daytime predation. It would be wild if it could be shown that two hosts in the *Ga* life cycle have their behavior modified in different ways to enhance host-to-host transmission. So the story is not complete and there are interesting things to be investigated. Nevertheless, with respect to the parasite in the snail host, this is a remarkable phenomenon. It may make adaptive sense, we can see *why* it arose in the parasite population, but how the hell would such a complicated thing be controlled by a parasite. For example, how would a parasite, inside an opaque snail shell, "know" whether it's day or night? Future biologists will have something more to figure out.

If we look further than just what happens in the time-frame of tides or days, there are more puzzles. Is there a season to host-parasite behavior, does it go on throughout a month and all year? I have some answers. In 1986, when I was having pretty severe foot problems, I didn't do much, but I did manage to keep track of snails frequenting a sandbar. Once a month from May through October I would visit the sandbar on consecutive day and night low tides. I had a 15 meter square area (circular this time) that I felt my way through carefully on each visit. I couldn't walk too well that summer so it was good to work on my hands and knees. Without going into the counts of snails, they were present at the end of May, most abundant in June, but present all through the summer, dwindled in September, and in late October there were zero snails visiting the sandbar peak. I did not take counts before or after this time, but more casual observations indicate that there are no or very few snails on the sandbar peaks early in spring, in very late fall, and during the winter. This suggests that the behavior is most manifest between about the time the sun is highest on the sky (summer solstice, 22 June) and the time the sun dips below a projection out into space of the equator (autumnal equinox, 22 September). Moreover, this period of time seems well-associated with the presence of active cercariae. If you dissect a *Ga*-infected snail from the field before or after this time, there tend not to be active cercariae in the snail, but virtually all have active cercariae during this time. Active cercariae apparently have something to do with generating the altered behavior. This makes sense because before late May and after late October, beach-hoppers would not be regularly active on beaches during nighttime low tides. The altered host behavior would be a waste of time at these times of the year, as there would be no next hosts around to infect. Apparently the parasite is adapted to avoid this seasonal waste of time, but not the waste of time associated with marching up and down sandbars. Their adapted state is not perfect.

What about within the period of summer solstice to autumnal equinox? What limitations are there then on the value of the behavior of host-parasites? During this period, late-day flooding tides, when host-parasites migrate up so they can be left stranded in the intertidal zone at night, occur at about a two week interval. These times roughly correlate with new and full moons, but this is an inconsistent correlation. At various estuarine locations the correlation is different and therefore I do not believe the altered behavior is run according to a lunar rhythm. In my view, but I must admit I don't really know, the behavior is timed by more immediate environmental cues. For example, a snail might anthropomorphically say, "I've got this pesky parasite onboard, it's late in the day, and I'm wet (the tide is flooding) so I will move upward." During summer when the behavior is manifest there are just six or seven series of days with late-day flood tides. The parasite had better get its transmission from first intermediate host (the snail) to second intermediate host (beach-hopper) for the year done then, as outside that period there would be little opportunity to do it.

As I have indicated, this work that I have been discussing [three published papers, Curtis (1987, 1990, and 1993)] was done mostly on Cape Henlopen, but also on a beach on the southern shore of Indian River Bay. The work itself spanned half a decade from 1983 to 1988. Getting the story told for the benefit of others and myself (published) that ran the time spent up to a full decade. I have mentioned a little about getting some of the papers published, but almost nothing about others. The point to be made here is that this aspect of doing scientific research is a key part of doing it at all. It's not an afterthought and anything but easy and automatic. Actually, with my temperament I would be rewarded by just finding things out. If discoveries never went any further than my own experience, I would be mightily entertained. However, this would not do for the enterprise of science. Whether information is used promptly by others or not, it needs to be out there in an organized form and available in libraries so future investigators can, if they want, have access to it. As I intimated previously, you just never know what use, if any, will be made of published work. It may be that, particularly in current times, the information contained in the published results of a study will be lost among the myriad of papers on a myriad of studies. But, it needs to be there so that investigators with whatever purpose or need in mind can use it. Some say it's a waste of time to publish obscure results in obscure venues that almost nobody reads. My friend Buck Cornell once said something to the effect that a flower of humankind passes through those pages. There's no intention to inflate myself too much, but I agree. Published scientific work is far from all perfect, but its tendrils reach far and deep. When Miller and Northup in 1926 published their paper on the trematodes found in *Ilyanassa* around Woods Hole, they could not have envisioned that use would be made of their work in the 1980's on Cape Henlopen. Nor could Stunkard, when in 1938 he published his works on the life cycles of *Himasthla quissetensis (Hq)* and *Zr*, have foreseen how important that publication would be to me some 30 or 40 years later. There are endless examples.

What does it take to publish a scientific paper? I am thinking in terms of a piece of field work in the biological sciences. Laboratory studies are often shorter, but they too require surmounting challenges all their own. A paper you might be able to read in 20 or 30 minutes from start to finish likely took years to put together in final form. To do a field study presumes that you first of all have observed something that looks worthy of further investigation. Once the issue is in hand, you have to come up with an investigative plan. As in Chapter 1, this might be the testing of a before-the-data-are-in hypothesis, or an unbiased data collection program for possibly generating an after-the-data-are-in hypothesis. Either way, much thought, library work, and time must go into this if the plan is going to work. Once you have figured out what you want to do the next thing is to do it, gather the data. This has its own problems. There are pieces of necessary equipment to gather, methods to try out and perfect, and the sheer physicality of going out to do the work. A field study is likely to entail months or a year or more of time just collecting data. Supposing you now have data in hand, the next challenge is to analyze it. Ideally, you have decided what statistical test(s) you will use on data before you gathered them, but, especially when you don't start with an hypothesis to test, it doesn't always work ideally. You may well have to figure out how to analyze data after you get them and this all by itself can take months or more. Once you have your analysis done you can begin to think about writing the manuscript. Journals all have their little idiosyncracies. There are usually specific rules for such things as titles, how referencing is to be done, keywords for internet searches, what sections are required, and in what order they are to be included. Supposedly you decide on the journal before you begin to write so there will be less to correct later on. I have seldom been so efficient. Papers that report on a piece of new scientific work, as opposed to papers that review other papers, generally have the following parts. The Title is best kept as short as possible and beautifully descriptive of the work. The Abstract is a summary of the work in typically a single paragraph. It is the toughest damn thing to write and I usually leave it to absolute last. Next comes the Introduction. This contains a statement of the problem to be addressed, a short review of the published literature about the problem, hypotheses to be tested, an account of the general plan of the work, what the work to be reported adds to the knowledge, and often the briefest indication of results. The next section, the Materials and Methods, explains what you did and how you did it. The goal is that, in principle, someone else could use your explanation to redo what you did. But, there's an old saying that you can't put your foot into the same river twice. The idea there is that you change the river when you put your foot in the first time. In the context of data collection and/or experimentation, especially in field biology, it is very difficult or impossible to recreate an exact set of circumstances, i.e. to exactly redo an experiment. A species of organisms is not made up of identical individuals and the "river" upon which you experiment will be at least slightly different. In this sense, because on a second run of the experiment and in addition to technique/equipment differences, different organisms will have been used. The hope is that a species is coherent enough, and the conclusions you reach about it are robust enough, that whatever outcome you got someone else can get by carrying out a similar procedure. An example would be the work on age and growth of *Ilyanassa* that Scheltema did around Woods Hole in the 1960's (Chapter 4). His conclusions were robust and when Larry Hurd and I analyzed the snail population on Cape Henlopen we got essentially the same result. In science you are supposed to be able to get the same results again and again if you do the same experiment. In fact, in many instances attempts to replicate field studies are never made simply because they are so involved and require so much effort. For example, who would choose to replicate almost any of my field studies on parasite-altered host behavior just to see if it would happen again? It's too much trouble and besides it took a bit of an obsessed idiot to do them in the first place. With all that said about the Methods section, the next part would be the Results. There are usually tables and figures (graphs, photographs) to prepare that will illustrate the results, which can be a major enterprise in itself. Here, the analyses of results are presented, presumably without comment. You don't try to discuss or evaluate results here. Ideally, you just present them without comment. The Discussion comes next. In this you can comment on the results, in effect spin them the way you think is appropriate. Here is where you say what you think results mean and how they fit in with previous work on the topic. This often is the most difficult section to write and where your creativity is most on display. I usually write the Material and Methods first, Results second, then the Introduction, and next to last the Discussion. Absolute last comes that pesky Abstract. Following the Discussion would come the Acknowledgments. Here you would pay homage to anyone who helped with the work in some significant way and to any funding agency that provided resources to do the work. I have often had people to thank, but only rarely a funding agency. I have no idea how much of my huge professor's salary I have spent supporting my work, but it was well worth it to have the freedom to pursue what I wanted. Applying for grants is incredibly time-consuming. You can easily spend a full year putting a big application together. And, in the end, most of the time you don't get funded. After submitting a few proposals in the early 1980's I made a decision to not waste my time with grant applications. I could get more done by basically funding my work out of my own pocket. Finally in the manuscript, and this is no small deal, you need to put together the Literature Cited section. All the papers you have cited in the manuscript, and no others, must be listed in proper form here. All this done, you now have a completed manuscript in hand.

The next step, after you have gone over the manuscript again and again and yet again, and changed and improved things, is to send it to the editor of the chosen journal. In most cases that editor will choose an associate editor who is particularly accomplished in the field at issue in your manuscript. The associate editor will select two or three, sometimes more, external reviewers for the manuscript. These are experts in the particular field involved and they are kept unaware of each other's involvement with the manuscript. These reviewers each go through the manuscript, write up an evaluation of it, and send it back to the associate editor. This individual then makes a decision about what he or she will recommend to the editor. This decision will be to reject the manuscript outright, accept it pending major or minor revision, or accept it as is. If accepted, there is almost always some revision to do. The editor will normally accept the associate editor's decision, but the editor has final say. In total this is the peer review process and all published scientific papers have gone through it. Whatever the final decision is it is communicated to the author(s). If it is not rejection, revisions will likely have to be made, which can be an extensive activity. You have to balance what the reviewers and editor want done with what you are willing to do. If you can't make the changes they want for some reason, maybe you don't agree or maybe you are asked to do something that brings up a personal integrity issue, in the extreme you may have to pull the manuscript from consideration. Assuming that doesn't happen, you make the revisions and write a detailed letter to the editor explaining what revisions you have made and not made. Then you send the revised manuscript and the letter back to the editor. At that point a final decision is likely made. The manuscript will be published, or not. If it is to be published, the only entity left to deal with is the publisher. Eventually, months or possibly more than a year later, the publisher will send you what are called galley proofs. These are the first cut of the paper in the form it which it will appear published in the journal. By this time you are normally on to something else and are no longer right on top of what you said in the paper. You re-read and it all comes back to you quickly. You are looking for errors made by the printer. Beyond that you avoid changing anything. After checking over the galleys carefully down to the minutest detail, usually within 48 hours of receiving them, you send them with any necessary corrections back to the publisher where it gets in line and is eventually published. Despite all these steps almost always some error seeps through. Nevertheless, the paper says what you want it to say. The only thing left is to pay page charges, which many journals have. These amount to something like 35 or 40 dollars per journal page. Usually these are paid for by your academic unit and sometimes I have had these paid by one department or another. However, I did not work in an academic unit that much valued research and many times I have had to pay these charges myself. I've spent money on very much less worthwhile things. These have been long paragraphs describing what goes into publishing a scientific paper and they should have been. Publishing the results of an investigation is a lengthy process and it should be appreciated as the essential and integral part of doing research that it is.

So, the work of the middle 1980's was over. It was fitting that it began with a romantic interest because it became a romance of its own in its execution. The activities involved affected many people, not least me. I had managed to do some fairly high profile scientific work with almost no support, except from the students who became involved. There was a key lesson in this. Quality work could be done with little financial or institutional backing. The University sort of exists in the background mostly ignoring, but tacitly accepting your activities. The real driving force behind work such as this, I would argue for all research, must be the people involved not the institution within which it takes place. The institution can if it chooses facilitate, but it cannot do. Before the 1980's I did not have that understanding. I thought that universities and other institutions could choose to do or not do. Maybe innately I knew better and it's a good thing I did because my institution chose not to give much support in my case. As it turned out it didn't matter. In fact, if there had been committed institutional support, it might well have screwed things up. It would likely have removed my underdog mentality and changed what I did. The circumstances under which my research was done, spoken of in Chapter 1, may well have mattered greatly. As it was I got to know some great students and work in a remarkable place. The Cape Henlopen Laboratory was an out of the way place. Even if kind of spooky at night, you could work there all night and all day with almost no one around to take up your time. For much of this time I had two dogs to pal around with me. Because there was no one else around much in that huge old mine storage building and minimal policing at the University campus in Lewes or in the Cape Henlopen State Park, I could almost always keep them with me. They would be under foot in the laboratory and go on trips out to the field. Kai and my dog Salt, a white Labrador Retriever that I acquired in 1984, were well behaved and marvelous partners in this work. Kai would even retrieve pill bottle samples and drop them in a bucket for me. They loved their life at that time. Regardless of what time or how often they had to get up and go they were always ready. In later days cops of various sorts became more persnickety about night work on the sandflat and dogs off the leash. It's not that there weren't problems and challenges to overcome, but overall it was an idyllic time pursuing the workings of the wiley *Ilyanassa* and its trematode parasites.

It is the fate of us all, but the other thing that came into focus for me during the 1980's was my own mortality. Before that, with a younger me and my affliction with diabetes more recent, I had the usual sense of indestructibility that comes with early life. That sense came to a screeching halt during this time. From here on I would have to adjust my activities around this vampire of a health problem. The days of 30-hour studies and weeks-long sample taking had to be over. The circumstances would be different from now on. I could do some things in the field, but now I would need help. I had started working with undergraduate students early in the 1980's, but now I began to look at the prospect of advising undergraduate research more seriously. Specifically, it should be possible to involve them in more than "grunt" work, taking samples, washing dishes, cleaning up, etc.. There was no reason why they could not be brought into and become an integral part of a research project, including publication. The idea of undergraduate research was around in the University at the time, but it was much less promoted than it is at present. In my academic unit, in particular, it certainly was not thought of in the sense of a student actually publishing something. It is not a simple matter taking a student through something they have never even envisioned before, conceiving, designing, executing, and publishing a piece of scientific work. By the late 1980's, partly because of my unstable health situation, I began to try to develop gifted undergraduate students with this in mind. This has turned out to be one of the most rewarding things I've done. It doesn't matter whether an undergraduate goes on into science or not. Some have, some haven't. Students are not in school for nothing. The ones I have advised in research were driven by different motives, but they all wanted to be involved and learn something. It's utterly amazing to watch the maturing of students as they go through the research process. They come to be the master of an area of knowledge. Their thinking, writing ability, powers of organization, attention to detail, and dedication to a job all markedly improve. They contribute something to the record of mankind's collective knowledge and are changed for life by the experience. Graduate students get this. Why not undergraduate students? Diabetes is a concern I'd rather not have, but insofar as it led me to start taking this approach I guess I should be thankful. The circumstances under which research is done rear their head once more and have their effect.

The Lewes Harbor site where nearly 100% of female snails had imposex in 1983-84.

Chapter 7

A SEXUALLY CONFUSED SNAIL

In the very first years of the 1980's a separate area of work was started that would not see its finish for me until 1998. There were a few papers published about the phenomenon of interest through the 1970's, but in 1979 a woman by the name of M. Jenner published a paper in *Science* about it occurring in *Ilyanassa*. She used the fifty-cent word "pseudohermaphroditism" in the title, which alone got my attention, but by that time there had been only two other papers about "my" snail in *Science* and that too grabbed me. The paper was about female *Ilyanassa* expressing male characteristics in certain habitats. An hermaphrodite is an individual organism that has both male and female features (ovary and testes) and the usage of this word usually entails that this is somehow abnormal for the species. A pseudohermaphrodite would be an organism having false hermaphroditism. When it is normal for individuals of a species to be both male and female, the species is said to be monoecious [mono - one, oecious (from oikos) - household]. Well, it's certainly abnormal in *Ilyanassa* where separate sexes are the rule. This snail normally has separate male and female individuals and it is therefore dioecious (di -two, eocious - households). So what's going on? Why would females take on male characteristics? In the early 1980's a series of papers by B. S. Smith made the point that females of this snail develop male features under the influence of a pollutant associated with marinas. Female snails found close to marinas would, in addition to the usual female features (capsule gland, oviduct, ovary), usually have a penis behind the right tentacle, a fold leading to the penis (like a male vas deferens that delivers sperm to the penis), and a convoluted oviduct (similar to a male's sperm storing organ, the seminal vesicle). In the case of *Ilyanassa*, such females were still basically females and could reproduce as such. They could not function as males. Females looked like male impostors and hence the term "imposex" was applied to the phenomenon.

You have to admit that, as a label, imposex is catchier than pseudohermaphroditism. I wouldn't want to have to talk about it very much under the longer rubric. It would eventually put you in a permanent pucker.

The pollutant that caused imposex in *Ilyanassa* was found to be tributyltin by B. S. Smith, TBT for short. This is an organic (carbon-containing) compound that contains the element tin. Such chemicals are known generally as organotins. A natural substance TBT is not. It was first produced in the 1940's, by the 1960's it was found to be useful as an antifouling agent in boat bottom paints. Fouling organisms on a boat bottom are a major problem, they drag in the water, slow down the craft, and result in greater fuel costs. An important advantage of TBT in paints over copper, which formerly had been used for this purpose, is that it lasts longer. With a TBT-containing paint you can keep fouling organisms (e.g., algae, barnacles) off a boat hull years longer than with copper in the paint. This meant that the boat would have to be hauled out of the water and scraped, an expensive operation, much less often. It follows that TBT paints were widely used. By the 1980's TBT had built up in the world's oceans to the point that it was a global pollutant. Imposex was observed in a variety of other snails in harbors all over the world, sometimes affecting females more seriously. In England females of a different snail with imposex have their ability to reproduce blocked. Females can even turn into sperm producing males. Imposex is brought on at very tiny TBT concentrations in the water. Two billionths of a gram (2 nanograms) in a liter (about a quart) of water will initiate it in *Ilyanassa*. [A gram is 0.035 of an ounce and a nanogram is one billionth (0.035/1,000,000,000) of that, a very tiny concentration indeed.] What happens is that TBT, which is meant to be toxic to fouling organisms, leaches out of the paint on boat bottoms and becomes free in the water. Imposex in female snails is one of its effects, but it affects many other organisms detrimentally and this is the main worry. People like me worry about snails, but screwing up a few or even many snails around the world is to most not a major concern. I mean, balance a little damage to some snails with having to haul boats out of the water more often. You can probably imagine where the weight lies in that contest. The main value of *Ilyanassa* and other snails in all this is as biological indicators of TBT pollution. It is possible to sensitively measure TBT concentrations in the water, but it requires fancy equipment. A snail on the other hand lives in the water all the time, absorbs TBT from it, and develops the visible features of imposex because of it. All you have to do is crack a snail out of its shell and look for features of imposex. That's easier and much cheaper and snails as a biological indicator of a globally distributed pollutant became a big deal. It certainly doesn't contain all that exist, but at present I have a computer file listing about a 1,000 published scientific papers on TBT and imposex. It was a big enough deal that in 1988 a law was passed in the U.S.A., the Organotin Antifouling Paint Control Act (as indicated, TBT is an organotin). This prohibited the use of TBT-containing paints on boats less than 25 meters (about 100') in length. Organotins were similarly banned in British waters a year before that. The idea was that there is collectively much less surface area on the bottoms of a few large boats, i.e., less surface to cover with paint and therefore less TBT would be used, than on the many smaller boats in use. With these introductory remarks about TBT and imposex under your belt it is now possible to relate the small contributions made by me and my students to this issue.

In the early 1980's Charles Jenner was traveling through Delaware and made it a point to look me up. Recall that he had done a number of pieces of work on *Ilyanassa* around Woods Hole starting in the 1950's. He was still involved with research on the snail, but at a university in North Carolina. I had begun to look for imposex in female snails after the paper by M. Jenner came out in 1979, but working where I did on Cape Henlopen it had not shown up yet. On that visit he had collected a few snails from the Lewes and Rehoboth Canal (hereafter the Canal) in Lewes and brought them with him to the biology laboratory where we met. I had never collected snails from that site. His basic point was that if I was going to be dissecting all these snails, I might as well learn to pick up on females with imposex features. How true. We dissected a few snails and I was amazed, amazed I say, to see females with a full-size penis. After he left I mentioned what I had been shown to a student, Ann Barse. She found the imposex thing interesting and soon we had decided to look into the Canal habitat with respect to imposex in female *Ilyanassa*. Also I started noting and recording its occurrence in snails associated with other work I was doing.

The research plan that we came up with was not complicated. There was no need of complication. We were going to explore (survey) an approximately 2,000 meter stretch of the Canal to find out how female *Ilyanassa* with imposex were distributed along it. As you might expect by now, there is a wonderful map of the area in the paper we published on our work (Curtis and Barse, 1990). The north end of the Canal is at Roosevelt Inlet in Lewes. This is in the southernmost reaches of Delaware Bay on its western shore. Tides are traded with Delaware Bay through Roosevelt Inlet. The Inlet's landward end is to the southwest with Delaware Bay to the northeast. From Inlet's landward end two waterways extend, the Broadkill River goes northwest, while the Canal goes to the southeast through the town of Lewes about a mile away. The Canal is a narrow (around 30 meters wide) waterway dug in the early 1900's. It's about two meters deep and extends southeast then south for about 15,000 meters (a little more than 9 miles) and connects with Rehoboth Bay. Because of its connection with Delaware Bay through Roosevelt Inlet, the Canal is tidal throughout its length. Tides from Rehoboth Bay on the south end are much less strong and there is a slow net movement of water in the Canal toward the south. There is vigorous tidal flushing from Delaware Bay at the north end, other than that the Canal is a relatively enclosed waterway. At the time our samples were taken, 1983 and 1984, boating was most intense in the northern end around the Inlet. We picked five points along the Canal between Roosevelt Inlet and Lewes. The most northward was at the mouth of the Broadkill River right next to the Inlet and they went south from there. The last sampling station was in Lewes Harbor next to the former public boat launching ramp there. A drawing of this site is shown in the frontispiece for this chapter. Stations were all on the western shore of the Canal.

At each station 25 snails were collected on seven occasions, in May 1983 and in June, August, and October in both 1983 and 1984. In total we examined 802 Canal *Ilyanassa* and 376 were females. Percentage of females showing some feature of imposex at the mouth of the Broadkill River was 43%. At another station close to the Inlet it was 59 %. Moving to the south it was 69%, then 96%, and finally 98% by the launch ramp in Lewes Harbor. Imposex was everywhere in the Canal, but there was a clear increase in percent females having imposex as you went south into the Canal toward Lewes Harbor. An indication of full-blown imposex is the presence of a full-size penis on a female. None of the females in the two stations right near the Inlet, where flushing action is greatest, had this, but many did at the three stations farther into the Canal. Apparently, from the greater proportion of females with imposex and its greater intensity of expression (the full-size penis), the concentration of TBT increased in the more poorly flushed waters south into the Canal. Put another way, imposex-inducing conditions were more intense there than nearer Roosevelt Inlet. Why might that be? Well, the waters were more enclosed and there were more marinas in there. Also, at the time (since closed) there had been a boat repair facility present in Lewes Harbor for some years. I remember it operating at least in the 1970's. Boats were hauled out of the water there on a ramp and their bottoms were scraped and painted. The scrapings were hosed down into the waters of the harbor. This was well before TBT paints on small boats were outlawed in 1988. No one has ever directly measured TBT concentrations in the Canal around Lewes Harbor, certainly not at the time we were doing this work, but I think we can reasonably presume that TBT-containing paints were in use and this was why imposex-inducing conditions were so strong there at the time. Ann and I also put together data from females collected and examined for imposex from other locations in Delaware. In summer 1983 Larry Hurd had sampled 747 *Ilyanassa* females with an undergraduate student, Barbara Conlin, from along the mosquito ditch in Canary Creek saltmarsh that we had earlier used to study crystalline style occurrence (Chapter 3). Canary Creek was directly off the Canal near the Inlet and they kindly allowed us to use their data on imposex in that ditch. From there, 19% of female sails demonstrated imposex. On the south shore of Indian River Bay (Chapter 6) of the 398 females examined from there, 5 % demonstrated imposex. Also, of 1,125 females already examined for imposex from Cape Henlopen, about 1% had demonstrated it. The hotspot for imposex in the early 1980's was obviously in the Canal, particularly near Lewes Harbor. This is supported by samples collected by other workers in the 1970's. In 1975 B. S. Smith found 88% of females from the Canal with imposex. Snails were also apparently collected from the Canal by M. Jenner in the late 1970's and she found 99% of females afflicted.

Studies of imposex thus far, and there were quite few by 1984, rightly emphasized the effects on females. Ann and I wanted to see if there were also effects on males. If TBT could cause females to produce a penis, could not the penis of a male also be affected? There were a few anecdotal indications of this, but no one had gone after it formally. Recall from Chapter 5 that after reproductive seasons males normally reabsorb their penis. This would happen in about late July in Lewes. Also recall that trematode-infected males usually had to be classed as m males because the parasites had castrated them and caused penis loss. First for uninfected males we determined how many had a penis present and how many were absent a penis under the influence of season and strength of imposex inducing conditions. For a reader taking an interest in statistical tests this is a three-way contingency analysis, as opposed to the two-way mentioned so far. We had two categories for penis condition (present/absent), three for season (spring/summer/fall), and two for imposex-inducing conditions (strong/weak). Strong was if the male came from the Canal, weak was if it came from one of the other habitats mentioned above. We found a highly significant effect of season on penis present versus penis absent. More uninfected males had a penis absent in summer whether from the Canal or not. This would be after the breeding season and is the expected penis loss event. Interestingly, there were too many males (more than statistically expected) with a penis in the Canal (strong imposex-inducing conditions) and too many without a penis in the other habitats (weak imposex-inducing conditions). The conditions that induced imposex in females also induced the retention of the penis by males. That is, there was no normal penis loss event in the Canal. The inference is that TBT causes penis retention in males as it causes penis development in females.

There was also an effect of strong imposex inducing conditions on parasitized males. The most frequent species parasitizing snails in the Canal were *Lepocreadium setiferoides (Ls)* and *Zoogonus rubellus (Zr)* and we restricted our analysis to infection with these species. In the Canal there were significantly more parasitized males with a penis present than outside the Canal. So, imposex inducing conditions were leading to both uninfected and infected males having a penis when they should not have one. That is, there were effects on males as well as females. Since that time more has been learned about what TBT does to snails, but at the time we didn't know what TBT did to cause this effect, we didn't know *how* TBT worked, but this was an important addition to the knowledge.

Work published some 20 years later, in 2005 (by E. Oberdorster and colleagues), provided an explanation for our observations on imposex in female *Ilyanassa* and the effects on male snails. In their work these authors were able to identify what amounts to a very small protein compound (only four amino acids long) that stimulates penis development in both male and female snails. The name of the compound is APGWamide, but don't worry yourself about its name. It turns out that snails exposed to TBT produce more of it. Greater amounts of this penis-generating compound under the influence of TBT goes a long way to explaining imposex in females and the effects on males we observed. In the Canal TBT concentrations were high. Therefore, more of this compound (APGWamide) was produced by snails there. Under these conditions females tended to develop an abnormal penis and males tended to retain a normal penis when they should not. It was a lot of fun for me reading this 2005 paper. The basis for our field observation, particularly with respect to male snails, could not be explained at the time we made it. Now we have a good explanation based on snail physiology for what we observed in the field. Such things happen, but it's pretty rare. Ain't science grand.

Ann and I did this work in 1983 and 1984 while she was still an undergraduate. When I first got to know her she was an unlikely student. She had been working as a plumber and as a mate on sport fishing boats that went offshore for marlin. She took one of my courses on a lark. In the 1980's (Chapter 6) I was involved with other things and Ann with completing her undergraduate degree in biology at the University of Delaware. The data sat quietly until the late 1980's. By that time I was slowed down by my affliction and Ann was pursuing graduate degrees at the University of Maryland working on parasites. Go Ann. We put data together and our paper was published in 1990. Before that we presented the paper at a scientific meeting. Ann gave the talk and she did a magnificent job. In fact, she won the best student paper award at the meeting. That was very satisfying to her, as well it should have been. I have given a number of papers at meetings. It's always a stressful time and it's good she came through it so well. The experience helped her with all sorts of future presentations. Since that time she has gotten her Ph.D. in biology and is currently a tenured university faculty member. For me, it was and still is most satisfying to see a student, that I had a role in introducing to biology some 10 years before, do and turn out so well.

After we had done our Canal work, but before our paper was published, I became involved with a group of people on another imposex study. In 1988 *Ilyanassa* from Sarah Creek off the York River near the south end of Chesapeake Bay were sampled. Two biologists from England, G. W. Bryan and P. E. Gibbs, were the main instigators for this work (Bryan, et al., 1989). They had worked extensively with imposex in a British snail. Imposex occurred in *Ilyanassa* in Sarah Creek and they wanted to tie the extent of it to TBT concentrations in the water and also in snail tissues. They knew I worked with *Ilyanassa* and sent snails collected in May 1988 from six sites in Sarah Creek (about 60 snails per site) for me to analyze for female imposex. Ann and I did the dissections. They took another set of snail samples and analyzed them at the Plymouth Marine Laboratory in England. Other people involved were able to measure concentrations of TBT in water samples and in snail tissues, something I did not have the equipment or knowledge to do in Lewes. All of the snails examined were around 25 mm or more in shell height, which meant that they had been living under Sarah Creek TBT conditions for several years. Concentrations of TBT in the water varied from 0-20 nanograms per liter, while in the tissues of snails it varied from 0-800 nanograms per gram of dry snail tissue weight. Obviously, snails were concentrating TBT in the tissues. In summary, where TBT was more concentrated in the water and in snail tissue, which depended on the amount of boat usage in Sarah Creek, imposex was more frequent among female snails. In the field imposex was initiated at about 2 nanograms per liter of water and by 20 nanograms per liter 100% of females were afflicted. This certainly was confirmation that TBT in the water caused imposex.

In 1992 I did a survey of snails on Cape Henlopen similar to the one Larry Hurd and I had done 12 years prior. There was a large pier at the southwest end of the sandflat then, which was at about the place where snails ceased to occur along shore. This pier used to be part of the "fish factory" facility, which by this time was no longer operating. It was a large part of the character of Cape Henlopen. Back in the 1940's -1950's there used to be a menhaden fishery in Delaware Bay and this was where the fish were landed, processed, and stored. Fish meal was stored in the compartments of a large cube of a tower. Bacteria would work on the stored meal and it would have to be mechanically moved around so it would not build up heat and explode. The tower was sturdily constructed with re-enforced concrete walls several feet thick. The factory had been abandoned when I first knew of it about 1970. There were a lot of old, large buildings on the site that had old equipment and junk left in them. They weren't locked up too well and I can remember roaming around in them on occasion in the 1970's and 1980's with my canine buddies Kai and Salt. It was all pretty wild territory in there. I remember encountering foxes hunting rabbits. I like stuff like that. It was an out of the way place with almost no one around and with a real history about it. I spent many interesting times there imagining what it must have been like when it was operating. In the 1990's the fish factory site was purchased to construct a "development." The place is now populated by expensive beach-front houses. I suppose it was inevitable, but I thought the land would have been better used as part of the Cape Henlopen State Park, or maybe the town of Lewes could have used it as a semi-natural preserve. Anyway, all the easy buildings were taken down and the site was cleared, except for the tower. A destruction company was hired to take it down with explosives. You know, like those casinos in Las Vegas. The charges were set and detonated. I remember hearing the very loud blast while standing over my stove at home a couple of miles away. The stubborn tower barely moved. The tower was a major feature of the skyline in the area and I can remember seeing it leaning over just slightly askew on the horizon. They set more charges and tried again. Another blast, and again the tower stood tall. A third try did no better. Some wag put up a sign: "Tower three, blasters zero." In the end it came down, but they had to take it apart piece by piece, which was a very expensive undertaking. The past dies hard sometimes. I was sorry to see the glorious old fish factory, its tower, and its pier go and houses come in, but I guess that's what goes for progress. There were ruins of another pier between the factory pier and the even now existing Cape Henlopen Fishing Pier. It was ordained that the pilings associated with those pier ruins had to go too. There was a pair ospreys that had built a nest on one of the pilings and they had at least one chick in the nest. I was sad the day the crane uprooted that piling and toppled the nest into the water. The parent ospreys were understandably all flustered, flying about, screeching. I never saw any bodies and I don't know for sure what happened to the chick or chicks, but I can well imagine.

In the 1992 survey, as in 1980, I used the pier at the southwestern end of the Henlopen sandflat as a beginning point for taking samples. The on- to offshore sampling transects were set up at about 100 meter intervals this time. They extended from the old factory pier at the southwestern end to about 1,300 meters northeast of the Cape Henlopen Fishing Pier. My assistant for this work was Karen Hubbard, who by this time had become an expert in matters of studying snails. There will be more about her in the next chapter. Also, there was even some money to support this work. A Biomedical Research Grant had been awarded from the Department of Biological Sciences and Karen was a paid assistant, not near what she was worth, but at least she did get paid. We did something important in this survey that had not been done in the earlier survey. Snails were sieved (2 mm sieve) from known amounts of sandflat area (0.36 square meters), rather than just picking up the first 40 snails seen around a sample point. If there were less than or equal to 40 snails in a sample they were all dissected. If more than 40 were included, a random 40 were chosen with the aid of random numbers for dissection. I plead guilty to writing the little computer program to generate the random numbers. It made things easier. Numbers of females with imposex, or numbers of females infected with certain trematode species, per 0.36 square meters were known. Samples were taken every 20 meters along transects and transects ran from the base of the beach face out to the offshore limit of snail distribution. A total of 44 samples along 14 transects were collected, with 1,686 snails being dissected (38% of all snails collected).

In 1980 Larry Hurd and I did not see any females with imposex from Cape Henlopen. Maybe we did not recognize it at that time, but it can safely be said that if it were common we would have picked up on it. It was at least very rare then if not absent. I pulled the data together for Henlopen snails observed for imposex in each year of the 1980's. From 1981 to 1984 it was found in less than 1% of females. In 1985 it was recognized in about 5%. In 1986 through 1991 the frequency was more like 10-20% and in snails collected in this 1992 survey almost 30% of females were afflicted. In other words imposex frequency of occurrence had increased some 30-fold over the intervening years between about 1980 and 1992. Observed cases of imposex on Henlopen were always definitely there, but intensity of expression was always low. Unlike the situation in the Canal, at most females would have only a stub of a penis, a vas deferens, and some convolution of the oviduct. The frequency of occurrence had increased, but not the intensity of expression. Cape Henlopen is on open water right at the mouth of Delaware Bay with a lot of tidal action (flushing) and there are no marinas there. The Cape May-Lewes Ferry docks about 2,000 meters to the east, but there is no concentrated boating activity there to serve as a source of TBT. Nevertheless, imposex became more common, strongly suggesting that TBT had gotten into the open waters of Delaware Bay and was not merely in enclosed waters such as in the Canal.

How was imposex distributed on Cape Henlopen? Well there was not much pattern to it in 1992. I calculated the percentage of occurrence on each of the 14 transects along the beach, southwest to northeast. It was highly variable. It ranged from about 10% of females to about 58%. It didn't start low or high and increase or decrease from one end of the sandflat to the other. The only pattern I was able to recognize was onshore to offshore. Females collected near the beach were somewhat more likely to have imposex. This made at least a little sense. In the literature it had been established that the highest TBT concentrations were in the tiny layer of water right at the surface. Even wave action would not break this TBT layer up. Snails close to the beach would be exposed to this layer, as the tides rose and fell, more than snails lower in the intertidal zone. Maybe that's why they express imposex more frequently. There was a side effect of this. Snails infected with *Ls* or *Zr* tended to occur away from shore. As consequence where these infection were frequent (offshore) there tended to be infrequent imposex. Other than that tenuous connection, parasitism and imposex were not connected. Being infected did not increase or decrease chances of a female expressing imposex. This observation did increase my wondering about how much snails move around over the course of a summer. Later.

There was one more investigation of imposex. Samples for this one were taken in 1995. That summer the situation with my left foot was pretty bad. There were a couple of bad sores on it and between daily soakings I had to keep it bandaged. I was still getting around on it, but badly. In the spring semester a student had shown up in one of my classes. Jeff Kinley was in his early 20's at the time and had an unsuccessful time at another university. He was living around Lewes and he made some use of the University of Delaware where I taught in an attempt to re-align his academic forces. We struck up a teacher-student relationship and we used to talk quite a bit. He expressed an interest in getting involved with some sort of undergraduate research. I told him about some of the work I had done and we ultimately agreed that a further study of imposex in *Ilyanassa* would be a great candidate for a start. I warned him that I viewed whatever project we came up with should be done with an eye to eventual publication. Jeff, never a guy to shun publicity, thought that was a great idea. A paper was ultimately published by the two of us (Curtis and Kinley, 1998).

In 1995 it was 11 or 12 years since Ann Barse and I had sampled the Canal for female snails with imposex. After that work, in 1988 a law had been passed prohibiting the use of TBT in antifouling paints on small boats. Jeff's and my initial idea was to take further samples to see if imposex had decreased in frequency since that law had come into effect. Also, there is a drawbridge across the Canal on the south end of Lewes Harbor and thus far only snail samples north of there had ever been taken. There were some eight miles of Canal south of the drawbridge that had not been sampled and there were only two very small marinas along there. We were curious about the imposex situation in the southern Canal. The plan was to take samples on both ends of the Canal stretch that Ann and I had covered in 1983 and 1984 and then continue sampling on down the Canal for the rest of its length. At the time Rehoboth Bay, off the south end of the Canal, was also un-sampled and we would also include a sample from there. Jeff submitted the research idea to the Undergraduate Research Office at the main campus of the University and they came up with some funds to help support him during the summer of 1995. To the University's credit, it was by this time well committed to undergraduate research. I was planning to go up for promotion to Full Professor in the coming academic year. Being pretty well crippled helped to keep me in my office/laboratory amassing the incredible amount of information needed for my dossier. If I succeeded, which I ultimately did, I would become one of two untenured Full Professors at the University, a dubious distinction afforded faculty in my academic unit. When summer started Jeff and I began taking samples. First I had to show him how to examine snails and get data out of them correctly. This always took time, but I was happy to see he learned quickly. As far as the actual getting of snail samples was concerned, I was utterly dependent on Jeff. My open foot wounds would not allow me to slop around in the deep mud along the Canal banks. You would sink down to your knees, or worse, in short order. However, I had a canoe and even if I couldn't walk I could still paddle strongly. Also, Jeff had a kayak. That was no good for me (too damn low), but he made good use of it. In the north of the Canal we could get to sites by car and on foot. In southern reaches we used one or the other of our crafts. Jeff was a young, agile guy who relished the field work, getting into the mud and collecting snails. In addition, the mud was a defense against biting greenhead flies, which would at times swarm around you once you had been at a sample site for a while. You couldn't swat enough to make the slightest difference.

Whereas Ann and I had merely picked up snails from a sampling area, by this time I generally was sieving snails from the sediment. This way the sample would not be biased toward larger snails. What Jeff and I did was to lay out a transect (a line) along the shoreline to be sampled. We would then take three samples along it at about equal length intervals. At each interval Jeff would place a one square meter frame on the surface of the mud. He would dig out the mud inside the frame down to about 10 cm (4") and pass it through a 4 mm mesh sieve. This would retain even quite small snails and give an unbiased sample, just what was there in the square meter. As in the 1992 survey above, if there were 40 or fewer snails present they were all dissected. If more than 40, a random 40 were chosen. By this time I had begun to be aware of how long this snail could live and we made it a point to return any unused snails to their habitat We took samples like this at five stations along the Canal. In Rehoboth Bay, at Savages Ditch, snails were very large and more sparse. They were collected as encountered along some 200 meters of shoreline.

Some of these sampling excursions were quite something. At the station about a mile south of the drawbridge in Lewes Harbor, there was nothing but saltmarsh around. We lovingly referred to the site as "Greenhead Marsh." I, of course, was confined to the canoe and Jeff was out there slogging through knee-deep mud getting the three samples along an 80 meter transect. There he was vigorously jogging through the mud, something I could not even imagine doing at the time. He was completely covered with mud. It kept some of the flies off him. For me sitting in the canoe there was no such complete muddy luxury, I got to use just what mud I could scoop up off the side of the canoe and apply. On another day Jeff sampled the next station using his kayak. It was about two or three miles farther down the Canal. It too was a very muddy situation. Somehow a photographer and the press got involved, Jeff gravitating toward publicity again. Shortly after that there appeared an above the fold spread in a local paper with pictures of Jeff in his kayak, covered with mud again, taking samples. There was a nice little write up along with it. As I recall they mentioned me just one time. That's about usual, but Jeff got a kick out of it.

So what did we learn? Among a total or 2,443 snails collected 675 were dissected and 369 were females. In terms of the frequency of females with imposex the simplest thing to do is just list what the results were. At the mouth of the Broadkill River 52% of females had imposex. At the station next to the launch ramp in Lewes Harbor 94% did. Next down the Canal at Greenhead Marsh the figure was 78%. The next station down it was 82% and at the inlet where the Canal opens into Rehoboth Bay it was 89%. At Savages Ditch in Rehoboth Bay 23% of females had imposex. In the area above the drawbridge this was not much different than results obtained 11 or 12 years before. In terms of frequency of occurrence imposex was still very frequent in the waters of the north canal, especially in Lewes Harbor. In addition it was very common in the whole southern Canal. There was no significant difference in a statistical test. Based on the one sample in Rehoboth Bay, it appears that imposex frequency is substantial there, but less than in the Canal. The thing that did change was the intensity of expression of imposex. In the earlier study there were many females with a full imposex penis. In this study this was never observed. Much smaller manifestations of imposex were always the case. So, after seven years of TBT paints being outlawed on small boats, frequency of any degree of imposex had not decreased, but snails expressed less intense degrees of it. The law was clearly having an effect, even if slowly.

Why would imposex still be so frequent if TBT had been banned for a number of years? During the 1970's and 1980's TBT had pretty obviously been abundant especially in the vicinity of Lewes Harbor. I mentioned previously that there is a net southward tidal drift in the Canal. A particle of water would slosh back and forth with the tides after it entered the Canal from Roosevelt Inlet, but it would slowly make its way to the south. There are no obvious major sources (marinas) of TBT south of the drawbridge. The most likely explanation is that TBT was carried south with the net tidal flow from the north end of the Canal. But, why would imposex be so common all these years later if there was no new TBT being introduced into the Canal? By this time it had been demonstrated that TBT can be bound up in the sediments and released as much as 20 years later. The snail is a deposit-feeder and it could be getting TBT that way. Also, the Canal is fairly shallow and power boats moving up and down it would stir up the bottom sediments and possibly release old TBT into the water. This would not be enough to cause full-blown imposex, but would perhaps be enough to maintain its frequent occurrence.

In 1989 Karen Hubbard slaving over a hot microscope checking for parasites.

There are many pollutants dispersed into the environment. This is an example of how pernicious their effects can be. A law is passed prohibiting use of the substance, but the effects don't just go away. They hang around for years after. And, recall that this and other snails involved are not themselves the prime concern in all of this. It is the multitude of other organisms (e.g. bacteria, algae, oysters, mussels, crustaceans, fish) caused problems by this foreign substance let loose into their environments.

Chapter 8

Sociable Parasites

Bear with me a moment. A few terms will need explaining, some of which we have touched on already. The reason is that this chapter seeks to deal with work on ecological relationships between and among species, parasite species in particular. Parasitologists with an ecological bent are interested in how parasite species get along when they occupy the same host individual. The interest here is in trematode parasites infecting *Ilyanassa*. It follows that a bit more information about those infections is needed. A definitive host (in this case usually a bird or fish) carries the adult stage of the parasite, the sexually reproducing parasitic worm. As an intermediate host, a snail infected with a trematode is infected with the larval (developmental) stages of the parasite. These stages reproduce only asexually, no fusion of sperm and egg involved. This is as opposed to sexual reproduction, where fertilization is involved. In short, *Ilyanassa* is infected by larval trematodes. There are three sorts of trematode larvae found in snail tissues, two of which I have mentioned before. Those two are sporocysts and cercariae. The sporocyst is a larval stage that lacks a mouth for food intake. It is structured like a simple sac. Across the surface of a sporocyst nutrients are absorbed from the host. The function of sporocysts in a parasite's life cycle is first to produce other sporocysts and ultimately cercariae (singular - cercaria, plural - cercariae). A cercaria, as already described, is typically a swimming larval stage. It breaks out of the sporocyst in which it was produced and gets emitted from the snail free into the water to find an appropriate next (second intermediate) host to infect. In a snail, an infection might amount to hundreds or thousands of sporocysts each containing many cercariae. The third type of trematode larva found in snails is the redia (singular - redia, plural - rediae). Infections with certain species consist of rediae rather than sporocysts. In these species it is rediae, rather than sporocysts, that produce cercariae. A redia is a different type of larva than a sporocyst. The main difference is that a redia has a mouth that serves to take in nutrition from the host snail. In a rough sense you can think of rediae as "chewing" their way about in the host snail's tissues. So a larval trematode infection in *Ilyanassa* will basically consist of all the sporocysts, or rediae, and the cercariae produced by them.

In biology, in the broadest sense, a population is made up of all the individuals of a species, e.g., at this writing the world's human population consists of 6.4 billion individuals. The term is often used in a more local sense. Parasite populations exist within and between the hosts in their life cycles and as a result subdivisions of their populations are required. All the individuals of a parasite species in an individual host make up a parasite infrapopulation. So, for example, if a snail is infected with *Himasthla quissetensis (Hq)*, the infrapopulation of that trematode would consist of all the larval stages (in this case rediae and cercariae) of that species present within the snail. There are host populations too and these have to be considered. For example, all the *Ilyanassa* individuals on Cape Henlopen make up the *Ilyanassa* population there. Some of those individuals are infected with *Hq*. The collection of rediae and cercariae of that species in all of the host snails make up a component population of that parasite species. In effect, an individual host contains a parasite infrapopulation and the collection of parasite infrapopulations in a host population is referred as a parasite component population.

In Chapters 5 and 6 I made the point that *Ilyanassa* individuals are often infected by more than one species of trematode at a time. Consider as an example a snail infected with *Hq* and also with *Zoogonus rubellus (Zr)*, a relatively common double infection. There would be an infrapopulation made up of rediae and cercariae of *Hq* coexisting in the same snail with an infrapopulation of sporocysts plus cercariae of *Zr*. These two infrapopulations are coexisting in a small space, the snail's interior, and drawing upon this single host as a source of sustenance. The question to be addressed here is: when larval trematode infrapopulations of different species find themselves in the same snail, how do they get along? Do they "fight" (compete) with each other for possession of the host? Might they somehow facilitate one another? Could rediae of one species, with their mouths, prey upon larvae of another species? When one trematode species is already there in a snail and another one comes along and sets up an infection in that same snail, is there an ultimate eviction of one or the other from the snail? Is it possible that they just ignore each other's presence? And, however they interact, on what basis do they do it? Specifically, have they become adapted, undergone natural selection, to be able to deal with one another? The questions come fast. There are at least nine species of larval trematodes involved with

Ilyanassa and therefore many possible interactions between species. On top of this, there are other many snail species in the world and many other sets of infecting trematodes to consider.

Let me introduce what was contributed to these issues by myself and one of my students in the early 1990's. If the interest is in larval trematodes infecting snails, there is one major problem: snail individuals infected by more than one species are generally quite scarce. You would generally have great difficulty in obtaining enough snails with multi-species infections to work with and find out what goes on. In Chapter 6 it was noted that on both the 1984 and 1985 Cape Henlopen sandbars multiply infected snails were unusually common. During the disaster beach experiment in 1985 I even collected one snail that had a combination of four species infecting it. In 1986 I found one more snail with this quadruple species combination infecting it. As far as I know, four is the greatest number of species ever encountered infecting one snail. In both snails all four species had active infections with mature, active cercariae in the infrapopulation. How four species got there is one question. How they managed to coexist is another. *Gynaecotyla adunca* (*Ga*) was one of the four species in those quadruple infections and in general it can coexist in a snail with a variety of other species. By 1987 I was well aware that snails carrying *Ga* tend to gather around sandbars and they were often multiply infected.

In the fall semester of that year an experience of a student showed up in a 100-level biology course meant for non-majors. At the time Karen Hubbard was maybe 10 years older than most of the other students. I didn't know it then, but for some time she had been enrolled in an associated technical college to improve herself. She had been taking a variety of courses in basic writing, mathematics, and criminal justice. Now she was taking on University course work. She always came to class, listened intently in lecture sections, and avidly tackled laboratory work. At first she just did the required work and did not say much. Then came the first examination. It is fair to say that my examinations are challenging, even in a course at this level. I choose never to use multiple choice questions. Students are often disgruntled, but at some future time anthropologists will discover that those damn things precipitated the downfall of humankind. I want students to learn to think and use information to argue logically for a point of view. To that end, even in large classes, I always included at least some essay questions on tests. Karen got an 86% on the test, which was one of the higher marks. What got my attention was that after getting it back she approached me to find out just what she had done "wrong" on the test. It wasn't that she was contesting anything; she just didn't want to make similar mistakes in the future. Her questioning was rigorous and to the point. I had to work to reconstruct what I was thinking when I graded her examination, but that's why they pay us the big dollars. I suggested that she come and ask if anything wasn't clear as we went along. She did. At the end of the course she had averaged out well over 90% and had earned a legitimate A. That was my introduction to Karen. Aside from having a unique intelligence, her main characteristic is that once she takes hold of something, no matter what else may be going on, there's no letting go until it is conquered. I liken her to an intelligent bulldog. In the spring semester of 1988 she took another 100-level course, this one about ecology, and in the fall semester of that year she started the two-course introductory sequence for biology majors. She aced all of these courses, her basic mode of operation as it later turned out. In the fall of 1988 I didn't know how long infections in snails lasted and began to question whether snails on Cape Henlopen were as often infected in the winter as was the case in the summer. I knew Karen pretty well by this time and asked her if she would like to help out with a winter survey. She was already thinking about doing an undergraduate project on snails and parasites and this would be a chance to get familiar with sampling and parasite identification. Much to her future discomfort she agreed. It was December when the fall semester was over and it turned out to be a real winter in Delaware, with major snow storms and all. I was raised in Connecticut where winter starts in November and ends in April. One makes it a point to dress for cold weather. Winters in Delaware aren't quite so definite and people seem to dress always the same, cold or warm, maybe a sweatshirt or wind breaker if it's cold. It's not cold for so long and people seem to just be cold, complain, and weather it. I had ideas of sampling along six transects distributed along the Cape Henlopen shore, similar to what Larry Hurd and I had done back in 1980. That winter the weather was so ugly and temperatures so cold that we only managed to do four transects and they required what might be called heroic effort on both of our parts. At that time of year snails pull well away from shore and distribute in the lower intertidal zone where they are mostly under water. Snails were buried in the sand and quiescent so we had to sieve snails out of the sediments. This meant scooping up sand and washing it through sieves, which meant carrying buckets of water to wherever a sample was being taken. Fortunately, my feet were in pretty good shape at that time. The sandflat was cold and windy and there was snow on the ground much of the time. Suffice it to say that getting these samples was cold, wet, and nasty work. As I would later learn, Karen was not one to expose herself to physical discomfort if she could avoid it. At the time she was still sort of in awe of Dr. Curtis and didn't complain about anything. That would soon come to an end, but not yet. I was not completely oblivious to how bad things were for her. I too was cold, but in my usual way the samples had to be gotten at all costs. At a later time she would vehemently explain to me that it was a warped way. There is a photo of Karen that says it all. When I look at it now a pang of guilt grabs me. There she is out on the sandflat on a cold, gray, windy day, hunched over a sieve box, bundled up to the point of anonymity in her hip boots and winter gear, and working wet sand through a sieve. It was later part of my education to learn how miserable she was. At the time, however, she found her way through it without complaint. I wasn't actually trying to put her through anything like an initiation, but she certainly earned her spurs. Once we got out of the field and into the relative comfort of the laboratory our respective senses of humor re-emerged and she quickly became expert at dissecting snails and identifying their trematode parasites. She became well-equipped to do her undergraduate research project the following summer. This ultimately led to an Undergraduate Thesis and a Biological Sciences Degree with Distinction at the University of Delaware, and two publications. Bulldogs have much to recommend them and I am much pleased at the contribution all this made to her future life.

What idea did we come up with for her project? First of all, if she worked around a Cape Henlopen sandbar, she would be likely to obtain a variety and a lot of multiply infected snails. In summer 1989 she selected a sandbar. Elevations on it relative to its peak were measured and a topographical contour map of it was drawn. The peak was a little more than half a meter above the sandflat. A nifty drawing is included in the published paper (Curtis and Hubbard, 1993). Random numbers were used to select 32 random distances and directions (locations) on/around the sandbar where snails would be collected. These locations are indicated on the drawing. It was important to choose these locations randomly to avoid getting too many snails from particular parts of the sandbar, which could affect the variety of multiple infections collected. Between 16 June and 17 August snails were collected from each of these random locations. In all Karen collected 379 snails and there was indeed quite a variety of infections. In all, five trematode species were observed in single infections plus one that occurred only in double infections. I won't list them infection by infection, but to give you an idea: in addition to 18 uninfected snails there were 162 infected with one of five different species, 134 snails infected with two species in 11 different species combinations, and 65 snails infected with three species in five different species combinations. Because the collection site was a sandbar where *Ga*-infected snails actively congregate, this batch of snails was purposely designed to contain many multiple infections. It is far from usual to collect so many and in such a variety.

These larval trematodes most intently inhabit a snail's gonad and digestive gland, but can occur anywhere in the snail's tissue with the exception of in the head and foot. One of our questions may be exemplified by: if species A tends to have a certain distribution within snails (it occurs in certain places in the tissues) when it is the only infecting species, is its within-snail distribution changed when it co-occurs in snails with species B? If species B causes the usual within-snail distribution of species A to be restricted, it would reasonably be concluded that B has a negative effect on A. Put more briefly, is species A is displaced in the presence of species B? This is the stuff of ecology. If there are such things as this going on in multiple infections within snails we would like to know it because they would be indicative of interactions between infrapopulations of co-infecting parasite species. Based on the literature, our expectation was that we would find such goings on.

How did we test for this? Karen took each of the 379 snails, removed it from its shell and stretched it out on a board with pins. With a razor she cut it into five sections of equal length, being scrupulously careful to not to contaminate any one snail section with material from any other. Each snail section was put in a small vial with some filtered baywater in it. The vial would then be sealed with a rubber cork and vigorously shaken to free up any contained parasite larval stages (sporocysts, or rediae, and cercariae) into the water. The water was quickly poured off into a dish and microscopically examined. At the end of this procedure it would be known which stages of which trematode species occurred in which sections of a snail. Then, for example, the distributions of stages of species A in snails singly infected only with species A could be compared with the distribution of A in snails doubly infected with A and B. If A is displaced when B is present, this would be the kind of thing we were expecting. The statistical test we used to detect such effects of one parasite species on others is prohibitively complicated. It certainly took us a while to master. It is explainable, but I won't attempt it in detail here. The test's name is a mouth full. It's a three-way contingency analysis using log-linear models. With this it was possible to analyze five of the six species of larval trematodes in our samples to see if they were affected in this way (displaced) by other species present in a snail. In short, none were. No species was displaced when co-occurring with other species. The most common situation was that trematodes of all species occurred throughout snails whether or not other species were present. So far then, it does not appear that there are interactions among infrapopulations of different trematode species when they inhabit the same *Ilyanassa* individual.

Perhaps interactions occurred, but on a different basis. For example, it could be that trematode A was not utterly displaced from an area of the snail when it co-occurred with trematode B. Maybe it still occurred everywhere it normally would in a snail, but in some snail sections one of its functions was cut down or eliminated. An infection in a snail has but one ultimate purpose in terms of the life cycle of the trematode involved and that is to produce cercariae. These fan out from the snail to infect a next host and continue on with the life cycle. That is, cercariae out equals reproductive potential out (babies out) for a trematode infrapopulation in a snail. Could it be that any of the trematodes in our snail samples had their cercarial production shut down in certain snail sections when co-occurring with other species? It is difficult to think clearly about multi-way contingency tests, but let me go a little deeper into ours: in our three-way contingency analyses the stages of parasite [whether there were 1. sporocysts (or rediae) and cercariae, or 2. just sporocysts, or rediae, or 3. just cercariae, or 4. none of the former three] present in snails sections was tabulated (organized) according to three factors, assemblage (which species infected a snail), section (which section of a snail), and stage (whether sporocysts, rediae, cercariae occurred). Whew! In this context the test was looking for a dependence of stage (specifically the absence of cercariae) on assemblage (which species were present) in sections of snails. Well, that was not found. Again using infecting species A and B as an example, one thing was that wherever A was distributed in snails when B was not present it was distributed that way when B was present. Furthermore, wherever A's cercariae were found when it was alone in a snail they were found there when A co-occurred with B. That is, species B was not completely shutting down the production of cercariae by species A anywhere in snails when B and A infrapopulations inhabited the same snail. Here was another indication that no consistent negative interactions among parasite species were going on. Parasite infrapopulations of different species occupying the same snail seemed to be ignoring one another.

So in snails one parasite species was not shutting down cercarial reproduction by any other. The question remained: was cercarial production merely being curtailed, as opposed to being shut down? To find out snails would have to be confined in cercarial release chambers such as the ones used on sandbars in 1987 and 1988. Recall that these chambers were glass jars with holes drilled in the lids. From the 379 snails taken from the sandbar 206 were placed in chambers in the field to release cercariae for 24 hours prior to being dissected (sectioned) as previously described. We wanted to avoid drift-in cercariae, as had happened when these chambers were placed directly on the sandbar surface. We put them where there were fewer snails around and suspended them off the sandflat. What would happen is that Karen would come back from her sandbar with snails from one or more of the random locations that had been identified there. We would place individual snails in individual cercarial release chambers and arrange them in a wire basket we had constructed. This was 1989 and the pier ruins (and the unfortunate ospreys) southwest of the Fishing Pier spoken of in Chapter 7 had not been destroyed yet in the name of "development." The packaged up basket, with snails in chambers contained, was taken out and hanged off a cross piece on those pier ruins so that the bottoms of cercarial release chambers were 20 cm (about 8 inches) off the sandflat. To keep the basket from swinging around in wave action at high tide the bottom of the basket had an anchor attached. We had jars without snails included (controls) and drift-ins were not a problem. In only one jar did a cercaria show up that was not potentially produced by the enclosed snail. We would hang a set of chambers on one low tide, let them be submerged on two high tides, and then come back (after about 24 hours) to retrieve the jars, their enclosed water, and the snails. In the laboratory the snails would later be sectioned and processed. So we would know what trematode species, if any, the snail in each jar carried. In the laboratory the water in jars was immediately adjusted to equal volumes and a random one third of it from each jar was examined for cercariae and they were identified and counted. The frontispiece for this chapter is a depiction of Karen working at this procedure. What the whole procedure ultimately meant was that for each snail it was known what trematode or trematodes infected it and how many cercariae of each kind were released during 24 hours in the field. We used 24 hours because cercariae are sometimes released at certain times of the day and we wanted to encompass all those times.

Results were interesting. What we could do, for example, was compare the number of *Hq* cercariae released from snails infected alone by that species with the number released by snails infected with that species plus different combinations of other species. In the case of *Hq*, there were four multiple-species combinations for which production of cercariae could be compared with singly infected snails. This was typical. For *Ls* there were five multiple-species combinations beyond those singly infected with *Ls*; for *Zr* there were seven; for *Austrobilharzia variglandis (Av)* there were four; and for *Ga* there were eight. So numbers of cercariae released could always be compared across several different multiple infections. First of all, snails would often release 100's or 1000's of cercariae in 24 hours. There was a lot of variation in numbers of cercariae released (sometimes only a few, sometimes very many), even from snails with only a single species infecting them. The number of cercariae released by parasite infrapopulations was often lower when other species were present, but not significantly so. The upshot was that we could statistically detect no consistent changes in cercarial output, up or down, for any of the above listed five trematodes, regardless of what other species co-occurred with them.

Single and multiple-species infections are more common in this snail than in most others. More data may lead to dethronement, but I think of *Ilyanassa* as the most infected snail on the planet. However, in sum thus far our tests indicate that larval trematodes infecting *Ilyanassa* do not displace one another in a snail's tissues and do not shut off, or even significantly reduce, cercarial output by co-infecting species. Of the several possible ways of getting along listed above (competition, predation, facilitation) these species seem to just ignore one another's presence; they seem unaware of each other. To say the least, when results came out this way we were taken aback. At the time, people had mostly investigated the ways in which species of adult trematodes interacted in definitive hosts and not many had looked into how larval trematodes interacted in snails. In cases where such an investigation had been done some intense interactions had been found. The cases of most import were those in which snails had been experimentally infected in the laboratory both singly and in combination with a pair of parasites. Results of the two species in the same snail could then be examined and compared with single infections. With only one small exception, laboratory infections of *Ilyanassa* have never been accomplished with any of its trematodes. Anyway, for other trematodes in other snails, cases of rediae apparently preying upon the sporocysts of another species had been noted. (I say apparently because it is not clear to me that the rediae were recognizing the sporocysts as prey, rather than just "chewing" their way around indiscriminately.) We thought evidence of this sort of thing would be discovered among trematodes in *Ilyanassa* as well. Well, it wasn't and we were left trying to explain how/why it might come out that way.

I outline below the four possible adaptive relationships among the species involved in a snail-trematode system that we included in our paper. Recall that for a species to be adapted to any aspect of its environment (here another species) that species would have to have undergone natural selection to deal with that aspect across previous generations. As an example, we might be talking about species A causing natural selection in species B. This is might be expressed as species A is acting as a selection pressure on species B. The end result would be that across its generations species B would become adapted to deal with species A. The four possible adaptive relationships in this system are as follows. (1) Parasite species are adapted to deal with the host. This is one possible adaptive relationship in the system and it surely must be in effect here because all these parasites have to deal with an *Ilyanassa* host in every iteration of their life cycles. Any genetic strain of parasite that could not do this successfully would quickly die off, leaving behind only those strains that could deal with the snail host. An ability to successfully use a host would be promoted in parasite populations by natural selection. So this is the minimal, required adaptive relationship in the system. Parasites are adapted to use their hosts. (2) The adaptive relationship parasite-to-host might not be the only one involved. The host might also be adapted to deal with the parasites. (3) It might be that parasites are adapted to the host and parasites are also adapted to deal with other parasite species. (4) Perhaps parasites are adapted to the host, the host to the parasites, and the parasites to each other. Everybody is mutually adapted. We used our knowledge of trematodes in *Ilyanassa* to support or refute these four models. As indicated the first model must be accepted: parasites must be adapted to use a host necessary in all their life cycles. Can the snail host be adapted to deal with these parasites? We concluded no and here's our reasoning. In its life cycle the snail has a planktonic larva that drifts at the mercy of water currents for two or three weeks. Let's suppose that a genetic change, a mutation, occurred in a reproductive (not parasitized) snail on Cape Henlopen that conferred the ability to resist infection by one or more trematode species. That snail would have an advantage on Cape Henlopen because it could resist infections with one or more species of trematode and the consequent sterilization. So that snail reproduces, while other snails not so "gifted" likely get infected and do not reproduce. Some fertilized eggs from that snail would likely possess that "gift" (the genetic change, or mutation). Those eggs would develop into veliger larvae that would spend weeks drifting around in the plankton. If any of these "gifted" larvae survive to settle and become juvenile snails there is no predicting into which *Ilyanassa* population they might settle. Many snail populations exist where there are few or no parasites to infect snails. Places like Cape Henlopen are more the exception than the rule. What I am saying is that there is a very small chance that offspring of this "gifted" snail will survive and settle in a snail population where parasites are prevalent. Thus, it is not likely to gain advantage from its genetic makeup over other snails in the habitat where it settles. There would likely be no or few parasites there to challenge it, i.e., no selection pressure to maintain the advantage. There are often metabolic maintenance costs for new mutations and this mutation may well be eliminated if there is usually no advantage to having it. In any case, the genes of the "gifted" offspring would be swamped by genes from snails not having (or needing) the ability to resist infections. Without a consistent selection pressure of parasitism generation after generation in snail populations, it is very difficult to imagine how snails could become adapted to deal specifically with trematodes. There's just too much opportunity for the snail host to exist free of parasite influences.

In principle this hypothesis could be tested in the field, but to do so would require Herculean effort. It would certainly be beyond my resources and indeed beyond the resources that are likely to be supplied from any source to test an idea of this sort. So, can we provide specific evidence that this is the way things are? No, but based on what is known about how snails reproduce and the distribution of these trematodes among snail populations, it is hard to see how *Ilyanassa* could be adapted to these parasites. The requirements of natural selection are not all there, especially consistent selection pressure, and the process of natural selection can't work to adapt this snail to deal with these parasites. If this is true, and we believe it is, then the second and fourth models outlined above must be eliminated from consideration because in the case of this system the host cannot be adapted to the parasites.

That leaves us with the third model to consider. What is required there is that parasite species be adapted to deal with one another. Is this possible? Again, we conclude no. First of all, these trematodes are spread among *Ilyanassa* populations by the travels of definitive hosts. Suppose an uninfected definitive host (a bird or fish) picks up an infection, say on Cape Henlopen, by eating an infected second intermediate host. Further, let's suppose that infection carries a "gift," a genetic change (a mutation), that aids in dealing with another parasite. "Gifted" adult worms develop in that definitive host. These adults produce fertilized eggs, some of which in turn carry the genetic "gift." These eggs get into the definitive host's digestive tract. Wherever that host defecates is where those "gifted" fertilized parasite eggs are released into the external environment. If they happen to land in a population of snails, "gifted" larvae from those eggs (called miracidia) develop and possibly infect snails. The infrapopulations that arise in snails from these "gifted" miracidia would be at an advantage because they would be able to combat that other infecting species. That other infecting species would not be able to infect the same snail as the genetic strain of the infecting species possessing the "gift." Such double infections would be blocked and this might be a real advantage for infrapopulations that develop from "gifted" miracidia. But, this process of spreading parasite eggs among snail populations via definitive hosts is comparable to the spreading of planktonic snail veliger larvae among populations. The problem is: infrapopulations of that other species may be too rare in the global *Ilyanassa* population. One of those may well be never again be met up with in a snail by the "gifted" strain of infection. The likelihood is that there would be no consistent selection pressure generation after generation and the mutation would be mostly useless and lost. The key issue here is the rarity of multiple infections. True, I have said they are more frequently found in *Ilyanassa* individuals than in most other snails, but still they are pretty rare, especially when you think of a specific pair of species. In our paper we used the following example. The most frequent species combination found among 4,870 *Ilyanassa* examined from Cape Henlopen, a hotbed for parasitism, was *Ls* co-infecting snails with *Ga*. This pair of parasites co-occurred in only 4% of snails examined. There's not much chance even these two species will meet up in a single snail. Other species combinations were less frequent, most of them occurring in minuscule proportions of snails. What all this indicates is that chances of infrapopulations of the two parasite species encountering one another within the same snails generation after generation are vanishingly small. Natural selection, operating to cause one species to become adapted to another is, to say the least, very unlikely. That is, there's no way selection pressure from parasite species A on parasite species B would be consistent enough generation after generation to produce an adaptation in species B to deal with species A. A genetic strain of species B might by mutation develop some way to deal with species A, but odds are it would never encounter species A again in an individual snail. It is hard to imagine how these trematodes could be naturally selected (adapted) to deal with each other. That is why we concluded no on this possibility.

The perceptive reader might say that maybe they meet often, as infecting larvae (miracidia), but one is adapted to quickly eliminate the other and the pair are never or seldom seen co-existing in the same snail. Just below there is some evidence presented that counters this possibility. Further, in future chapters evidence will be presented indicating that miracidia infrequently get to snails.

In the final analysis we were, and are, in the same situation with regard to this hypothesis as we were with the hypothesis that the host is not adapted to deal with the parasites. It would be very difficult to gather evidence in the field to test it. Nevertheless, it is hard to see how one parasite could be adapted to another. If the third model above is not operative, we are left with the first model, the parasites are adapted to use the snail as a host and that is all. This is a key point in our thinking about how this large snail-trematode system works. Whatever adaptive effects these parasites have in this system it is because of how they are adapted to use the host. We can't assume that members of the snail population are adapted to resist trematode parasitism. Natural selection couldn't have worked to produce that. Further, we can't assume that one species of trematode is adapted to "fight off" another to take over a snail host. If two parasites can or can't coexist in a single snail, it has to do with how one or both are adapted to use the host, not fundamentally with any parasite adaptation to deal with the other parasite. We are not saying this applies to all snail-trematode systems. They all have their own peculiarities. We are suggesting that this is the way we should think about *Ilyanassa* and the larval trematodes that infect it along the many miles of the eastern seaboard of North America. Ideas to the contrary about interactions of species in the system should not be assumed true without evidence to support them. It is always possible that we are wrong. That's the way it is when you go beyond the observable data in science, but therein lies the adventure. In any case, the process of natural selection requires certain conditions be in place for it to work and they don't seem to be there to produce adaptative relationships of host-to-parasite, or parasite-to-parasite in this system.

I suggested just above that there is evidence that parasite larvae (miracidia) of different species don't meet up very often in the process of infecting snails. There will be more on this later, but at this time evidence to this effect was limited. To discuss the issue we will need to introduce a couple of extensions of terms used previously. Any assemblage of more than one species can be referred to as a community. People who study assemblages of species are called community ecologists. When the infrapopulations of two or more species of trematodes exist in a single snail, that assemblage is called a parasite infracommunity. (Even when only one infrapopulation is present in a snail, to maintain the usage, that is sometimes spoken of as a (species-poor) infracommunity, even though only one species is present.) If a parasite component population is all the infrapopulations of a parasite species in a population of hosts, then all the infrapopulations of all parasite species in that host population becomes a component community. These added terms are necessary because there is the possibility that what goes on in infracommunities (e.g., such negative species interactions as competition or predation) will have an effect on what is observed at the component community level. For example, if species of redial infrapopulations prey on sporocyst infrapopulations often enough in infracommunities, that will affect component community composition. Put another way, if rediae prey on sporocysts often enough in infracommunities, redial and sporocyst infrapopulations would seldom or never co-occur in single snails in the component community. This is the kind of thing we were expecting to see with *Ilyanassa* trematodes.

If we suppose that different species of trematode larvae (miracidia) make frequent attempts to infect snails, encounter one another often in the process, and "fight it out" (compete) for possession of snail hosts, a definite phenomenon should show up in the component community. Specifically, when you examine young snails there should be mostly single infections because miracidia of all sorts initially colonize snails. When you collect and examine somewhat older snails there should be lots of multi-species infections. The number of multiple infections in them should be higher because parasite species have assembled in snails and are in the process of "fighting it out" for possession of snails. When yet older snails are collected, the hypothesis goes, there should be a reduction in multiple infections because "battles" have been settled and evictions of certain species have occurred.

Was anything like this apparent in the component community of trematodes in *Ilyanassa* on Cape Henlopen? We knew something about the ages of snails of different sizes. It turned out later that we were underestimating the age of snails, but this doesn't make the thrust of our analysis wrong. We looked at the parasite component community in two time frames. One was over the summer of 1989.

Another was over a period of years by looking at different size-classes of snails. Over the summer, was there indication that single infections were more abundant in snails early in summer, multiple infections more common in midsummer, and then in late summer snails had mostly single infections again? There was no such trend. As one example, we observed a greater proportion of triple-species infections in snails collected in late summer. This is counter to what would be expected if different species colonized snails, "fought it out," and only the winners remained in snails at the end of summer. What about in the longer time frame of years? Again, there was no such trend. Among larger (older) snails there was a larger proportion of triple infection. In either time frame this is not what should have been observed in the component community if what was postulated above to be going on was in fact going on. The indication was that infecting trematode miracidia of different species were not frequently getting to snails and "fighting it out" for possession of those snails.

Many of these larval trematode species seemed quite able to coexist in individual *Ilyanassa*. Some coexisting pairs were rediae-sporocyst combinations and many were sporocyst-sporocyst combinations. There was one particular pair that serves to illustrate our interpretation of what was going on. That pair was *Hq* and *Ls*. A paper in *Science* in 1969 by W. B. Vernberg, F. J. Vernberg, and F. W. Beckerdite had identified these two species as not co-occurring in the same snail in their sample of 5,025 *Ilyanassa* from Beaufort, North Carolina. Indeed, in my previous samples from Delaware I had never seen them together in a snail. Therefore we were surprised, and very careful about identifications, when *Hq-Ls* (*HL*) double infections showed up in the more detailed examination of snails in this work. Both species' infrapopulations consist of rediae and cercariae, but even a novice observer would not confuse these two larval trematodes. Rediae and cercariae of the two species are markedly different. I remember Karen, by this time equal to me at identifying these trematodes, asking that I confirm what she had found when she ran into these double infections. In all, *HL* infections showed up in four snails. In all four it was observed that there were many *Ls* rediae and cercariae, but only a few of *Hq*. The *Hq* rediae appeared sluggish or moribund. In the one snail that was subjected to 24 hours of cercarial release before dissection only 18 *Hq* cercariae were shed, while 567 *Ls* cercariae were. From all appearances *Ls* was doing better in co-infected snails than *Hq*. This left us trying to explain why. What's the basis for it? Many investigators of this sort of situation would conclude that these trematodes were competing for the snail and *Ls* was winning. (Incidentally, that would be at variance from what would be expected because the family to which *Hq* belongs is usually considered to be made up of superior competitors.) According to our conclusions these and other trematode species cannot be adapted to deal with each other. We thought that adaptive relationships should be most prized because they would produce the most consistent interspecific interaction. If they are not adaptively based, they could easily be some sort of accidental relationship between species. The apparent conflict between these two species must be an accident, a consequence of the way one or both happen to be adapted to use the snail host. Species compete with one another for some resource in short supply. If anything like this is going on it is indirect and operates through the way these two parasites are adapted to use the host, not the way they are adapted to co-exist or deal with each other. This, it struck us, is an important thing to understand.

An angle on this work resulted in another paper (Curtis and Hubbard, 1990). Many workers with snail-trematode systems depend on released cercariae from snails to judge whether a snail is infected or not and with what species. This is especially true of freshwater snails and in fact the method is more reliable there. As mentioned above 206 snails were put in cercarial release chambers for 24 hours. Released cercariae were identified and counted and later each snail was examined for infections. This gave us a chance to answer the question: if we had depended solely on shed cercariae, how often would we have been wrong about what species of trematode(s) infected a snail? Results were instructive. Overall, if we had depended solely on shed cercariae we would have determined parasitism in snails wrongly in 59% of cases. That's bad enough, but where snails were doubly or triply infected we would have been wrong in 80 and 93% of cases, respectively. In future work I would have to rely on shed cercariae to assess state of infection of snails, but I had to do it judiciously because of this analysis. If a snail sheds a species of cercariae it is definitely infected with that species, but failure to release a species of cercariae does not guarantee that the snail is free of that parasite. The only way to be absolutely sure is to dissect snails. If you need the snail alive for an experiment, obviously dissection is not an option.

From the winter of 1988-1989 through summer of 1993, despite everything else that was going on with her, Karen was nearly as involved as I with research on *Ilyanassa* and its trematodes. She showed a lot of grit and was a very valuable colleague. I don't say assistant because she was more than that. She, as we all do, needed rounding out in matters of biological science, but once facts were in hand her thoughts were penetrating. She had a way of cutting through the details to the heart of a matter. In 1991 we signed on to give a paper on our work at a scientific meeting. Karen gave the talk. My job was to organize and prepare the visual aids for it. They were inspired if I do say so myself. There were some pretty complicated things to present in a 15 minute talk, such as three-way contingency analyses and log-linear models, to reach what was not a popular conclusion. All of the student-presented papers were given the day before when Karen couldn't be there. She had to work. Karen gave our paper the next day. I was more nervous than she was, which from all appearances was not at all. I think it is fair to say she blew the gathered audience away. There were several questions following the talk and after its time was over and the meeting ended a number of people came up to ask questions and make comments. A couple of people were actually quite agitated that she had limited adaptation in the system to the parasites being adapted to the host. To the degree that people had thought about it they thought that our fourth model, everybody mutually adapted, should be the way things were. It happens, but it's not often you can generate that much reaction at a scientific meeting. Karen handled herself well and dealt with compliments, questions, and agitation with cool and accuracy. It can't be known, but I was there for the other student papers and I think she would have won the best student paper award had she given the paper the day before. As her professor I did have a role in the success, but the credit mostly goes to Karen. When students get to the point where they can do something like this before an audience containing unknown Ph.D. and other accomplished biologists it demonstrates a sense of confidence that can't really be obtained any other way. It went into the future with her.

Karen had yet to finish her undergraduate degree. A challenge yet to come was the official writing of her Undergraduate Thesis. Writing a thesis is always difficult. There is, of course, the subject matter to get across correctly. That's no small deal. On top of that there are numerous University deadlines and rules and regulations to follow. Not least there is the defense to go through, essentially like one a Masters or Ph.D. student would have to go through. You have a committee of faculty members, four in this case, who have read your thesis. After giving a talk summarizing your work, you have to stand before them and answer questions for 45 minutes or an hour. If you have screwed up writing the thesis, or badly enough during the defense, your thesis may not be accepted. It's a tense time. Karen is pretty feisty and was confident. I had faith in her, but as usual was nervous. As I might have expected she did very well and her Thesis was accepted.

Karen was tied to southern Delaware and not able to go to the main campus to finish her course work for her degree there. Here was another intellectual challenge set out for her. She had to convince the University administration that according to their rules there was a different way to finish a degree at the University. She had exhausted the available University course work in the unit where I taught. Another nearby university was used to meet her degree requirements. As usual she got A's in every course. In fact, during her college work, which was very extensive considering where she started, she never got anything but A's. In 1993 Karen left the University of Delaware at the very top of her graduating class of 1,000's of students, Summa Cum Laude, with a Degree with Distinction (because of the Thesis) in Biological Science. How to go Karen. As for me, I was lucky to have been a part of it.

100's of snail cups to be searched for parasite stages in 1991 to weed out infected snails

Chapter 9

Old Timer

During the time that Karen was doing her work I became interested in some other questions. One of them was: how long does an *Ilyanassa* individual stand to live? The answer to this question fundamentally changed my thinking about the ecology of this snail. Think back to Chapter 4. There I made the point that it takes three summers for a snail to grow into the largest size-class, which contains snails 18 mm and up in shell height. That "and up" is important. I have found snails on Cape Henlopen that measured as much as 30 mm. They are very rare, but there they are. Snails up to about 25 mm can regularly be found. It must have taken more than three years to attain that size, but how much more?

When you attempt to answer a question the first thing to do is always to find out what is already known. A fairly comprehensive study on the ages attained by various animals was in a book written by A. Comfort and published in 1964. It was noted therein that snails could be up to 15 years old and three years were allotted to *Ilyanassa*. By the early 1990's a few snail species had been found that lived for as much as 30 years. When Larry Hurd and I finished our work on snail growth in the early 1980's we thought the snail must live to be 4-5 years old, maybe more. A study by some people I have referred to previously, C. Jenner and M. Jenner, concluded that *Ilyanassa* could live to be seven years old. So, in 1991 when I began this work I was aware that snails in general might live to be some 30 years old and the maximum estimate was seven years for "my" snail. That's what Karen and I thought and based some of our reasoning on in our 1993 paper.

It's all well and good to ask a question, but how is it going to be answered? The problem is this. The age of snails during the first three years of life can be determined by means of size-frequency analysis. In Chapter 4 I described how this method could be used to follow a cohort of snails on the sandflat from the year they settled out of the plankton as juveniles until they grow to about 18 mm and merge into the largest size group. As noted, that took three years. But, and it's a big but, how can age be measured in snails in that largest group? That group would be made up of snails three years old and older. How much older is possible and how could that be found out? Some species of snails have what are called "growth checks" on their shells or elsewhere on their bodies (sort of like annual rings on a tree trunk), but no such thing was, or is, known to exist on *Ilyanassa*. So that is not a promising avenue. Another approach would be to measure the growth rate, the rate of shell height increase, in snails in the largest size group. To do this would require tracking the growth of snails in the laboratory and/or in the field. My predilection would be in the field because it is there that snails normally live, feed, reproduce, and do all the other things associated with being a snail. In the laboratory it is impossible to feed snails a natural diet for a very long time and moreover they are couch (aquarium) potatoes. The laboratory environment is likely to affect growth in unnatural ways. So the field was chosen. The Cape Henlopen sandflat was handy and a logical place to do the work. A snail loose on the sandflat would gradually add to its shell height. If I could measure a snail's size before release onto the sandflat (no problem) and then recapture that snail at some later time (big problem) and re-measure its size, by dividing shell height change (ending size minus starting size) by the amount of time between measurements, I would have a growth rate for that snail. One problem is, when I later find "that" snail how can I be sure of its identity so I would know which initial size is to be subtracted from which ending size? Another is, these snails would not be infected with *Ga* and therefore not "glued" to a particular sandbar. They would be free to roam where they would on the sandflat. How far do they roam anyway and how the hell am I going to track them down? In 1985 I had some experience with marking snails with individually numbered "bee" tags, but those experiments ran at most for about a week and for appreciable growth snails would have to be free on the sandflat for a much longer time. Anyway, in principle individual marks would handle the snail identification problem. I could be sure of making the correct subtraction and getting an actual growth rate. The other problem involved a real crap-shoot. I had no idea of how far snails wandered on the sandflat. Released snails could maybe be seen around the release point for a few days, but then utterly disappear into the huge snail population on the expansive Cape Henlopen sandflat. For one thing, and as a hedge on this possibility, I had to release a lot of individually marked snails. In terms of effort marking snails, releasing them, and later finding them this was a huge gamble that simply had to be taken. Gloomy, dismal failure of this enterprise was very possible.

This was the general setup. In June of 1991 snails were marked and released onto the Cape Henlopen sandflat. Because trematode infection could affect growth, I wanted to release uninfected snails to measure growth rates. There was a secondary reason. Snails were to be released in the high-parasite zone of Cape Henlopen (Chapter 4) and I also wanted to keep track of the colonization of released snails by trematodes. How quickly does that happen? Many native snails in the chosen release area (about 80%) were already infected and I expected the uninfected snails I released to be quite literally assaulted by parasites. There will be more about that later. Cape Henlopen is not a good place to collect uninfected snails because many are already parasitized. Back in the days of crystalline style studies, now more than a decade ago, we had examined many snails by dissection from a saltmarsh ditch in Canary Creek (Chapter 3). Those snails were seldom infected. In the first half of June 1991 snails were collected from that ditch as a source of potentially uninfected snails. I wanted to be surer that these snails were actually uninfected than would come with merely collecting them from a site where in the past snails tended to be uninfected. I employed what I call cercarial release tests to achieve this. In the laboratory these marsh snails were placed individually in plastic cups of filtered seawater and confined there for a minimum of 24 hours. This procedure took up essentially the whole laboratory. There was a three by seven foot table in the middle of the room that on three occasions was covered with hundreds of cups containing water and a snail. I have a photograph and there is the frontispiece for this chapter. Snails would often make escape attempts. We confined each one in its cup by laying porch screening over the cups. After the 24 hours was up the water in all the cups had to be microscopically examined for the presence of cercariae. The tedious hours of work were absolutely thrilling. If cercariae were found, the snail in that cup was known to be infected and it was excluded from the experiment, i.e., not marked and released. It was later returned to the marsh. If no cercariae were found, the snail from the cup was considered putatively uninfected. It would not be correct to call it certainly uninfected because infected snails sometimes do not release cercariae, as Karen and I had recently found out. All putatively uninfected snails went into a pool of snails that were ultimately marked, measured for initial shell height, and released onto Cape Henlopen. By 22 June 1991 a full 1,200 of them, yes 1,200, had been released. Needless to say they represented a lot of laboratory work. Thanks go to Karen for being an integral part of laboratory and field work throughout this whole project. I was lucky to have such a quality colleague in the effort. About 1,200-1,300 meters alongshore northeast of the Cape Henlopen Fishing Pier and about 10 meters offshore from the base of the beach face there was a metal stake driven into the sand that had been there for years. In 1991 the stake was on the northeastern watery edge of a large tide pool right next to the beach. There was a prominent gully at low tide that drained this tide pool in the offshore direction. We chose that stake as our release point because it would likely be a stable reference point. If snails were found, their positions on the sandflat could be determined relative to that stake. That way it would be possible to gain some knowledge of how quickly and how far individual snails roam. That is, if any could be found after release. We could only hope.

Once I had my putatively uninfected snails amassed the next job was to get them marked and measured. Marking 1,200 snails individually is not something easily done. Let me describe the procedure. Marks had to not hinder snails in any way and be put on securely, carefully, and without damaging the snail. After all, to accomplish the purpose, a mark had to remain on a marked snail in the field for months or longer. The bee tags used were small plastic disks (2.5 mm diameter by about half a mm thick). They weighed almost nothing and would not burden a snail. As mentioned in Chapter 6 tags come in sets of 500. Each set has tags of five colors, 100 of each. Tags of each color are numbered from 1 to 99 with a blank included for 100. When it comes down to applying a tag to a snail the snail first must be prepared. When they come from the field (or an aquarium) they are wet and typically have a layer of biological film on the shell, living algae with associated mud. There is a wonderful German word, "aufwuchs," for a biological film growing on the surface of an organism. Anyway, a tag would not stay stuck on such a surface. So each snail's shell had to be dried and a patch on it cleaned up to receive the tag. Drying snails was relatively easy, but scraping the aufwuchs off a patch was difficult. I went out and bought a Dremel Mototool and a bunch of wire brushes for it. I had an old Black & Decker Work Mate around that I pressed into service. It had a vice capability on it that could hold the Mototool securely. When the Mototool was turned on you would have a small, rapidly turning wire brush in front of you. You could then put on some eye protection, grab a snail with your fingers, and use the brush to clean a patch on the upper shell surface where the tag would ultimately go. It is surely unlikely that a snail ever experienced such treatment. We took to calling the buffing apparatus the "orgasmitron." The trick was to clean the shell without gouging into it too much. If the shell was weakened, the snail would not likely survive the rigors of its natural environment, which would defeat the purpose. Quite often the snail would be wrenched out of your fingers by the wire brush. The snail would go flying across the room and likely be lost to dry up and die somewhere in all the paraphernalia resident in the laboratory. To combat this problem we erected a large piece of cloth like a sail to intercept a snail's trajectory and provided a cushion for it to fall down upon. From there it could be retrieved. After the shell was dried and buffed the tag could be affixed. The snail was placed in a spring clothespin so it could be stood up in a small vice and worked on. A small dab of clear fingernail varnish was placed in the cleaned patch. Then using small forceps one of those pesky, tiny, easy to lose, and hard to control tags was placed in the dab of varnish on the shell and oriented so the base of the number was toward the shell apex. Orientation was important. Later on we wouldn't want to mistake, for example, a six for a nine. After the tag was oriented another coat of varnish was put over the tag's surface. The tagged snail in its clothespin was put aside to dry a bit. At this stage tag edges tended to curl up away from the shell surface. It was necessary to inspect each tag, poke it back down if necessary, and perhaps put on another coat of varnish. After all this a snail would be left in its clothespin for several hours to dry thoroughly. When all snails in a batch had been marked each one had to be carefully measured for shell height. Each snail's identifying number (e.g., Red 56, or Yellow 25) was recorded in a computer spreadsheet (no paper anymore) along with where and when it was collected, its initial shell height, and any special features (e.g. a notch in the shell, or a less than perfect orientation of the number). Karen and I collected the first snails from the marsh on 4 June and they were tested to remove infected individuals, marked, measured, and 500 were set free at the release stake on 9 June. Would they ever be seen again? On 12 June more snails were collected from the marsh. I got 500 putatively uninfected snails together and they were processed and freed by 18 June. Snails in this batch were distinguished from the first 500 by having the pointy end of their shells (apexes, "tails") painted with red fingernail varnish. So, for instance, Red 56 in this batch became Red Red 56. The third batch of 200 snails had their tails painted with white varnish (giving a White Red 56) and they were released on 22 June. The ultimate product was that as of that date we had 1,200 snails, almost surely uninfected and individually identifiable, with shell heights 18 mm or greater loose on the Cape Henlopen sandflat. Now, in the future, the job would be to find them, re-measure them, and see how much growth had taken place.

Between 22 June and 26 October in 1991 I made 33 trips out to the release stake area to search for marked snails. Karen helped on most of these occasions. Two sets of eyes are much better than one. By this time Cape Henlopen was getting more crowded. Park rangers had motorized three-wheelers and were more likely to travel to remote beach areas such as the release area and check up on what you were doing. Dogs were allowed, but had to be on leash. My dog Salt was a constant companion. Most of the time there was no one else around at the site and I confess to leaving her loose much of the time. She was older now and pretty well behaved. She got to be very good at appearing to be tied up to a piece of driftwood on the beach. Anyway, we got checked out by rangers a number of times, but never got arrested.

Typically I (mostly we) would get to the release area about two hours before low tide. That would be about the time that the water was receding off the peaks of sandbars. We would have time to get set up. On note cards or a tape recorder, air and water temperature and water salinity were noted as well as comments about wind and weather conditions. At some point photographs of the area might be taken. Searching was begun and would go on until maybe two hours after low tide. A search was tiresome and a bit of an ordeal. Our puny 1,200 snails were swallowed up in the native population. A searcher would walk about widely in the release area with no particular pattern in mind looking for marked snails. They were not easy to spot. A marked snail had to be unburied and positioned so that the mark or tail paint was visible. The water couldn't be more than a few inches deep and couldn't be cloudy with suspended sediment. When a mark was found the snail was placed in a clothespin. We carried a quiver full of stiff wire stakes, about 18 inches long, with us as we searched. In its clothespin a found snail would be hanged off a stake, which was stuck in the sand where the snail was discovered. If we were lucky enough to find some number of snails we could then go around to each wire stake and read the snail's mark number, measure its distance from the release stake with a tape measure or optical rangefinder, determine its direction (degrees from north) from the release stake with a lensatic compass, re-measure the snail with calipers, and carefully record all these data. If the snail was not to be recollected, it would be re-released to wander, grow on, and maybe be found again. The stake and clothespin were retrieved and recycled for the next find. It was a bit of a surprise, but we had pretty good luck finding marked snails. In 1991 there were 998 sightings of marked snails, one a full 180 meters from the release point. That was certainly a fortuitous find.

Doing the marking and the searching and keeping all the data straight kept me busy through the summer and fall of 1991. By this time I was having trouble standing in one place without constantly moving my feet to stay in balance. My balance was all right. That is, I knew it when I was losing it. But, the nerves and muscles controlling my lower legs, ankles, and feet were not able to make the small adjustments required that allow you to just stand still in one place. This problem aside my feet were O.K.. At least all former wounds were healed up. I had to use a staff as a prop to keep me steady when doing the field work. That was a new thing for me and it was a real pain in the ass, but once I sort of got the problem under control (by using a prop) this project was one of the most enjoyable I had undertaken. I got down to a routine the getting of people, dogs. and equipment ready to go out to the site. It was a mile or so walk out there from the laboratory and it took something like half an hour to do it. It was a commitment. A constant challenge was gaging tides and weather. Tides could expose a lot of sandflat or very little depending on phases of the moon and wind strength and direction. A fairly strong northwest or west wind would drive water out of the Bay and make the tide go lower, exposing more sandflat to search. Sometimes the tide would not recede sufficiently and a search trip would be a bust. Even though not doing it easily, the walk was invariably pleasant for me. There was always a risk of thunder and lightning storms developing during the three or four hours that we were on site and searching. If that were to happen, and on occasion it did, there was no shelter anywhere even remotely handy and a long walk back to be made in the rain and lightning. In such an environment you (especially me) are the tallest thing around and that made the lightning a scary proposition. While searching you had to keep an eye to the west looking for dark, billowing thunder head clouds and keep your ears open for distant thunder. Many times clouds or thunder would be seen or heard, but would pass north or south of the area. You didn't want to bail out on the enterprise too early, or too late. So you had constantly to make decisions, decisions with consequences. Which are: you either get to continue raking in data, or you have a wet, nervous walk back. Of course, most of the time the tide was at least reasonably low and the weather not threatening. The sandflat can be inspirational. There is something about wandering around on it, with a little wind to keep biting insects at bay, eyes expectantly down, as if held there by a martingale, searching with the sun preferably at your back so you can easily pick out a marked snail should there be one among the myriad of snails in front of you, seagulls chattering and carrying on about you, maybe a black skimmer flying along fishing with its lower beak in shallow water, and the encompassing, profound solitude. I recall those moments poised between the past and the future with great affection.

Of course I was still teaching and that took up much of my time. I worked my way through the academic year of 1991-1992. During that year I arranged for my first sabbatical since starting as an Assistant Professor in 1975. I should have gotten on that sooner. These are wonderful breaks that typically come along every seventh year of university teaching. Some professors take a whole year at half pay. I chose to take half a year in the fall semester of 1992 at full pay. A university awards sabbaticals to revitalize and rejuvenate faculty interests, to prevent or stave off staleness. It doesn't always work, but that's the idea. Faculty on sabbatical often visit another university for purposes of research and/or teaching and to collaborate with a different set of colleagues. My own goals were to catch up on getting papers written for publication, do some reading, perform a survey of snail imposex on Cape Henlopen (Chapter 7), and to look for snails marked and released in summer 1991. Aside from reading materials associated with two papers I was writing, which came out in 1993, I was afforded time to read a large book entitled “The Growth of Biological Thought, Diversity, Evolution, and Inheritance” by Ernst Mayr, an influential and wise biologist. The author and book were quite famous. The book was full of content put there by a clever man. Reading that book cover-to-cover was informative and indeed a revitalization. It influenced what I taught for years to come and was time well spent.

As far as trying to find last years snails out in the field, I was in for a disappointment. I, with Karen’s help, spent a lot of time searching the 1991 release area in summer and fall of 1992. Further, I expended substantial effort sieving snails out of square meter plots of sediment. I individually looked at nearly 8,000 snails to see if I could find a marked one. None were found with much effort expended looking for them and that surprised me. What had happened? Had the marks of all snails not recollected last year fallen off over the winter? That was certainly a possibility, but some with evidence of tail paint remnants would have been expected. Maybe the snails were killed, or had simply moved away. I can’t be sure, but in all likelihood it was basically the latter. I (we) didn’t much recognize it at the time, but the area had changed rather dramatically since the previous year. It all had to do with moving sandbars. On low tides in 1992 the stake was well out of the tide pool most of the time. In 1991 the pool had often encompassed the stake and the pool itself had changed its configuration. There was still a gully draining the pool to the northeast, but as I later determined it was not the same gully. Years later, after I had a chance to compare photographs taken between 1991 and 1992, what happened could be reconstructed. Once again (Chapter 6) the ecological wisdom that “you can’t put your foot in the same river twice” raises its ubiquitous head. I was not aware at the time, but the whole landscape had changed. It looked similar, but it wasn’t; it was a different system. In 1991 most marks had been found somewhere in the main 1991 gully. The 1992 gully looked about the same, but between 1991 and 1992 a sandbar had swept northeast to southwest across the old 1991 gully and obliterated it. A new gully had formed and replaced it. I could see it all happening in the series of photographs I had taken from 1991 through 1992. Any snails in the 1991 gully would have been forced to move along or be buried. I imagine they moved along and any that still had their marks would be located well to the southwest of the 1991 release area. Basically, I was looking in the wrong place. Sometimes people look under the lamp post because it’s lighted there, not because the sought item was lost there. That’s sort of what I was doing and probably why no marked snails could be found. Any still around would have been elsewhere.

Not one to be daunted by failure, after my sabbatical I went back to teaching in the spring semester of 1993. Along with other responsibilities I started to plan for releasing more snails the following summer. I wanted to release more uninfected snails, but this time I also wanted to release infected snails so their growth and movement on the sandflat could be studied too. So, with Karen's help (she was graduated now, but still hooked) and using the same procedures as in 1991, 200 putatively uninfected saltmarsh snails were individually marked, measured, and transplanted to the Cape Henlopen release stake as of 2 June. That stake was indeed a fixture on the sandflat. Another student, one Chris Morrisey, helped out regularly with the work in this year. It was a new experience for him and he deserves many thanks. On the day we released the 200 marsh snails we collected 680 native Cape Henlopen snails from throughout a pool next to shore adjacent to the release stake. Nearly all were in excess of 17 mm in shell height (at least three years old). From these a random selection of 324 was made. Of these, 300 were tested for infections, cleaned up, marked, measured, and released on 7 or 10 June. All 1993 snails were given red tails because there were no 1991 snails still around with which to confuse them. As expected many Cape Henlopen snails revealed infections in cercarial release tests. *Himasthla quissetensis (Hq)* cercariae were shed by 66, *Lepocreadium setiferoides (Ls)* by 27, and 50 shed *Zoogonus rubellus (Zr)*. Bearing in mind the results Karen and I got when we tested the effectiveness of cercarial release tests, it is likely, and so it ultimately proved out, that many more of these native snails were infected. The positions of sandbars were much different now and they allowed wider and quicker dispersal of marked snails than in 1991. Anyway, between 7 June and 14 November 38 trips were made to look for marked snails. The summer and fall of 1993 were just as inspiring and adventurous as the idyllic summer and fall of 1991. We were getting better at this. Marked snails were sighted a total of 948 times. We did about as well with these 500 as we had done with the 1,200 released in 1991.

With respect to growth in that largest size group of snails, what was learned from these three years of work? It's all officially recorded in Curtis (1995). First, I did not depend on cercarial release tests to conclude the infection status of any snail. In this context that would be asking for errors. Rather, infection status of snails was determined by dissection after they were recollected from the sandflat. Growth was measured as shell height change during the time a snail was free on the sandflat. My analysis of growth was based on snails that had been free on the sandflat for at least 40 days. The reason for this amount of time is that it turned out most snail growth occurred within 40 days of release. Little or no growth seemed to occur after that. This indicates that growth is episodic, occurs in restricted portions of the year. There was an initial issue to be settled. Of the total

1,700 snails released 1,400 originally came from the marsh and 300 were native to the Cape Henlopen sandflat. All of them did their growing while on the sandflat. Did marsh snails grow differently from sandflat snails on the sandflat? To analyze this in a table I delineated the seven groups of recollected snails and compared their growth characteristics. A summary of this table is shown below. Abbreviations are: NI = not infected, Hen = Henlopen, *Hq* = infected with *Himasthla quissentensis*, *Ls* = with *Lepocreadium setiferoides*, *Zr* = with *Zoogonus rubellus*, *HZ* = snails doubly infected with *Hq* and *Zr*. For instance, the individual growths of 130 recovered snails (all with initial shell heights 18 mm more) that came from the marsh in 1991 are subsumed under the first group. Only the average growth of groups of snails are indicated.

Group	1991 MarshNI	1993 MarshNI	1993 HenNI	1993 Hen*Hq*	1993 Hen*Ls*	1993 Hen*Zr*	1993 Hen*HZ*
Number of snails	130	31	11	24	7	7	11
Average growth (mm)	1.3	1.1	1.2	0.4	0.2	0.2	0.1

I analyzed these data in a statistical test (something called the Kruskal-Wallis test) that compared the growths of snails in the seven groups. All groups grew quite slowly, but the key message is that, in terms of growth on the sandflat, it did not matter where a snail came from. What mattered was whether or not a snail was parasitized. A parasitized snail, no matter with which species, grew much less than snails that were not carrying a parasite. I didn't leave it there, but this is an important result. In some snails there is the phenomenon called "gigantism." When they get parasitized with a trematode their growth rate increases and they become giants. The notion is that nutrition no longer has to be channeled to reproduction (the snail is sterilized by the parasite) and it can now be channeled to growth. As a result they get larger than if they were having to "feed" reproduction. Gigantism caused by being parasitized clearly does not happen in *Ilyanassa*. Snails get big because they have lived a long time, not because of being parasitized.

Released snails were not all of the same initial shell height. Uninfected released snails varied from just over 17 mm up to just under 24 mm. The question arises: did initial snail size affect the amount of snail growth? Among the 173 recovered uninfected snails the answer is a guarded yes. Growth data led to a mathematical formula that allowed prediction of how much a snail would be expected to grow based on shell height at release. Initially smaller snails grew a bit more rapidly, while initially larger snails grew a little more slowly. But, there was lots of variation in growth. Uninfected snails grew as much as just over 3 mm and as little as zilch, nada during their up to 130 days of freedom.

With regard to infected snails, the 49 recovered ranged in initial shell height from about 19.5 to 26.5 mm. How much they grew did not depend on their initial shell heights. There was variation in growth (maximum = 0.8 mm, minimum = 0.0), but less than for uninfected snails. The average increase in shell height for parasitized snails was 0.2 mm during time free on the sandflat. They were growing something like five times more slowly than uninfected snails.

I wanted to know how old an uninfected snail might get to be. I won't go into detail about how I did this, but recognizing the large inherent amount of growth variability and the formula predicting the amount of growth expected based on initial size it was possible to make minimal and maximal estimates of snail age as snail size increased. Minimal age assumes that a snail adds shell height as fast as possible each year and maximal age that a snail adds shell height as slowly as possible each year. It is assumed that a snail is three years old at 18 mm. If it grew at the fastest possible rate, by the time it was 20 mm it would be about four years old; at the slowest rate it would be about 10 years old. At 22 mm it would be between five and 28 years old. At 24 mm a snail would be between seven and 40 years old. It is not reasonable to presume that a snail would consistently grow at the fastest or slowest rate every year. In different years it would grow different amounts within the approximate limits that the data indicate. In any case, it seems clear that seven years as a maximum age for *Ilyanassa* woefully misses the mark. A 22 mm uninfected snail is likely to be somewhere between 10 and 20 years old and there are lots of them on Cape Henlopen.

Parasitized snails are another issue. Among them average growth was about 0.2 mm/year and growth rate was not related to initial size. They all grew about the same rate regardless of how big they were initially. Let's consider a 27 mm snail. These are found on the sandflat. In our work on parasites (Chapter 4) Larry Hurd and I had found that 98% of 23 mm snails from the general area where snails were released were parasitized. At least we can assume that it is highly likely that a 23 mm snail on Cape Henlopen will be parasitized. If it grew as fast as possible as an uninfected snail until it reached that size, it would only be five or six years old. At 23 mm, possibly before that, it would in all likelihood be infected. Just using the average growth rate of infected snails, to grow the 4 mm to 27 mm at the rate of 0.2 mm/year (five years to grow 1 mm) would take some 20 more years. Giving a total age of something like 25-26 years. If it remained uninfected and grew as slowly as possible to 23 mm, it would then be about 37 years old. Four millimeters of growth at 0.2 mm/ year would take it to 27 mm and

57 years of age. We are on pretty good grounds saying that a 27 mm parasitized snail would be between about 25 and 57 years of age. As indicated, stating the age of a snail based on its size is not an exact exercise, but this puts brackets around it. Snails can get to be several decades old. That was a major surprise!

But, is it reasonable to believe that a parasite infection in a snail can be so long lived. The general concept afoot was that trematode infections weaken and debilitate host snails. A number of counter indications came to mind. Larry Hurd and I (Chapter 4) got the suggestion from the our size-prevalence curve that once a snail gets infected it stays that way for life. At that time we had no correct idea of how long that life might be, but it appeared that snails didn't lose infections. The work with movement of marked snails on the 1985 sandbar showed that parasitized snails moved around as much as uninfected snails, more in the case of *Gynaecotyla adunca (Ga)* infections. Infected snails may be castrated, which is certainly serious harm, but they aren't impaired as far as moving around in the habitat goes. Besides, two snails survived with infections from one year to the next. Also, I have previously mentioned Miriam Rothschild. In 1940 she and a colleague (E. Meyerhof) reported in *Nature* on a trematode-infected marine snail that they kept in the laboratory for five years, all the while infected and putting out cercariae, in total millions of them. Isolated from its natural environment, it wouldn't have lost one infection and gained another. After all, where would new infecting miracidia come from? A final thing, if snails in certain areas of the sandflat (such as the release area) almost never get beyond 23 mm without taking on a parasite and snails routinely get to 25 mm or more, that must mean that snails are living in the infected condition for 10 years or more. The possibility that this might be the result of a series of infections, rather than a single, persistent infecting infrapopulation, would require that snails lose infections (indications are otherwise) and are routinely and often colonized by new infections. This turns out not to be the case, as I will describe later. For now, it is clear that snails can live a long time and, if infected, apparently their parasites live right along with them. All of a sudden we have to think of at least infected snails, and probably uninfected ones as well, living for something like four decades.

As I said above, this changed my whole way of thinking about the ecology of *Ilyanassa* and its trematode parasites. Most of the snails I had read about in the parasitology literature were from freshwater and had life spans of only a year or two. When host snails die any parasites they have picked up must die with them. Parasites must re-establish themselves in snail populations every year or two. I knew that marine snails often had longer lives, maybe a decade or two, but I was not ready for 40 years. There is apparently a strategy at work here. Karen and I had drawn the conclusion (Chapter 8) that the parasites are adapted to deal with the snail host and that is all. If a host stands to live long, it stands to reason that its parasite would also be adapted to live long, especially since chances of it making its way through the life cycle to another snail are slim. In effect parasites are "saying:" "If I manage to successfully acquire a snail host, I am not going to harm, debilitate, or kill it. Rather, I'll hang onto a snail host, and farm it for resources as long as I possibly can because my progeny are not likely to get to another." One thing this means is that when I collect a batch of snails at one time and do dissections to determine the prevalence of trematode infections among them, as I have done many times (Chapters 4, 5, 6, and 8), what I am looking at is a snapshot of the infections that have been accumulated in snails in the population over a very long time. I think of this as my "Charles Darwin moment." In the 1830's, when Darwin thought the earth was only a few thousand years old he was puzzled how evolution could have produced all the adaptations and diversity of species he observed around him. Then geological science made the case that the earth is actually millions upon millions of years old. [The current estimate is 4.5 billion years (that's 4.5 thousand million years) for earth's age and living things have been around on it for something like 3.5 billion years.] Evolution thereby was given an extremely long time to work and all of a sudden it all made sense to him. Insofar as I had a similar experience with the age this snail and its parasites, I take the liberty of aligning my experience with Darwin's. The system of snails and parasites that I had been observing was the product of multiple decades rather than just a few years. An infection observed today might have arisen 10, 20, or even 30 or more years ago. There's a lot of history involved and my whole image of the system was changed. For one major thing, I no longer felt comfortable collecting and sacrificing by dissection large numbers of snails. I had effectively done myself out of a long-standing research technique. When I thought that snails only stood to live at most maybe a few years I wasn't much bothered by this. Individual snails were fleeting. Although this too may have its problems, it was sort of like killing ants, fleas or some other abundant, short-lived organism. Now, when I knew *Ilyanassa* individuals to live maybe 40 years or more, it becomes a different matter. When especially a large, probably parasitized, snail is killed there is a lot of history eliminated from the earth. That individual is not easily replaced. I would have to change my approach.

There is a related piece of work, put forth in Curtis (1996), to which I have already briefly referred, that goes along with the work described so far in this chapter. It is a short paper, a so-called note, but absolutely pregnant with implications. I have had occasion to mention miracidia previously in Chapter 8 and in this one. A miracidium (miracidium is singular, miracidia is plural) is the trematode larval stage that infects a snail. Miracidia arise from fertilized eggs produced by the adult trematodes that parasitize definitive hosts in the life cycle. Adults live in the digestive tract of a vertebrate, a fish, amphibian, reptile, bird, or a mammal. There adult worms copulate and produce fertilized eggs, which are packaged in capsules. Capsules exit with the host's fecal material. If capsules are deposited into an environment where snails are present (e.g. onto Cape Henlopen), one of two things might potentially happen. A snail might come along and eat the capsule, the miracidium in the capsule hatches within the snail, and the snail becomes infected. Or, the capsule might sit there for a while in the habitat and the fertilized egg in it eventually develops into a swimming miracidium that hatches from the capsule and seeks a snail to penetrate and infect. Infection, by whichever mechanism, amounts to a mother sporocyst developing from the miracidium and this stage, depending on the species, then produces daughter sporocysts, alternatively rediae, and ultimately cercariae. Little is known about miracidia in the life cycles of *Ilyanassa* trematodes. The two species of miracidia that have been observed swam to snails rather than capsules being ingested by snails. Here's the question: however it is done, how often (at what rate) do miracidia colonize snails in the field? It has always been amazing to me how little this question has been addressed by parasitologists. To understand how a trematode life cycle actually works it should be a critical concern. Anyway, that is the concern of the work to be described here.

As described in the foregoing, in springs of 1991 and 1993 snails were collected from a low-prevalence site in a saltmarsh. These were screened for the presence of trematode infections and 1,400 of them were determined to be trematode-free. They were released to roam where they would at a stake on the Cape Henlopen sandflat. Such snails may be termed sentinels because they were almost surely (putatively) uninfected. They, like a sentinel lamb exposed to lure in a predator, would be exposed to lure in trematode parasites.

A further measure was taken to see if any of the snails, tested to weed out infected snails, managed to get through the test with an infection. In all 565 saltmarsh snails that had tested as uninfected were dissected to see if they nevertheless had infections. This was an experimental control. Six of them were infected. Therefore, when saltmarsh snails were tested to detect uninfected snails the test was 98.9% effective, but 1.1% of snails that showed no parasite (released no cercariae) actually had one. Put another way, of the 1,400 putatively uninfected snails released, 1.1% (about 15 of them) were probably infected. That's important to know. It is possible that a sentinel recovered with an infection from the field did not acquire it in the field. It had it when released. This will serve as a correction for that possibility.

Another question was, into what sort of infection environment were the putatively uninfected sentinels released? This was the high parasite zone that L^2 had found more than a decade earlier. Interestingly, now that I knew how old snails and their infections can get to be, there would be snails and parasite infrapopulations alive here now that might have been collected alive back in 1980. In 1991 samples were taken to confirm that prevalence there was still high. It is important to understand (reiterate) that snails do not infect each other. They are infected by miracidia that arise from adult worms in definitive hosts. So these data simply indicate the proportion of snails that had become infected in the area in the past (over perhaps decades), not the chances of a released snail getting infected in any given year. It turned out that 80-90% of snails in the release area were already infected with five different species of trematodes. Infections included four two-species combinations plus two triples. What this means is that the miracidia of several species had been active and successful in the area in the past. The expectation was that miracidia of all these species were probably still active in the area and released sentinels too stood to be infected.

So what happened? A total of 71 excursions to search for marked snails were made in 1991 and 1993. Snails were free on the sandflat for from 17 to 793 days before being recovered and dissected to see if they had become infected. Snails recovered early on were few and taken because I was nervous that they would be lost forever. My colleagues and I were able to find and recover (from the 1,400 released) 185 sentinels. Five of the 185 (2.7%) carried infections. In 1991 it was five infections among 153 recovered and in 1993 it was zero infected out of 32 recovered. It is of interest that the snail free for 793 days remained uninfected the whole time. Two were infected with *Hq* (free for 81 and 103 days); one had *Ls* (free for 103 days); and two had *Zr* (free for 98 and 103 days). So 2.7% of sentinels free on the sandflat for months during two years were recovered infected. Recall that there is a 1.1% chance that sentinels failed to release cercariae when tested initially and sneaked through the cercarial release tests with an infection. Consequently, it may be said that overall 2.7% - 1.1% = 1.6% of sentinels probably picked up infections while free on the sandflat.

This low colonization rate was a surprise, especially because so many snails native in the area were already infected. It was generally thought that a high parasite prevalence and a high rate of colonization by parasites would go hand in hand, but these results indicate not in this case. We have a snail that lives a long time and a population where prevalence of trematodes is very high, but at the same time snails in the population accumulate, are colonized by, trematodes very slowly. How can this be? Let's accept for the moment that 1.6% of snails get infected by a trematode each year. Of course, this is an estimate and subject to error. Had I done only the 1991 work the estimate would have been 3.3%; in 1993 alone it would have been 0.0%. Surprisingly, there are almost no other estimates of this sort for any snail. No estimate is necessarily correct in detail, but we are on good grounds that this one reflects a truly low rate of colonization. A matter of a few percent of snails get parasitized each year. Each year many uninfected juveniles are recruited to Cape Henlopen from the plankton (Chapter 4). These snails have a 1.6% chance of picking up an infection each year. [A few snails with shell heights less than 10 mm (one or two years old) have been observed infected.] If a snail does not get infected in one year, in the next it stands another 1.6% chance. If it escapes infection in that year, in the next it has another 1.6% chance. By the time a snail is, say, 20 or 25 years old these 1.6% chances add up and the snail is almost surely infected. This is the simplest explanation for why a size-prevalence curve (Chapter 4) shows that few small (younger) snails are infected, but almost all large (older) snails are. I have intimated this before, but this means that when you observe a high prevalence of trematodes in an *Ilyanassa* population, say on 18 July 1980, you are looking at infections that have built up in the population at a slow rate over a long period of time. Decades are involved. Few people are likely concerned with the fate of snail populations and even fewer with the associated component communities of parasites. If I have been at all successful here, maybe a reader has been given cause to pause and think about this a little. Let me return you to Chapter 4 where I mentioned *Ilyanassa* as a keystone species because of all the trematodes that depend on it to complete their life cycles. Many other species populations, a variety of invertebrates, fish, reptiles, birds, and mammals, serve as 2nd intermediate and definitive hosts for these trematodes. What I was suggesting in Chapter 4 was that there is an extensive network of effects on the whole estuarine system brought on by these parasites. The role of *Ilyanassa* in all their life cycles makes the snail a central player in the system. Human activities of all sorts are very intense in estuaries. That we have at least one component of estuaries, this snail, that is laden with historical content stretching back for decades increases the stakes considerably. Were a snail population and the parasites in it somehow destroyed, say in the name of profits from coastal real estate, a system that has built to its present state over this kind of time would be removed and far from easily replaced. With the present state of knowledge I can't predict exactly what the consequences would be, but they would surely go beyond the removal of a few snails. Were the whole snail species to be somehow driven to extinction, it is difficult to imagine the extent of the consequences.

This work in the early 1990's essentially brought my investigations of this snail and its parasites on Cape Henlopen to an end. The Henlopen State Park was changing. It was never exactly wild and untouched in my experience, but by the 1990's many more people were using the Park's facilities. Down on the ocean beach there used to be a modest, unpaved parking lot. There was a very small building containing crude showers and a tiny concession where in summer you could get a soda, potato chips, and maybe a hot dog or hamburger and french fries. Now, a more elaborate and in comparison huge building sitting on pilings serves the same purpose. More people meant more aggressive (now armed) rangers with more problems with which to deal. Now the Park had an entrance fee. Most important as far as this research goes is that the sandflat, which had formerly been closed to it, had been opened to clamming. On every low tide there would be anywhere from several to a hoard of people out there trying to dig up clams. The effect was that as you looked over the sandflat a very prominent feature would be the piles of sand they would leave behind as they gouged about with their rakes. That put me off. My former idyllic, lonesome days on the Cape Henlopen sandflat were now over and they were not likely to return. Neither was my younger self. I was fortunate to have had that rewarding circumstance while it lasted. All my research and small adventures went on without much recognition, caring, or support by the powers that be in the University or the State Park. That is probably just as well. The students involved certainly made up for it. I like to think I made good use of the situation, but all circumstances and these too must pass. I decided that Cape Henlopen was no longer a good site to pursue research.

In 1995 I began exploring for *Ilyanassa* along the eastern shore of another estuary in Delaware, Rehoboth Bay. This Bay is about 10 miles south of Delaware Bay and I was sort of barely aware that there were some very large *Ilyanassa* along there. By the end of that summer I was limping around and pretty well disabled. I was using crutches to help keep weight off my feet. Open wounds had developed especially on my left foot and I was soon going to have a major problem with that situation. But not yet, and in August of that summer Buck Cornell and I, both lovers of classy beers and ales, decided to check out a much ballyhooed establishment in Ocean City, Maryland. In summer, that populous town is a famous seaside vacation site in this area. Almost 40 years ago now I took a plane ride and was impressed with how precariously Ocean City sits on an exposed sand spit. It has been damaged on occasion, but has so far withstood all weather events. In its presently populated state, with vertical evacuation up the high rise buildings along the coast recommended, a major event now could be disastrous. The town is about 30 miles south of Lewes straight down Route 1. That road goes right by Rehoboth Bay on a piece of the sand spit, actually a barrier island, about a quarter to maybe half a mile wide. The Bay is on the west side and the Atlantic Ocean on the east side of the barrier island. Buck and I stopped off at several places along the Bay to look for snails. We found some at a place called Savages Ditch. Moving along shore there we, mostly Buck, as crutches weren't too amenable to the intertidal sand, collected 40 snails that day. Compared to Cape Henlopen snails they were all giants, 25 mm and greater. One was by me an unheard of 34 mm. I later found that every last one of them was infected, almost all with *Hq*. Hmmm. When we finally got to the beer joint the beer was unique and excellent and the atmosphere was jovial, especially a couple of women that happened to be there. They thought a pair of biologists were interesting for a while. That was fine, but I was mulling over a major question in which our new-found beer compatriots weren't too interested: why were snails at Savages Ditch so large? Did they grow faster there or were they older? And later, why all the *Hq* there? Here we go again.

South shore of Indian River Bay, a 1984 view west from Gray's Point along Blackwater Beach

Chapter 10

A 1981-1995 CAPSTONE

This will be a briefer chapter. Toward the end of 1995 my foot problems were getting quite severe. Teaching was hard enough and field work was out of the question. I decided to put some time into bringing together data on all the snails dissected from 1981 on. Snail data collected prior to 1981, mainly from Cape Henlopen, by Larry Hurd and myself had already been reported upon and would not be used in this. Also involved in this demarcation of data was that by 1981 I had become accomplished at identifying the infections I encountered. The idea was to see what those data would tell me about *Ilyanassa* and its trematode parasites in Delaware. While this enterprise was indeed easier on my feet, it was hard on my sit down and proved to be anything but a simple undertaking. Between 1981 and 1995, from nine different Delaware sites, 11,774 snails had been collected from the field and dissected in the service of various surveys and experiments. Each snail had been measured for size and at least examined for sex and trematode infection. Most snails (7,500 of them) had come from Cape Henlopen because that is where most of the work was done. Just for starters, all these data had to be put in order. That took some doing. I began putting together a summary manuscript that would ultimately be published in *The Journal of Parasitology* under the title "*Ilyanassa obsoleta* (Gastropoda) as a host for trematodes in Delaware estuaries" (Curtis, 1997).

Recall that parasite populations and species assemblages (communities) must be grouped or nested within host individuals and host populations. To reiterate the groupings, a parasite population within an individual host is referred to as an infrapopulation. In some host individuals more than one species of parasite is present and that species assemblage is termed a parasite infracommunity. In the context of a host population (e.g., *Ilyanassa* on Cape Henlopen), all the infrapopulations of one parasite species residing in snails in the population is a parasite component population. By extension all the parasite species infecting snails in a population, in whatever combinations, would be a parasite component community. Having to nest populations and communities this way is a complication, but a necessary one when dealing with ecological parasitology. My manuscript would have to deal somewhat with all these levels of parasite groupings. For trematodes in marine snails there were some, but not all that many studies that did that this. So a capstone study would be a useful addition to the knowledge and I took that as sufficient justification for the project.

All the *Ilyanassa* in the native range from the Bay of Fundy in Canada to about northern Florida constitute what can be thought of as the global population of the species, at least on the east coast. The snail [apparently with only one of its parasites, the schistosome *Austrobilharzia variglandis (Av)*] was introduced on the west coast about a century ago. It was supposedly delivered there with shipments of oysters from the east coast. It is now abundant, for example, in San Francisco Bay. The global *Ilyanassa* population is made up of spatially separated local populations, such as the one on Cape Henlopen, that are interconnected to various degrees by means of planktonic veliger larvae that drift around among them. There is no natural way veligers could be exchanged between the two coasts so those two populations are permanently isolated. My manuscript was intended to deal with the native east coast global snail population, but my resources (data, money, and materials) dictated that it would have to be done by way of a large sample of snails and their trematodes from a regional subset of the global population, the one in the mid-Atlantic around Delaware. A now "ancient" genetic study (1972) with the aforementioned B. S. Smith (Chapter 7) as one of the authors had shown that *Ilyanassa* populations were genetically similar (had the same genes) along the Atlantic coast. They were similar most likely because snail planktonic larvae (veligers) were exchanged among them. It has never been done and would require quite an effort, but carefully and systematically sampling snails all along the east coast would lead to a much better understanding of the way populations of *Ilyanassa* and their trematode component communities are organized and related.

As they say, if wishes were fishes. For the present, the scientific world will have to be assuaged with a regional study. Things have not changed much in the decade or so since my paper came out. What did my capstone study show? First, I had anything but a random collection of snails, a point I was at pains to make in my manuscript, and that meant I had to be careful how I used the data. When you are analyzing component populations or communities of parasites in certain ways it is important that a random, representative sample be used. Let me hold off on explaining why for a bit and for now just indicate how I did use the data. I have said that *Ilyanassa* is probably the most infected snail on the planet and these data well bear this notion out. As already indicated 11,774 snails were dissected. They came from nine sites as much as about 50 km (30 miles) apart. With the exception of Bowers Beach, I have had occasion to mention all these sites in previous chapters. Larry Hurd worked with an undergraduate student, Barbara Conlin, now with the Army Core of Engineers, who helped him analyze snails and parasites (for instance in that ditch in the Canary Creek saltmarsh). Her family had a cottage at Bowers Beach. It was in the old style, small and quaint. That site is well up inside Delaware Bay near the Murderkill River entrance. In 1983 she and Larry Hurd collected and examined some snails from there. As with their meticulous work in Canary Creek, they kindly made the data available to me. Vastly different numbers of snails came from different sites. Prevalence (% infected) and variety of infections also varied greatly. The best way to convey the variety and richness of the data is to present an abbreviated version of a table presented in the published paper. The North End, "Greenhead" Marsh, North Shores, and Thompson Island are all sites along the Lewes and Rehoboth Canal spoken of in Chapter 7. If, for some reason, you would like to know exactly where all these sites are, there is a simply gorgeous map in the publication. To the right in the table is recorded the number of different parasite species combinations found in snails. For example, in Canary Creek among 1,738 snails dissected, eight different single species infections, only one double, and no triple or quadruple combinations were found, giving a total prevalence of 10.9%. A view of the Indian River Bay site is shown in this chapter's frontispiece.

	No.of different species combinations					
Site	No.snails	%infected	Singles	Doubles	Triples	Quadruples
Bowerseach	185	8.7	4	0	0	0
CapeHenlopen	7,500	62.3	8	15	6	1
CanaryCreek	1,738	10.9	8	1	0	0
CanalNorthEnd	1,112	23.4	8	1	0	0
"Greenhead"Marsh	120	70.0	4	2	0	0
NorthShores	120	37.5	4	2	0	0
ThompsonIsland	115	8.7	3	1	0	0
SavagesDitch	80	100.0	4	3	1	0
IndianRiverBay	804	78.9	7	11	4	0

The most widespread parasite infections in snails were *Zoogonus rubellus (Zr)* and *Lepocreadium setiferoides (Ls)*, both using fish as definitive hosts. Both species were present at all nine sites. *Himasthla quissetensis (Hq)*, a seagull parasite, occurred in snails at seven of the nine sites. The other species were less widespread, some being quite restricted in their distribution. One, *Diplostomum nassa (Dn)*, is not often encountered. It uses shorebird definitive hosts, and was restricted to places where they gather (Cape Henlopen, and Gray's Point in Indian River Bay). When you open a snail infected with this species the effect is quite striking. My former student, Ann Barse, even today expresses being startled by the experience, which for a person who has worked on marlin guts full of tapeworms is saying something. The sporocysts of the species are very unsporocyst-like. They are very elongate, highly mobile, and active. Bear in mind you are looking through a microscope where things are magnified perhaps 10 or 20 times. The snail is out of its shell and when you probe or cut the body sporocysts explode forth into your field of view. You are presented with a boiling, writhing mass of activity, sort of like you had just cut open a bag full of snakes. Ann was glad this species is rare. I think its kind of cool.

In all, 51.0% of snails collected were infected with one or more trematode species. That's a pretty amazing percentage. For most snails that figure would be closer to 10%. Arbitrarily, I separated parasites into two groups. Those that occurred in 1% or more of snails were labeled "core" species; those that occurred less often were labeled "satellite" species. There were nine trematode species identified in the total sample. Five of them were core species including, in order of prevalence among the 11,774 snails in the sample, *Zr* (23.1%), *Gynaecotyla adunca (Ga)* (15.4%), *Ls* (13.6%), *Hq* (10.7%), and *Av* (1.8%). The other four species were satellite species occurring in well under 1% of snails. I won't list them here, but just before the bibliography at the end there is a table in which they are all included.

With nine species of trematodes there are a conceivable 36 combinations of two species that might occur in snails. For example you might have a species, say *Zr*, infecting a snail together with *Hq*, or *Ls*, or *Ga*, or some other species, and you end up with 36 conceivably possible double combinations. [You can just take my word for this, or if you are interested there is a formula for determining it, which I always forget as soon as I've used it. Your head can get all muddled up retaining too many such things. Many calculators will do the operation for you. The formula is ${}_9C_2 = 9!/2!(9-2)! = 36$. That is, if you have nine items (parasite species in this case) and take them 2 at a time (as in double infections), there are 36 possible combinations. The ! in the formula means factorial, e.g. 2! = 2X1 = 2; 9! = 9X8X7X6X5X4X3X2X1 = 362,880.] Of the 36 conceivable double infections, 16 were observed and all involved one or two of the five core species. There is a lesson here. If two parasites rarely occur as single infections in snails, they are not likely to show up in double infections. If one core and one satellite species occurred together in a snail, the core species was always either *Zr*, *Av*, or *Ga*. These are all sporocyst infections in a snail and they are the most likely to be combined with other species in multiple infections. I had occasion in Chapter 6 to mention *Ga* in this regard. Two double infections, involving core species, are very seldom encountered. One, *Hq* and *Ls* (abbreviated *HL*), was mentioned in Chapter 8. The other is *Hq* and *Av (HA)*. Only one snail among the 11,774 was infected with the *HA* pair.

As to triple infections, there are 84 conceivable triple combinations that might be found in snails. Of these seven were observed. The seven triple combinations mostly involved two or three core species. There's that lesson again, rare (satellite) species tend to not show up together in multiple infections. Of the 84 conceivable triple combinations, 72 that did not occur in any examined snail would have had to include a satellite species. Interestingly, there were core species that might have been expected to co-occur in certain triple combinations, but did not. If a triple involved *Hq* and *Ls*, or *Av*, those triples did not occur. There appears to be some sort of conflict, as discussed in Chapter 8, between certain parasite species when they try to infect the same snail. In some way *Hq* is the likely "trouble maker," as the other two species (*Ls* and *Av*) can and do co-occur in snails. To hammer home this point, look at the one quadruple combination that was encountered on Cape Henlopen. You may recall that two snails bearing infections with four trematode species were found, one during the rainy, but fun-filled, experiment on the beach in 1985, and another in 1986. Those snails were infected by four of the five core species (*Ls, Zr, Av,* and *Ga*) and all species were simultaneously producing mature cercariae. The absent one was *Hq*. So all the others could clearly coexist, but for some reason not the absent one. That reason, as indicated in Chapter 8, most likely has to do with the way *Hq* is adapted to use the snail as a host.

In the 11,774 snails the prevalence (51%) and variety of multiple infections (12.6% of snails were multiply infected with a total of 24 species combinations) was very great. This is not often matched in other snails. As Karen and I had discovered in the early 1990's (Chapter 8) these parasites are for the most part pretty sociable. I did a three-way contingency test (a log-linear analysis) to see what influenced the likelihood of a snail being infected and with what. The three factors considered were infection category [including uninfected and infected with various parasite species (and combinations thereof)], size category (shell heights under 10 mm, 11-15 mm, 16-20 mm, 21-25 mm, and 26 mm or greater), and snail sex (male or female). In the test these factors must be considered alone as single factors and interactively in combination with one another. What this test told me was that, if we were going to try to predict whether a snail in the sample of 11,774 was going to be infected and with what, as single factors all factors (infection category, size, and sex) were all important to know (were significant) because all three were highly variable among samples. Additionally, the interaction between infection category and size is important to know because if you have a smaller snail it less likely infected. This was in agreement with the size-prevalence curve produced by Larry Hurd and me (Chapter 4): larger snails tended to be infected; smaller snails tended not to be. So size mattered. Finally, the interaction between sex and size was also important because males had some tendency to be smaller. Of interest was that there was no importance of the interaction of infection category and snail sex. If you had a female or a male snail, it was equally likely to be infected, so you would not need to know a snail's sex to predict whether a snail was infected or not. A shorter statement would be: lots of snails were infected with a variety infections, but if as an exercise you wanted to predict whether a snail was likely infected, the infection category you predicted and snail size mattered, but snail sex did not. The result was the same whether the whole sample of snails was tested, or just the snails from Cape Henlopen. So it was applicable to both the wider and the more restricted area.

Let me take some time to discuss something that has been a "bee-in-my-bonnet" for a number of years. It has to do with the analysis of parasite component communities and using the results to make inferences about what goes on between species in parasite infracommunities. Let me state again, that the component community is made up of the parasite species (all infections) present in a host population, an infracommunity is the species present in a host individual. In principle the analysis is perfectly valid, but it's not so easy to apply in practice. The analysis revolves around a question of probability. A probability is a number between zero and one, 0 is no chance of an event occurring, 1 is a guaranteed chance. Probability can be expressed as a decimal or a percentage. That is, the probability of an event occurring might be expressed as 0.4, or 40%. Specifically, the question here is: what is the probability of a chain of events all occurring. If you have a single and honest (not loaded) die, what is the probability of rolling two sixes in a row? This is governed by what is called the product law of probability. To arrive at the probability of rolling two sixes in a row, the job is to multiply together the individual probabilities of the component events. Chance of rolling one six is 1/6, or 1chance out of 6. The die is a cube with six faces and only one of them has six dots. Let's say you have rolled one six. What's the chance of rolling two of them in a row? You pick up the die and roll it again. As with the first roll, chance of getting a six is 1/6. The chance of rolling two sixes in a row is 1/6 X 1/6 = 1/36 (= 0.028 or 2.8%); the two individual probabilities are multiplied together. Chance of rolling three sixes in a row is 1/6 X 1/6 X 1/6 =

1/216. This works in many venues. For example, what is the probability of drawing an Ace of Clubs and then an Ace of Spades from a well-shuffled, standard deck of 52 cards? It's 1/52 X 1/51 = 1/2,652. If you divide 2,652 into one, the probability comes out to 0.00038, or 0.038%. The second term in the calculation is 1/51 because after you draw one card there are only 51 cards left in the deck and only one of them is the card you want. This reasoning can be transferred over to parasites. If in a component community, species A parasitizes 25% of the snails and species B parasitizes 25%, then ideally 0.25 X 0.25 = 0.0625 (6.25%) is the random chance that a snail will carry both parasites.

The logic of this is important to get. To this end let me set up a sort of standard "straw man" so I can later majestically knock him down. Imagine a billiard table with 100 billiard balls on it. There are 25 red, 25 white, and 50 black balls. The 25/100 reds represent infection probability with parasite species A (= 0.25), the 25/100 whites infection probability with B (= 0.25), and the 50/100 blacks the probability of being infected with some other species or uninfected (= 0.5). Note that a ball must be red, white, or black and the probabilities (0.25 + 0.25 + 0.5) add to one. Now imagine these 100 balls bouncing around on the table top randomly. They would be bouncing off one another and the sides of the table. Randomly means that there are no preferred directions or distances for ball travel. The question is: what's the probability of a collision between a red and white ball, a double infection? As above, 0.25 X 0.25 = 0.0625 (6.25%). There are six and a quarter chances out of a hundred that a collision will be a red and white ball. The rest of collisions will be red-red or red-black, or white-white or white-black. If you increase the number (actually the probability) of red balls and/or the number of white balls, you increase the probability of a red-white collision. Let's say there are now 40 red, 35 white, and 25 black. Now the probability of a red-white collision is 0.4 X 0.35 = 0.14 (14%). In a perfectly sampled parasite component community you would expect 14% of snails to be doubly infected with species A and B. Take note of the descriptor "perfectly sampled."

As far as I know, in 1937 W. W. Cort and colleagues were the first to use such an approach to analyze the component community of larval trematodes in a population of snails. It was a freshwater gastropod in Michigan. It was an important study reporting nearly 63% infected with 17 trematodes among 7,259 snails examined and very clever because it was first to analyze things this way. There was even a quadruple infection observed. This is the only one in any snail that I am aware of other than the two I have seen in *Ilyanassa*. As described in the previous paragraph, what these researchers did was to take the observed prevalences (proportions infected) for pairs of parasites, multiply them together, and thereby predict what proportion of snails would be expected at random to be doubly infected with both. Since then many authors have used this same basic approach. Once you have the expected proportion of doubly infected snails you can multiply that proportion by the total number of snails examined to get the number of doubly-infected snails you would expect to see. If, say, the expected proportion of snails doubly infected with a pair was 5%, then the number of snails doubly infected with that pair expected to be seen in Cort's particular sample was 0.05 X 7,259 = 363. So far so good, but here comes the potential problem. For any given pair in any given study, the next step often taken is to compare the observed number, the number of doubly infected snails actually observed during dissections, with the expected number. Three possible outcomes exist. The observed number could be less than, equal to, or greater than the expected number. Let's use some pretty clear-cut numbers to illustrate. Suppose you collect (observe) only 30 doubly infected (AB) snails, but you expect 363. You might be led to infer that there is some sort of barrier to co-infection of snails by the two species. Maybe the parasites prey on (eat) one another or compete for resources (e.g., for nutrition or space) in a snail so fervently that one of the two is out competed and evicted from the snail. Or, there's another possibility. Maybe some factor external to the snail prevents co-infection. Maybe miracidia of the two species for some reason (e.g., they occur in different sections of the sampled habitat) seldom manage to get to a snail individual to infect it. Suppose instead that you observe 350 doubles and expect 363 in your sample. The inference from this might be that the two species are combining at random and there is no barrier to co-infection of a snail. The third possibility is that perhaps you observe 553 doubly infected snails, but expect 363. This could lead to another inference: perhaps the presence of one of species in a snail promotes or facilitates infection by the other.

The map is not the territory. For one thing parasites pursuing their life cycles are not billiard balls bouncing around on a two dimensional billiard table surface. Snails don't infect each other and the image of two billiard balls colliding is not exactly appropriate. It is debatable whether there is any even vaguely comparable situation on, for example, Cape Henlopen. Snails are infected by miracidia coming from definitive hosts. If we wanted to figure the expected probability of two parasites infecting the same snail, we should most directly multiply the probabilities of both kinds of miracidia getting to the snail. Add to this that contact by the miracidia of the two species at the snail (closer to a "collision") might not be in the same year and possibly separated by decades. Things get quickly muddled up. The history of infection with two species in a habitat [the individual prevalences (probabilities of occurrence) of the two parasite species in snails] is only remotely associated with the probability of a double infection coming about. I do recognize an association, but far from a direct one. Multiplying two prevalences together to arrive at the randomly expected chance of a double infection may not be the best way to go about things.

Even if we put the previous paragraph aside there is yet another problem. It has to do with that "perfectly sampled" descriptor I noted above. The design with which a component community of trematodes in a population of snails is sampled is all important. Keeping this at the level of one pair of parasite species, at base what is required are accurate estimates of the prevalences of both parasites in the component community in the snail population. This is not so easily accomplished. Ever since the 1983 *Ecology* paper that Larry Hurd and I published, I have been impressed with the amount of spatial heterogeneity inherent in populations of snails, and more important, in the spatial distributions of snails infected with different trematodes. Let's look at some already covered examples: in our paper Larry Hurd and I noted that *Hq* was frequently observed near shore, close to the base of the beach; snails infected with *Zr* in contrast tended to be concentrated away from shore, near the extreme low water line; in later work (Chapter 5) I found that *Ls*-infected snails tended to be aggregated on carrion sources; and in the small vertical aspect of the sandflat there is the whole business of *Ga*-infected snails. If you collect snails from sandbar peaks they are almost surely infected with that species. If you collect lower down on the sandflat, that infection may be present, but pretty infrequent. There are more examples of this sort of thing, but they have to do with later work at Savages Ditch. My point is that maybe just now, after all the previous work, I know enough to design a sampling program that would result in accurate estimates of the prevalences of trematodes in the component community in a snail population. Just maybe. (And, it is fair to ask: are prevalences in the *whole* component community actually relevant to the prediction.) Were someone to happen to just collect snails near the extreme low water line, the prevalence of *Zr* would be elevated, near the beach it would be *Hq*, and snails collected from a sandbar would be a different story. The design of sampling would have to pointedly include all these areas. Here's the point: without careful, knowledgeable sampling design, if the goal were to accurately estimate parasite prevalences in the overall trematode component community in *Ilyanassa* on Cape Henlopen with these snail samples, they would all be seriously in error. A case in point, as extensive as the sampling by Larry Hurd and myself was in the late 1970's and in 1980, we still utterly missed snails infected with *Ga* because we didn't know to sample on sandbars or the beach. What this means is that any attempt to estimate expected numbers of double infections, by multiplying together a pair of (likely erroneous) prevalences, would result in nonsense. Unless sampling is done with a lot of knowledge and completeness (that is, lots of effort), I question such calculations with any snail species and its trematode parasites. I pointedly did not make such calculations in this work. The data are there in the paper to make them, but the samples are anything but random and representative measures of parasite prevalences. They would most likely lead us astray.

The author lolls about at the operation base point (release stake) at Savages Ditch in 1996.

In extreme cases, such as multiple infections with *Hq* and *Ls*, or *Av (HL* or *HA*), the scarcity of multiples certainly suggest a heads-up. There is something going on that prevents co-infected snails from being observed. Whether it's some sort of interaction going on between parasites internally within the snail (probably having to do with how one or both is adapted to use the host), or something external to the snail, is unknown. Any suggestions we might get from less extreme cases would have to be used cautiously. If two parasites are fairly often found together in snails, it means they can coexist there long enough to be observed, but not much else can be safely inferred. Maybe they get to combine, but in the future one will be eliminated. Maybe they combine freely. Maybe one somehow facilitates the other. We won't get a clue by multiplying prevalences together, unless the prevalences are purposely very accurate. The point I made in my paper was that if you want to study how species get along in infracommunities, look directly at infracommunities. That's the way to understand what goes on there. Karen and I did that and so have a few other investigators. To really get at chances of co-infection for any species pair would require learning about the ecology of the miracidia. So far only two species of miracidia (*Zr* and *Av*) among the ones that infect *Ilyanassa* have been observed even in the laboratory and one (*Av*) was observed infecting a different snail in Hawaii. This would be a very difficult larval stage to study in a field setting. This is open to future ecological parasitologists and they will have much to investigate. As usual, a scientific investigation has produced more questions than answers. To me that's a comforting thought. Does it give you any ideas?

Chapter 11

New Place, New People

I was getting to be a new people. Near the end of the fall semester in 1995 a very nearly deadly event occurred. About three weeks before the fall semester's end my left foot, which had been afflicted with open wounds for some time, almost literally blew up on me. I couldn't get in to teach my classes for a few days, something I never allowed to happen. Before this occasion I don't think I had missed a single class in some 20 years. I was homebound with especially my left foot, but also my right, hideously swollen. I had a high fever. Karen caught up with the fact that I hadn't been around for a few days and came by to check on me. She took one look at me, got me into her car, and took me to the emergency room at the local hospital. I owe her for her decisiveness, a debt I can probably never repay. With some assertiveness on Karen's part they fairly quickly got to me and took some basic measurements. My blood sugar was ridiculously high and the fever was serious. I soon passed out, was admitted, and put on insulin and intravenous antibiotics. Fortunately I did wake up that night, but I was in pretty bad shape. The doctor who ended up with my case (one M. Saberi) deserves credit for probably preserving my life. The rest of the semester had to be written off. When they x-rayed my feet they were both infected, probably my whole system. In 1961, just before I went off to college, I had stupidly managed to shoot myself in the right foot with a .22 caliber pistol. I spent the first 13 weeks of college in Maine on crutches. With the snow and ice, that was interesting. There was still some lead in there and I can recall explaining to numerous doctors and nurses that it had been there for more than 30 years. Well, at least I was able to do that. Days later, on a laptop, I was able to construct final exams for students in my classes. I'm sure the students, one and all, were most grateful for that effort. They kept me in the hospital for a long stay all the while pumping me full of burning antibiotics. At some point they put in a port, through which antibiotics could be run into my system. The port was in my left arm and from it a tube ran via a blood vessel up and over into my chest area. The idea was to release the antibiotics into my central circulation. They would better get to my feet to fight the infection that way. At the end of about two weeks they sent me home with a battery driven pump and bags of antibiotic. I would have to keep taking them for about three months. For that time I would have to carry a satchel around with me so that antibiotics could be continuously administered. It was such fun teaching that way. Very slowly the infection started to clear up a little and eventually I was freed from my "succubus" shoulder bag with all the equipment in it. It was oral antibiotics after that. You may recall that after the ill-fated beach experiments back in 1985 I had a similar episode. That one was troublesome, but less serious. At that time the satchel system was not available and I had to hang bags of antibiotics from the ceiling with a tube running to my arm. Air bubbles entering my circulation were a constant worry. Difficult to live with as it was, I liked the new system much better. This event put me very close to being a goner. For me the point was made that tomorrow is guaranteed to no man. I was a very lucky guy.

Before this disruption I had been talking to two students about doing research. Jeff Kinley I have mentioned before. He was involved in the 1995 study of imposex along the length of the Lewes and Rehoboth Canal that was published in 1998 (Chapter 7). He was a quality student of known abilities. At the time he lived in Lewes, but during the school year he had to be about 90 miles away on the main campus taking courses for his degree. We had been discussing the possibility of him doing an undergraduate thesis. I had recently found those comparatively giant snails at Savages Ditch and the central question of his thesis was to be whether they got that big by growing faster, or for a longer time. We had comparison growth data from Cape Henlopen and it was important to know if growth rate in different habitats was different. The other student was Nathan Tanner. He was definitely an unknown quantity. He was clearly a bright guy interested a variety of things, including biology. It is fair to say that he was sort of inconstant in applying himself to academic pursuits. On the other hand his talents were very great. His artistic talents allowed him to illustrate this volume. He new his stuff, could learn easily, and could write effectively. The ability to write well is a good window into the organization of a mind. Nathan had demonstrated on tests that he could write about matters of biology, but it went well beyond that. He wrote essays, short stories, and further had a facility with poetry. You can't do that without the benefit of an organized mind capable of acquiring and using knowledge. His might not be the "standard" organization for a mind, whatever that is, but it worked for him. If he represented a bit of a gamble, it was one well worth taking.

I conceived of the project at Savages Ditch in Rehoboth Bay this way. Because Jeff and I wanted to measure growth rate of snails over 18 mm in shell height it would be necessary, as on Cape Henlopen (Chapter 9), to mark, release, later find, and re-measure snails in the Savages Ditch habitat. Nathan's problem would be different. Prevalence of trematodes in snails at this site was very high. In the samples I collected during summer 1995 in this habitat it was 100%. His concern would basically be this: infection prevalence at Savages Ditch seemed to be even higher than on Cape Henlopen, is the probability of an uninfected snail becoming infected there also higher? Also, does the trematode species composition infecting a snail often change? Releasing sentinel marked snails, as I had done on Cape Henlopen in 1991 and 1993, would be required for both these projects. In addition snails with known infections would have to be marked and released. My own, solitary, interest was going to be the movements of marked snails. I already had data on snail movements from Cape Henlopen, though I had not yet analyzed them. Data from snails at Savages Ditch would contribute further to an understanding of snail movements. This will be covered later. Here I want to focus on Nathan's and Jeff's work. Forget me not. They couldn't have done their projects without my handicapped help.

Because similar work with sentinel snails was already covered in Chapter 9, and further because Curtis and Tanner (1999) was published first, I will describe Nathan's project first. In February when the spring semester of 1996 began I was able to teach my classes, but it was done around all sorts of medical activities. I was sort of up and around. My left foot had to be soaked twice a day. Between soaks it was bulbously bandaged and could not take any weight. Crutches were my means of getting around and of course there was my succubus satchel. If Jeff and Nathan were going to get their projects started in a timely fashion, my physical state of affairs would be a major impediment. As the spring semester proceeded I made it a point to stay in contact with my two aspiring researchers. I knew that if the project was to be started at all, I would not be able to actually do most of the work required. It was important to get Jeff and Nathan to commit to the project and understand that they would have to be the prime doers of the things that had to be done to get it going and see that data were collected. I knew what had to be done and how to do it, but couldn't be much help in the execution of it. My role would be truly advisory. They both agreed to take the project on. Little did they know then what was in front of them. They signed up for research credits, for which they ultimately earned deserved A's. As usual I taught these credits free to the University. The University would pay some piddling little amount for teaching such credits, but relative to the work involved it seemed like an insult and I always declined. The Undergraduate Research Office provided some support for each of their efforts. That certainly helped, but as is usual such support does not begin to compensate for the hours, days, weeks, months, and indeed years that would be put in.

The first problem for all of us was that the laboratory was in utter disarray. During 1995 and so far in 1996, with me being in bad shape, it had been used only a little. It hadn't been altogether unused, but I had not spent a lot of time keeping things in order. That survey of the Lewes and Rehoboth Canal with Jeff had been executed and I had been in the midst of preparing a promotion dossier. I was going up for Full Professor in 1996 and the dossier supporting that effort is a time-consuming process. Every last thing done in teaching, research, and service since in my case promotion to Associate Professor 12 or 13 years ago had to be dredged up, written down, and intelligently presented. Anyway, because I couldn't do it, Jeff and Nathan would have to put the laboratory in order so this 1996 project could be launched. They, of course, were not familiar with all the clever places I had put things so they would be readily available to me should the need arise. As I recall they spent a long day getting things cleaned up and in workable order. Later there were a few things I could not find without extensive searching, but that was a small price to pay for the excellent job they did. Now work could be started. We would need some filtered baywater to work with during the summer. Between them Nathan and Jeff collected 60 gallons of it (27 ppt salinity) from off the Cape Henlopen Fishing Pier, filtered it to remove any cercariae that might be present, and hauled it back to the laboratory for storage and use. Having done this myself a few times I can attest that this is not a small job. In overview, our plan was to release 500 marked snails at Savages Ditch and snails had to be collected, marked, measured, and deployed. The first snails to be released were to be sentinels. They had to be collected, tested to remove demonstrably infected snails, marked, and measured. It would have been desirable to collect these at Savages Ditch because then they would not be transplanted snails introduced to a new environment. However, because previous evidence indicated that most or all Savages Ditch snails were already infected (later it proved more complicated than that), we decided that sentinel snails should be collected where infected snails were not so likely. A logical place to collect potentially uninfected snails was the ditch in Canary Creek saltmarsh, where historically few snails were infected. In the rain on 3 June Nathan, Jeff, and Josh (Jeff's unfortunate cousin) made a trip to the saltmarsh. I could not go because the marsh was too sloppy underfoot, crutches would not work, and there was all too good a chance of getting my bandaged foot muddy and wet. They collected about 800 snails, of which 450 were placed individually in cups of filtered baywater. (I will not specifically say so in every instance, but as I have expressed before I was now concerned about taking snail lives without very good reason and it became my general practice to return unused snails alive to the habitat where they were collected.) Because I was the only one on this team with experience in identifying cercariae I made it into the laboratory to begin introducing my compatriots to this aspect of the work. I just sat in one place at the microscope with my foot up while the others brought cups to be examined. When I would find cercariae they would come over and take a look to develop skill at identification. History does not always well predict the future. When we examined the cups for cercariae to detect infected snails it turned out that about half were infected. Compared to past experiences there was obviously a more infected population of snails inhabiting the site in 1996 compared to previous years. We were able to sequester 140 putatively uninfected snails from this batch, 40 of which were later dissected to see if any had gotten through the test with an infection (three did). The high frequency of infected snails from the saltmarsh concerned us and we sought out another place to collect potentially uninfected snails. We settled on a location along the Lewes and Rehoboth Canal next to the public launch ramp in Lewes. In the past few snails from there had been infected. Jeff and Nathan went and collected 306 snails. We put them in cups and were able to acquire 210 putatively uninfected snails from among them. Sixty of these were dissected to see what proportion had gotten through the test with an infection (none did). In summary so far, what this means is that we obtained 100 putatively uninfected snails from the marsh plus 150 from the launch ramp. We dissected 100 supposedly uninfected snails that had been through the test and 3% had gotten through with infections. This suggested that perhaps 3% of the sentinels to be released carried infections, but the overwhelming majority were uninfected.

It was now time for another first for my colleagues. We had to set up the marking apparatus and mark snails. Jeff and Nathan went to local stores for fingernail varnish and to clean them out of Mototool wire brushes. We needed clear varnish to affix the tag to a snail and red varnish to color its tail (shell apex). The red on the tail had proved the easiest to pick out in the field during Cape Henlopen studies. The orgasmitron came back into being again. We set up the snail-catching sail, got out all the clothespins, and assembled the other paraphernalia associated with this job. The orgasmitron evoked considerable humorous banter. Before long I had them both indoctrinated in the intricacies of snail marking and they became accomplished at it. I was able to impress upon them that this had to be done carefully. There could be no mark edges sticking up, proper orientation of the mark number was essential, and harming the snail with overzealous buffing was asking for failure. The mark had to stick and be readable. You never knew how long a snail, a tag, might have to exist in the wild before it was found again. At this point they could not yet anticipate the vast rewards this seemingly silly work would produce for them. They did it well, but there was a lot of joking around. The room was alive with it and that was fine with me. After all, for a couple of young men putting little individual numbers oh so carefully on snails at the behest of an older, crippled up professor was a strange thing to be doing on an early summer day.

At the end of all this we had 250 marked and measured sentinel snails to release, but during measurement one was found to be dead. We made arrangements to take the sentinels down to Savages Ditch and release them. The day was to be 8 June 1996. I will never forget that day. We gathered up the 249 living sentinels in a container. Also, a galvanized metal pipe about six feet long and a heavy hammer were commandeered into service. Nathan had a pickup truck at his disposal as well as an old canoe, which was loaded on the truck at a jaunty angle. It seemed pretty secure. The truck was supposedly sound, but there was some question about the brakes continuing to work. Nathan, Jeff, and I got the gear loaded and we piled into the cab. I was sitting in the middle and we started down the road. To get to Savages Ditch meant travel down Route 1 to the south. From my house it's a total of about 13-14 miles. You had to go through the outskirts of Rehoboth Beach and right through Dewey Beach. There was a lot of traffic. Before Dewey Beach the brakes failed. They would work a little and then go away leaving us with very uncertain stopping capabilities. Being a supremely sensible guy, in my advancing years, and currently in a crippled state I probably would have abandoned the mission for a time. Well, maybe not. Anyway, I was with a couple of young men of more adventurous temperament than me. They never thought about stopping. So I steeled up and got with the program. There were some anxious moments, but with the brakes working a little sometimes, gearing down as necessary, and with judicious use of the emergency brake Nathan worked our way through Dewey Beach traffic and ultimately onto the open road south of there. There was no need to stop now and we soon made it to Savages Ditch. The place was desolate. We would worry about the trip home later. The Savages Ditch site is on the eastern shore of Rehoboth Bay. The Ditch itself runs nearly east-west and is about 10 meters wide and 300 meters long. Nathan and Jeff portaged the canoe from the truck down to the Ditch and put the pipe, hammer, snails, and other gear into it. I was pretty useless, but at least I had been there before and had some specific idea of what had to be done. On my crutches I made my way through the sand and grass to the canoe and managed to get myself seated in it. It was on land when I did this because my bandaged foot could not get wet. I am more than two hundred pounds, but Nathan and Jeff nevertheless easily slid the canoe into the Ditch's water. I just sat there and went along for the short ride. It was great having "helpers" who could do such things. The guys climbed aboard and paddled us about a 100 meters west down the Ditch. This was more like what they both were built to do. At that distance the Ditch opens into a wider area that is bordered to the south by a long and narrow saltmarsh island, maybe 100 meters long and 10 meters wide. To the north of this wider area the border is saltmarsh on the strip of land (a barrier island) that separates Rehoboth Bay from the Atlantic Ocean. The wider area at its eastern end is as wide as the Ditch. About 150 meters away on its western end it is about 130 meters wide where it opens into Rehoboth Bay. There are no official names for the features of this habitat. The wider area, because of its shape, we began calling the "Funnel." The Funnel is not drained of water on every low tide, but on this day much of the Funnel's bottom was exposed. There were *Ilyanassa* visible all around. We had to go along the northern side of the Funnel where there was more water. Even so, to preserve my foot my two "engines" had to get out and skid the canoe with me in it along over the sandy-mud bottom. We came to a site near the western terminus of the Funnel on the northern side about a meter away from shore. We decided to release snails there. Nathan and Jeff got out the hammer and pipe and drove the pipe deep into the sediment, leaving about a foot of it above the sediment surface. In the frontispiece for this chapter, as usual there I am, a loafer near the release stake. We then took some notes about weather conditions, measured temperature and salinity of the water, and air temperature. With great ceremony the 249 sentinel snails were dumped onto the sediment in a pile at the base of the pipe. It could now be dubbed the release stake. That stake would last there for some eight years. We watched the snails become active and begin to disperse away from where they had been deposited. It didn't take long before they were quite spread out. While we were waiting and watching Jeff and Nathan haphazardly collected 250 native snails from the Funnel area. These were expected to be infected. They would be marked and released as native snails to see how they would grow, whether they changed infecting species, and where they would go. There was nothing left to do at Savages Ditch on this day. We made our way back along the Ditch and grounded the canoe where we had put in. Sitting for so long, Nathan and Jeff had to literally lift me out of the canoe and get me upright on my crutches. They got the canoe back on the truck and off we went back up the road to Lewes. Now experienced, Nathan was quite proficient at handling the faulty-braked truck. For some reason the brakes had even come back a little while the truck was sitting. Anyway, there was not so much call for a heart in the mouth on the way back. I was dropped off at home and off they went. What happened with them and the truck later that day I don't know. Whatever adventure they encountered it couldn't have been too perilous because they were both back marking native snails the next day.

For the sentinels, from which infected individuals supposedly had been screened out, we had developed a computer spreadsheet listing each snail with its data. Native snails were to be added to the file. Because we expected the native snails to be infected, the idea was that we would first put marks on them and then check to see what species of cercariae each one released. That way we could readily use the number to list a particular snail with its infection and other data. Attesting to their newly acquired competence, Jeff and Nathan came in the next day and marked the first 125 natives without me. I was resting. The following day I did make it in and we put the first batch of natives in cups for cercarial release and we (mostly them) marked the rest. On 11 June Jeff was otherwise engaged, but Nathan and I were in and checked the first batch of natives in cups. Most of the snails, which again were from the Funnel, released cercariae. Some did not, but as it later proved out virtually all were actually infected. The most frequent parasite was *Himasthla quissetensis (Hq)*; next was *Zoogonus rubellus (Zr)* with the most infrequent being *Lepocreadium setiferoides (Ls)*. There was a goodly number of double infections with the first two listed species and fewer with the second two. After completing this job we put the second batch of 125 natives in cups. When we checked those cups the next day the infection story was similar to the first 125. As we examined a cup we would also use calipers to measure the snail from the cup and enter its data into our spreadsheet. In the end the spreadsheet listed each of the 500 snails with its origin (marsh, launch ramp, native), mark number and color, initial size, cercariae released (if any), date set free, and any identifying unique features. In the future, that spreadsheet would grow mightily as data came in from snail sightings in the field. While measuring snails several appeared suspiciously inactive. Aside from the smell, you could tell if a snail was dead with a simple test. On death they rot quickly and if you insert a bit of paper towel into the shell it will come out dirty if the snail is dead. If it comes out clean (no black specs), it is still alive. It turned out that 19 of the 250 were dead when measured, possibly victims of too much buffing. The 231 live natives were to be released the next day, on 13 June 1996.

Jeff was tied up on that day, but Nathan and I made it down to Savages Ditch in the afternoon to release the marked native snails. His truck was repaired and it was a relatively uneventful trip. Nathan is a very experienced canoeist and he rapidly got the canoe down by the water, me seated in it, and underway down the Ditch. I was perhaps a bit more useful, but still confined to the canoe. By 3:15 PM the natives were piled at the release stake and free to roam where they would. They didn't disperse quite as quickly as the sentinels had. Perhaps they "knew" they were in their usual habitat. We took some environmental data. It was a wonderful day to be there. The air was 29°C (about 84°F), the water was also 29°C, water salinity was 28 ppt, and it was sunny. We (Nathan) began looking for sentinels released on the previous visit. I sat in the canoe right next to the release stake while Nathan walked about looking for snails. My main job, among all the other complicated duties I had within the confines of the canoe, was to write down data. We found 61 of the 249 sentinels. Most of them were quite close to the release stake, but some had gotten as much as 50 meters away in the six days since release. At this stage of our searches when Nathan would find a sentinel he would loudly and clearly yell out its tag number (e.g. BR1, blue red 1). I would call it back to confirm I had gotten it right. We had to take care that the old man's ears didn't mishear. Good grief! I never imagined I would live so long. He used a tape to measure the snail's distance from the release stake. I would use a lensatic compass to get its direction from the release stake. Thus we would have polar coordinates (a distance and a direction) for the snail's position at that time. It was about this time that I realized what a gem Nathan was for this project. Now in his early twenties, he had grown up on a large farm in Delaware passing many hours alone searching for arrowheads, fossils, and other such small items. His eyes were incredibly sharp. Later, when I was somewhat able to walk, in a given amount of time I might find 10 marked snails while he raked in perhaps 25 or 30. He had the technique, patience, and eyesight that made him the best I could hope for in a snail searcher. Among the snails he found that day were four that had been killed since release. They were all at least several meters from the stake so likely they had been alive to move away when released. On one the shell had been severely damaged and in another the snail had been evicted and the shell taken over by a hermit crab. There would be more dead snails found later. This is something I had not seen much of on Cape Henlopen. There, only one sentinel had been found dead. Whatever the cause, Savages Ditch must be a tougher habitat in which to live. Nathan kept doing his job and I mine for some four hours and about 7:00 PM when the water became too deep we packed up an left. After sitting all that time it was a bitch getting upright and moving on my crutches once again. But that's O.K.. It was a good day and thanks to Jeff, Nathan (and me) the project was now nicely underway and pretty much on schedule.

Compared to the sandflat on Cape Henlopen, the Savages Ditch area is a feature-filled habitat. I had not worked in and around Rehoboth Bay very much, but in the fall of 1995 I had spent some time exploring the immediate Savages Ditch area and had a pretty good idea of what features were where. The habitat layout plays a large role in this work and it is useful to describe it. Most of the features did not have official names and, as was done with the "Funnel," we gradually made up labels for them as we went along. This is a much smaller area than the Henlopen sandflat and there are fewer snails. Consequently, the likelihood of finding marked snails was much greater. That is, if they don't just vacate the area.

As already mentioned the mouth of Savages Ditch itself opens up toward the west into a wider area we named the Funnel. On the immediate left at the mouth of the Ditch is a waterway about 25 meters wide that leads off to the south into a wide ditch (later called "Big Snail Ditch"). In the fall of 1995 I had an experience with this waterway that led to the name I applied to it. Recall that at that time, just before my hospital stay, I was walking very badly. I was exploring the Funnel, moving slowly eastward along the north shore of that long, narrow island that forms its southern border. It was already pretty muddy going. As I got closer to this waterway I thought I would like to see what the bottom in it was like, how deep it got, and whether there were any snails there. So I incautiously waded into it a bit. Bad mistake! It got deeper very quickly, not just the water but the mud on the bottom too. I was using two seven foot long poles to help me get around and keep myself steady. As I went down one of those got away from me and floated away south. The mud was very soft and on my knees I quickly found myself stuck in it to hip level. It only took a little thrashing about to get into this situation. Snails, smarter than me, did not inhabit that bottom. Where the mud gets too soupy they avoid the area. Over the bottom there was about a meter of water. As they say, so there I was, nobody else anywhere around, not walking well under even good conditions, up to my shoulders in water, and my legs utterly stuck in the mud. Fortunately, I had not lost both poles. Trapped as they were my legs were useless, but my upper body strength was still intact and useful. I used the one pole I yet had flat against the mud as resistance and ever so laboriously extricated myself from the mire. I crawled free and rested my muddy self on the muddy shore. I thought: "That was incredibly stupid. Make it a point not to do that ever again. No leaping before you see the other edge. Test before you venture forth."

This was an experience from hell and I decided to name the evil waterway "Hell's Gate." The long, narrow island just west of Hell's Gate, on the south edge of the Funnel, obviously had to become "Devil Island." At the western end of Devil Island, where the Funnel terminates, there is another waterway. This one is some 40 meters wide. It connects to an officially named major channel called Stockley Gut, which trends off toward the south. In this waterway the bottom is firmer and the water shallower. It is a friendlier place. This became "Heaven's Gate." On the west of Heaven's Gate there is another island, smaller than Devil Island, which I gave the name "Angel Island." Recall that the release stake is near the western terminus of the Funnel on the northern shore. The shoreline on the northern edge of the Funnel runs about northwest to southeast back toward the entrance to Savages Ditch itself. The stake is now uprooted and gone, but its position was sort of directly across the Funnel mouth (to the north) from the western end of Devil Island. As the crow flies from the stake to the western end of Angel Island is about 140 meters (about a football field and a half). About 10 meters west of the position of the release stake the shoreline abruptly turns north and extends that way for several hundred meters. Beyond that on the eastern shore of Rehoboth Bay is large area of saltmarsh islands and waterways called the Rehoboth Marsh. Offshore of this north-south shoreline and marsh is open Bay. That is the basic layout of the area. You can find a drawing of the area in Nathan's and my paper. There are photographs depicting the area in later publications about Savages Ditch. One of them can be found in Chapter 16. Or, should you choose, Google Earth on the internet [Rehoboth Bay (release stake) at about 38° 37' 42" N, 75° 04' 20" W] could be used to get a view of the area.

All the marked snails were now out. The next day we did a search was three days after releasing the natives, nine days after releasing sentinels. All three of us went down to the site and, as usual, we brought a canoe so I could keep dry feet. I hadn't yet learned how the tides behaved at this site. We got there about 2:30 PM and the water was still pretty deep. It was still above the exposed foot of the release stake, too deep to see any snails, marked or not. The estimates for the time of low and high tides were given for Indian River Inlet. That was only about a mile away, but tides were delayed by landmasses between there and Savages Ditch. It turned out that the delay was something like an hour and a half to two hours. Not only that, but tides were heavily affected by winds. Depending on direction they could lead to a very low tide or no low tide at all. I had to put up with some fairly severe groaning and ribbing from my colleagues for my error in timing. I got the tides better figured out, but never got perfect at timing them. By about 4:00 PM the falling tide began to expose the Funnel. With both Jeff and Nathan searching with sharp eyes and me sitting in the canoe recording data (and, I hasten to add, finding a few marked individuals near the stake), we found 117 snails that day. We had released 480 and found four dead the last time out. Not much time had passed, but 117 finds out of 476 snails was a red-letter day, as mark-recapture work generally goes. It spoke well for the future success of this project. Sentinels had been out longer and were the most dispersed, one of them 56 meters from release. In their shorter time, natives had also become quite well dispersed. Two of them, both infected with *Zr*, had moved 50 meters from release. That agrees with the 1985 results from Cape Henlopen: infected snails move about just as much as uninfected snails.

In the years between 1996 and 1999 (four spring-summer-fall periods) in various combinations we made 100's of trips to Savages Ditch to search for snails. More than 1,000 snail sightings were made and the data recorded. Some trips went pleasurably, some not so much so. Savages Ditch was at that time a very quiet out of the way place. Most of the time we would be alone (no tourists or park rangers) to do our work, which was always good. There was often (not always) a breeze blowing that kept biting insects from flying around easily and having at you. It could get very hot (at other times very cold) and thunder storms might boil across Rehoboth Bay from the west and spoil your day, but by and large it was a pleasant place to be, especially when snails were being found. To illustrate the sorts of things that transpired, let me recall one day. We had made an agreement to meet at the site for a fairly early low tide. It seems that both of my partners had been out the night before misbehaving. They both got there nevertheless, but hangovers were in full effect. Death warmed over would be a charitable assessment of their conditions. Of course I had figured the time of low tide a bit incorrectly. We were there early and the water was a little too deep to work effectively, hangover or no. After directing some probably well-deserved comments at me for my ill-timing of the tide, both of them found some shallow water and lay down in it on their backs with their toes and noses pointed skyward. From the canoe I can remember observing their quiet, still, half submerged prone profiles emergent from the cool, calm water as we waited for the water to recede. It must have been soothing because when the time came they got up and searched, long and effectively. Many snails were found that day. Such were the circumstances under which this phase of the work was accomplished.

In any scientific project data must be kept in good order. Data for the spatial positions and sizes of a large number of snails is a lot to get and record accurately. This is especially true when a searcher has to shout it across some distance in the wind and amid all the wisecracks and exclamations that go with finding snails. For each snail found, the mark number, distance from the release stake, direction from release stake, and the size of the snail had to be accurately recorded. It was at this time too soon to expect snails to have grown, but later on changes in the size of snails would be critical to Jeff's project. Frankly, I can't recall whether we had yet developed the procedure I will describe by this second day of searching. Anyway we soon did. I was mostly confined to the canoe all summer, although as I got stronger and better I began to venture out a little on "dry" intertidal ground. Later on, when snail finds got fewer and when the water was deep enough, I would pole my way around in the canoe looking over the side for marked snails. The wind was a constant challenge when doing this and made for pretty strenuous exercise, but I did find some marked snails. In the canoe I would receive and record all the incoming data. We used walkie-talkies to communicate between me and a distant searcher. Tongue in cheek, an exchange would go something like this with a tape recorder running. Like military folks, we would say "over" following each transmission. "This is Distant Searcher 1 calling lazy-ass Sedentary Base. This is Sedentary Base. We have snail number BR1 here. Base reads BR 1. How big is it? 23.2 mm. Base reads two three point two. Let me get distance and direction (some time passes). That's 55 meters at 247°. O.K., Distant Searcher 1 out. Sedentary base out." Jeff and Nathan would walk about sniffing out snails. Each of them would have a section of PVC pipe strapped on (Jeff's creation) that served as a quiver and was filled with wire-like metal stakes. When a snail was found they would mark its location with one of these. The snail was clamped in a clothespin. If needed, its mark was touched up or repaired and the snail was hanged on the stake in its clothespin to let the mark dry as long as possible before letting it go again. The mark number and snail size would be relayed to me, I would sight with a compass the direction to the snails position and get its distance from the release stake with a tape (if fairly close) or with an optical rangefinder. This instrument was sighted on a vertical staff held at the snail's position by the searcher. (In later years we replaced the optical device with a sonic distance measurer and ultimately with a laser rangefinder.) Great care was taken to say numbers clearly, repeating them often, so no errors would be made. Thus we would have recorded where a particular snail was (its polar coordinates) and its size at a specific time. Then it was on to the next snail. At the end of the day's searching all the data would be amassed on audio tape, which would be protected in a sealed plastic bag against any mishap. Dropping things in the salt water was an ever present and sometimes realized risk. Over the years I have destroyed quite a number of tape recorders, prize digital calipers, and other equipment that way. Oh well.

Back at the laboratory a substantial effort went into getting field data accurately into the spreadsheet. I developed a Quick Basic computer program that would take the data in the order acquired in the field and sort it into the order in which snails were entered in the spreadsheet. What would happen is that the tape would be played and as snails came up on the tape their data would be typed into my program, one snail after the other. When all snails were in, the program would sort the snails together with their data and a list would be printed out with snails in the order they would be encountered in the spreadsheet. From that list data would be entered into the spreadsheet. This kept you from having to skip around among 500 snails in the spreadsheet, depending on which snail was found when in the field. If you routinely had to do that, it would lead to entry mistakes. Besides, it was quicker this way. For Jeff , changes in size of individual snails over time was critical. For my personel and ultimate interest changes in spatial locations were key. I was stretching my programming ability pretty far to develop programs to plot the positions where snails were found on an overlay of the habitat. My programming skills were all learned "on the street" and this was anything but easy for me, but when accomplished it proved essential and most interesting. For students learning about how research is done, this business of keeping and managing data is among the most important lessons they learn. If you are not ready for it, it can be overwhelming. I'm sure, for example, that both Jeff and Nathan had handled small amounts of data from laboratory exercises, but they had never had so much to keep in order that was so important to the success of such an effort-filled project. This was not just another course undertaking.

With regard to Nathan's and my interest, the accumulation of trematode infections by populations of *Ilyanassa*, we used data collected in 1996, 1997, and 1998. Also, because they were related, Nathan and I also made use of data from native snails released on Cape Henlopen in 1993. In those years Nathan, Jeff, and I spent many months worth of days searching for snails. A single search could last anywhere from one hour to over five hours, depending on the tide. This does not include time preparing, on the road, cleaning up, and entering data. Naturally we found more snails at first and fewer later on. In 1996 alone there were 976 snail sightings. After that my colleagues had to be on the main campus in Newark working toward their degrees. They couldn't be so intimately involved with the project, but each continued to come and search for snails as he could. If it does eventually get accepted, it is still likely to be one or two years before a paper actually sees publication. There is value to a student in getting a paper published while still in college, or at least shortly after graduating. With this in mind Nathan and I decided to get on with writing ours in 1998.

Let me first deal with what happened in terms of the sentinel snails released at Savages Ditch. The idea when we released the sentinels on 8 June 1996 was that they would be allowed to roam the habitat until fall, at which time they would be recollected as found and dissected to see if they had gained infections during their time on the loose. The place where snails could be most consistently searched for and where most sightings were made was the Funnel. It is a confined area and the water either leaves or gets shallow on most low tides. Each month in 1996 and in the whole summers of 1997 and 1998 we always sighted more natives than sentinels. We released more sentinels than natives and the reason for this likely is that they vacated the Funnel fairly soon after release and simply were not sighted. A reasonable answer to the why-they-left question would come from another experiment that I later did. For now they had just disappeared, but not altogether and not instantly. We made 388 sightings of sentinel snails during the summer of

1996. We were going to start collecting them in September. Of course by that time they were all but not to be found. In all we were able to recollect 16 sentinels. In 1996 we got 12, two of which carried infections. Good old BR16 (blue red 16) infected with *Hq* had been free for 104 days and from the infection's state of development (immature) it was likely produced by a miracidium that infected the snail at Savages Ditch. Yellow red 83 (YR83) carried *Zr* and had been free 119 days. That infection had likely come with the snail from the Lewes launch ramp because it already was mature (producing cercariae) and rife throughout the snail's tissues. In 1997 three sentinels were recovered, one of which (YR30) was infected with *Gynaecotyla adunca (Ga)*. This infection could have been transplanted with the snail, but we concluded it wasn't because *Ga* is extremely rare in the source habitat, making it very unlikely that it was transplanted. Finally, one more sentinel (BR28) was collected in 1998 after it had been free over two winters and for 776 days. It was uninfected at recovery. The upshot is that two of 16 sentinels appear to have gained new infections while free in the Savages Ditch habitat, which comes to a 12.5% colonization rate. However, it was over two years so the estimate becomes 6.25% of uninfected snails become colonized by trematodes per year. If we are wrong about that *Zr* infection, the estimate is 9.4% per year. This is a good bit higher than the 1.6% I earlier found on Cape Henlopen (Chapter 9), but is still quite low. Besides, the Savages Ditch estimate would have to be used cautiously because it is based on so few recovered sentinels. Anyway there it is, the second field estimate for the rate at which uninfected marine snails become infected. It still surprises me that there are not more such estimates in the literature.

A brief note about marked snails found dead in the field is warranted. As I said before not much of this occurred on Cape Henlopen suggesting that Savages Ditch is a riskier place for a snail to live. Could this explain why fewer sentinels were always sighted? During the summer in 1996 we found 10 sentinels and eight natives either as empty shells, or shells taken over by hermit crabs. These shells all had mark numbers on them so we knew "who" they were before just dying or being killed. Statistically speaking, the same number of identifiable sentinels and natives were found dead. In addition to these carcasses we also found several that had indications (e.g. red on the tail) of once being marked, but no number and we couldn't be sure whether to put them with sentinels or natives. Shells without living snails inside were often severely damaged. Whatever happened to them was not pleasant, often involving extensive mauling. It's a mystery what killed them. There are a number of suspects (hermit crabs, blue crabs, skates), but it is unsolved. Clearly sentinels did not preferentially stay in the Funnel. That is the major reason why they tended not to be sighted or recaptured. More on this later (mainly Chapter 17), but they just moved out into other areas.

Now we will enter the more murky (but consequential) world of already infected snails and whether they change their infections as time proceeds. If you want to make repeated observations on a snail over time, it has to remain alive in the field and be found again and again. This is a murky area because there can be no dissection to be absolutely sure what species of larval trematode(s) infect the snail, except maybe when it is collected for the final time. With present techniques, if dissection is not possible one is dependent on cercarial release to determine what trematode or trematodes are present in a snail. This makes the analysis of results far from straight forward and cut and dried. The uncertainty of cercarial release discussed previously (Chapter 8) is the root of the problem. If a snail sheds cercariae of a species, the snail is definitely infected with that species, but failure to shed a species of cercariae does not guarantee that the snail is not infected with that trematode. In short, a positive is a definite positive, but a negative leaves questions.

Let's consider the Savages Ditch natives first. Nathan and I released 231 native snails that had been tested to see what cercariae they would release. Most of them demonstrated infections, but 28 did not. The question always is, were there infections present where none was demonstrated and were there other infections present besides those demonstrated? Sorting this out requires dealing with some uncertainty and making informed judgements. In 1996 from September on we recovered 26 natives and dissected them to determine infection status. Twenty-two of them had the same infection that was revealed by cercarial release at the beginning of the summer. That is, no change in infection status was indicated for these 22 over three or four months of freedom. Among the other four one released no cercariae initially, but had *Ls* when dissected. All indications are that a new infection takes many weeks to develop to maturity. Also, recall that infected snails grow very slowly. This snail was 27.3 mm in shell height when released, was free for 114 days, and grew only 0.1 mm in that time. That infection was likely there all the time, but it failed to release cercariae when initially tested. The other three revealed only *Hq* initially, but also had *Zr* when dissected. This is a common double infection at Savages Ditch. In that 1995 sample 25% of the snails dissected had it. Further, that double often fails to kick out both kinds of cercariae when tested. So we were suspicious. These three snails had been free for from 99 to 114 days. In two of them both infections were producing cercariae (mature) and extensive throughout the snail's tissues. Based on this our judgement was that those infections were present at release in June and not new. However, in the third snail the *Zr* infection was immature and quite limited in extent in the snail's tissues. Our conclusion was that this was a case of *Zr* newly adding itself to *Hq* to form a doubly infected snail (*HZ*) sometime during the summer of 1996. Of course, we are guessing (we think in an informed way) about these scenarios, but it appears that a miracidium made it to one out of 26 snails (3.8%) and changed its infection status. That change involved the formation of a double infection. Well, *HZ* double infections are quite common and they must form sometime. We were pleased to have been on the scene and doing an experiment in the field that allowed the "witnessing" of one coming into being.

Most of the marked native snails escaped recapture and being dissected in 1996. We were able to find 35 of these during 1997 and 1998, in their second and third years after release. We even found 10 of these snails three successive years running. Let me summarize what happened to infections that these snails carried. Of 35 snails free in the Savages Ditch habitat from June of 1996 into the succeeding year or two, 30 gave no indication of having changed infection status and we accepted that there was no change in infecting species composition in those snails during that time. So at least 85.7% of them had experienced no change. Two of these 30 snails were dissected. Both shed *Hq* cercariae in June of 1996. They were found in June and August of 1997 and both revealed *Hq* infections when dissected. The 33 others of the 35 were collected as they were found, but not dissected. We thought ourselves very lucky to find them. Snails free in the wild for that long and then recaptured are a rarity. Rather than put these guys under the knife we brought them back to the laboratory and subjected each one to another cercarial release test. They would be re-measured, the mark repaired, and within a day or two brought back to the habitat and re-released where found. Maybe we would find them again and many we indeed did. Anyway, that leaves five snails to discuss that appeared to change infections while in the field. Two of these five released no cercariae when initially tested in June 1996. One of these was not seen in 1997, but released *Ls* cercariae in June of 1998. It had been free for 720 days and showed only 1 mm of growth over that time. The other shed *Zr* cercariae when tested in June of 1997. It had been free for 367 days and had grown only 0.7 mm. During their time free these two had clearly grown as parasitized snails do (slowly). These infections were most probably not newly acquired after release in 1996. Both were likely present at release, but failed to release cercariae when tested at that time. Two others definitely had *Hq* in June of 1996, but later in addition revealed *Zr*, one in May 1997, the other in July 1998. It is an educated guess, but both species were probably present in both snails at release. They just did not reveal both infections when tested in June 1996. The last snail of the five shed both *Hq* and *Zr* in 1996 and 1997, but when recaptured and tested in March of 1998 it only shed *Zr*. This suggests that one species (*Hq*) had been eliminated from the snail. I wish we could have found that snail again to retest it, but no such luck. There will be more on this, but it is unlikely that *Hq* was evicted by *Zr* from the snail. It is more likely that both infections were still present in 1998, but only one revealed itself.

If we take these results at face value, five out of 35 snails (14.3%) possibly experienced infection status changes. However, as described above it seems likely that most or all of these apparent changes may reasonably be discounted as not real. So, we concluded that zero out of these 35 snails actually changed infection status. In the final analysis data from Savages Ditch indicate that changes of infection status (from uninfected to infected, or additions/eliminations of infecting species) are pretty rare events. The nature of the system requires some guesswork, but it certainly looks like infection changes occured in well under 10% of snails per year.

What about the native snails released onto Cape Henlopen in 1993? This gets even murkier. Perhaps you recall that 300 native snails from the tide pool near the release stake there were tested by cercarial release in early June 1993. They were marked, measured, and ultimately released on 10 June. In that year from 10 September on they were collected and dissected to assess infection status. There is one exception. One snail was found after 439 days and retested by cercarial release. For science I probably should have, but I couldn't bring myself to kill it. Nathan and I made comparisons between initial infections detected and final infection statuses. It is useful to bear in mind that these 300 natives were released along with 200 uninfected sentinel snails (Chapter 9). Sentinels were released in the same area, were out during the same time, and 32 were recaptured in fall 1996 and dissected. None of the 32 had picked up new infections.

Let me first dispense with the snails that did not appear to change infection status. Of the 300, Karen and I were able to recollect 62 snails after 92 to 439 days of freedom. Of these, 37 (59.7%) gave no indication of having changed infection status. The one free for 439 days was one of these. It demonstrated *Hq* initially and also when recollected.

There is no escaping that 25 snails (40.3%) gave some indication of a change in infection status. It seemed strange to Nathan and me that trematodes had colonized none of the recovered sentinels, but there appeared to be so many infection changes among native snails. At first we thought that maybe there was a parasite susceptibility difference (a genetic difference) between Henlopen and Canary Creek snails (the origin of sentinels). But, that's very unlikely because indications are that planktonic snail larvae (veligers) spread out to genetically homogenize the global snail population.

Growth rate of a snail helps to tell if it is parasitized or not. Growth rate of these native snails was measured and we used that to give us a clue whether a snail that originally tested as uninfected was actually so. We also used Karen's and my work on the likelihood that a multiple infection will release all cercariae when tested as a source of clues. Of the 25 natives giving some evidence that a change of infection status had occurred, 15 of them were snails that did not release cercariae initially, but had infections when recollected and dissected. Growth rate over the summer would be a help with these 15 in determining whether snails were in reality originally uninfected. We developed a criterion. It was: if over the summer the snail grew 0.6 mm or more, that indicated the snail was not infected. All but two grew very little, were clearly infected for the whole summer. In all likelihood they were infected initially. So 13 of the 15 apparently new infections can be reasonably discounted. The other two not releasing cercariae initially showed *Hq* infections when dissected on recapture. Over 92 days of freedom they grew 0.8 and 1.9 mm. We accepted these as new infections because the snails had grown substantially and they must have been uninfected for at least part of their time free. The other 10 of the 25 snails demonstrated an infections initially, but had a different infection when later dissected. What about them? How likely is it that they in reality changed infection status? In seven of these 10 cases double infections with that common pair, *Hq* and *Zr*, were involved. In all of these one of the species showed up in the initial test, but on later dissection both were present. Control samples taken in 1993 to measure how frequent different snail infections were in the release area showed that 9.9% of snails in the area had *HZ* double infections. Moreover, Karen and I had found that this double infection fails to reveal both species by cercarial release 70% of the time. It is possible that among these seven snails one or two may have acquired new *HZ* double infections during the summer of 1993, but is more likely that these snails were doubly infected with *HZ* from the start. Among the remaining three snails two released cercariae at the beginning of summer identified as *Zr*, but when recovered and dissected at summer's end they had a different species infecting them, *Hq*. The indication is that *Hq* evicted *Zr* from these two snails. For one thing these two species can coexist in a snail without one evicting the other. At Savages Ditch there are two cases (snails) where the double infection was observed to long persist, for two and three summers. These apparent changes were probably caused by an initial misidentification of cercariae. It's my fault. During the initial cercarial release test I left snails in the release cups for too long a time (at least 66 hours) and when I looked cercariae had encysted as metacercariae on the walls of the container. At that stage in my experience with metacercariae it is all too probable that I mistook encysted *Hq* for encysted *Zr*. In other words, these snails were probably always infected with *Hq*. The final snail of these 10 emitted *Ls* cercariae at the start of summer and on dissection after 115 days of freedom had *Austrobilharzia variglandis (Av)*. Cercariae of the former species have an unforked tail with bristles on it, while cercariae of the later species have a forked tail and no bristles. There is no way to mistake one cercariae for the other and we accepted this as a real change. In Australia there is a congener (a species belonging to the same genus) of *Av* that has been demonstrated to invade and eventually evict other trematode species. Likely that is what happened here. So, in summary among the 25 snails that appeared to change infection status, our conclusion is that there were two uninfected snails that had become infected with *Hq* and one that had a replacement of one parasite species (*Ls*) by another (*Av*). The rest of the cases probably can be discounted. Maybe one or two should not be, but that won't change things much. At face value data indicate that 40.3% of snails changed infection status. There is some guesswork involved, but it is more likely that only three out of the 62 recovered natives (9.7%) actually did.

Why does it matter whether parasite species in a host often get changed around or not? Well, it's about the very nature of parasite infracommunities, or species assemblages, and ecologists are interested in what goes on there. Does species composition change frequently? How dynamic are they anyway? In the literature two extremes have been identified, interactive and isolationist communities. If previously uninfected hosts are often infected and already infected hosts routinely experience new parasite species trying to infect them, and perhaps the resident parasites being evicted, that would be called an interactive parasite community. You can imagine a host living in an environment in which the host is bombarded by parasite infective stages. A parasite in such a community often has to interact with other parasites. The key feature of such a community is a high rate of colonization. This sort of situation has been found to be the case, for instance, in the communities of adult helminths (parasitic worms) in the intestines of vertebrates. At the other extreme is the situation wherein a parasite infecting a host is very unlikely to have to deal with other parasite species. Colonization of hosts by parasites is a rare event and it is likely that a species infecting a host will get to exist in isolation from other parasite species. A low rate of colonization is the key feature and such a parasite infracommunity is termed isolationist. Whether larval trematode infracommunities in snails tend to be interactive or isolationist has been considered by only a few authors. Most of the attention has been directed at adult helminths in vertebrate definitive hosts. To distinguish these two sorts of communities some sort of break point should be decided upon. That is, if colonization rate is above this level then the community is interactive, if below it's isolationist. But, no such demarcating level of colonization exists. Some have looked at snail-trematode systems and, though there are few measurements of colonization rate, concluded that infracommunities are interactive. Certainly, with the low colonization rate of uninfected snails in evidence and the few changes in infecting species composition in snails that seem to occur, the indication is that trematode infracommunities in *Ilyanassa* are near the isolationist end of the spectrum. Single infections are most common and parasite species mostly exist in snails isolated from all other parasite species.

Let's try to sum this all up. Sentinel snails provide a good bit of the evidence that trematode colonization rate of *Ilyanassa* is low, probably under 5% of snails each year. If we summarize results from both sites combined with regard to native snails, taken at face value 34 of 123 (27.6%) native snails re-evaluated for infection status after being free in a habitat for at least one summer appear to have undergone some change. This would be substantial if it were actually the case, but as explained above most of these cases can be reasonably discounted. On Cape Henlopen, let's say for the sake of argument that two of the seven apparently new *HZ* doubles were actually new. Beyond this, there were also two cases where previously uninfected native snails became infected with *Hq*. And, there was one snail that exchanged a *Ls* infection for an *Av* infection. At Savages Ditch there was pretty definitely the formation of one new *HZ* double infection. Maybe a few more, or possibly fewer, changes were real, but we end up concluding that six of 123 snails (4.9%, rather than 27.9%) actually changed infection status. At least at Savages Ditch, this was partly over multiple years. So it would therefore not be correct to say that probability of an infection change is 4.9% per year. It is less than that. We recognize the guesswork involved in this rough estimate, but even with more certain knowledge it is not likely to climb much above this. When we were writing our paper Nathan came up with a way of phrasing the situation that is clear and has been influential. It has been cited by other authors several times. It goes as follows. Suppose we accept as an estimate that 10% of snails experience an infection change each year. This means that on average 10 years would go by before a change in infecting species would occur. Such parasite infracommunities would almost have to be seen as isolationist and far from interactive. Our estimate is closer to 5%, which would mean that on average 20 years would go by before an infection composition change took place. As far as the larval trematodes of *Ilyanassa* are concerned, we came down proposing that parasite (infra-) communities are on the isolationist end of the spectrum.

We are not done with Nathan, as he co-authored another publication with Jeff and me, but let me sort of close him up here. Nathan is a contented explorer at heart whether it is big territory or small. In 1996 and 1997 I owe him for exploring Rehoboth Bay with his crippled professor. I never could have done it without him and the process really helped me to understand Rehoboth Bay and the snail population in it. It was all done by canoe, sometimes with misadventure a constant possibility. After using Nathan's canoe and truck a couple of times we began using my canoe and truck. It was an aluminum Grumman canoe about 18 feet long that I had purchased back in the early 1980's, a very sturdy and capable craft weighing about 75 pounds. My truck at that time was a 1987 Bronco. With its V8 engine and four on the floor, I really liked that vehicle. On some days we would go exploring to kill time when the water was yet too high to search for snails, on others we would go expressly for the purpose of exploring. One of our first outings was to look into southeastern Rehoboth Bay. This is an area of saltmarsh islands and channels just south of Savages Ditch, maybe one mile north-south by three quarters of a mile east-west. There is a huge snail population among the many saltmarsh islands there (millions upon millions of snails). Before our canoe excursions into the area, I was totally unaware of the layout and the snail population there. The explorations were instructive. On one of our trips in that area we happened upon a dead seal washed up on a small beach. That was a surprise. It's not often you see that around here. North of our research site, we also explored Rehoboth Marsh. There were few snails there. I can remember going down some very narrow ditches, barely wider than the canoe, hoping beyond hope that we would not run out of water under our keel on the falling tide. If we got stuck, I sure couldn't walk out of there. Turning around would be a major undertaking if we could do it all. Sitting there waiting for the tide to come back up, biting flies would show us no mercy.

In July of 1996 we took a run down the western side of Rehoboth Bay. That was an adventurous trip to say the least. The Bay is about five miles long north to south. We put in at the north end and paddled along the shore to the south, our faces and the canoe working against a brisk southwest wind. When we would look along shore we could find few snails there, although we did find an isolated population at one protected site near a place called Camp Arrowhead. The snails there were all very old (big). I wondered how long it had been since *Ilyanassa* veliger larvae had made it to them and contributed new juveniles to the population. It could have been 20 or 30 years, possibly more. The southernmost point of the day's journey, about four miles down the Bay, was eventually achieved. We looked around for a while and then started back. We were in for a faster, thrilling trip. If we went offshore a bit, it would be a pretty straight line back to where we had put in. Heading north the wind was strongly at our back and it eventually pushed us perhaps a mile offshore, a little farther than we had intended. The water was choppy and there were wind-driven swells, higher than the gunnels of the canoe, overtaking us from the stern. That's where my paddle was working and I was expecting to get swamped at any moment. It was a wet and wild ride. Up in front Nathan was absolutely soaked. My foot, or anything else, didn't stay dry that day. With the adverse wind it had taken us about four and a half hours to get south, but the trip back only took 40 minutes. It was a fun experience, in retrospect, but we were both happy to get back to our landing. We would have been able to handle it, but had the waves filled the canoe with water it would have been a major deal getting us and the canoe back toward shore and afloat again. These explorations involved more than just the events covered, but overall it was quite an experience and you get the idea. Seeing the world from a canoe gives you an intimate perspective.

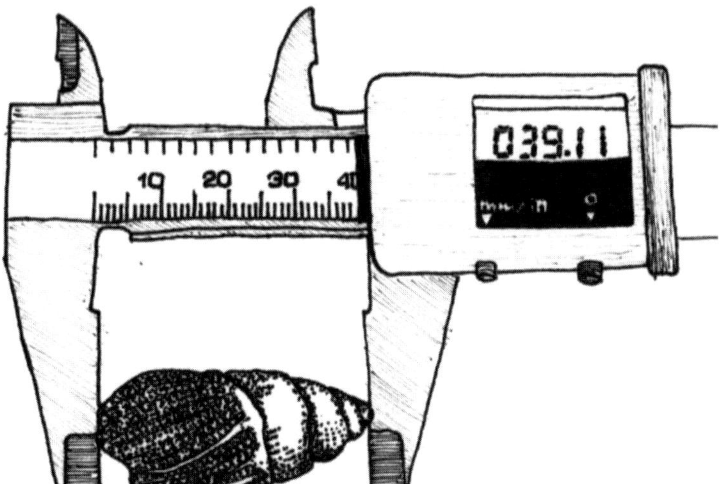

Record *Ilyanassa obsoleta* from Savages Ditch in 2003, at this size (39.1 mm) ~75 years old.

Ultimately, Nathan finished his degree in biology at the University of Delaware with two undergraduate publications. Having been a co-author on both of those I can only say that his contributions in fieldwork, thinking, and writing were essential. He may have been a novice biologist at the beginning of his intense affair with research, but by the end he had mastered the subject matter and was able to add substantially to the ultimate product. A lot of his time over about four years went into it, not to mention interest and ability. He is to be congratulated on his undergraduate career. Nathan is an interesting and talented guy. He has become and remains a valued friend to this day. I am honored by that. I am also thankful for the illustrations he contributed herein. I think they are pretty cool. He did not go on into officially doing biological science, although he certainly could have. Nevertheless, biology is still a part of making him what he is. After graduation such real adventures as hiking the Appalachian Trail from Georgia to Maine and a major kayak trip came to take up his time. The kayak trip was north along the Mackenzie River in the Northwest Territories of Canada. That river is about a 1,000 miles long and flows north to the Arctic Ocean. He paddled the whole way. Wow! He's into different things at present, but at that stage of his life it couldn't have been any other way.

Chapter 12

REALLY OLD GIANTS

Now onto what Jeff Kinley's project told us about the snails and trematodes, specifically at Savages Ditch, but in a more general sense too. This is first told in Jeff's 1998 undergraduate thesis "Growth Rate and Inferred Life-span of Giant, Parasitized *Ilyanassa obsoleta* at Savages Ditch in Rehoboth Bay, Delaware." We later further developed and published his thesis in Curtis, Kinley and Tanner (2000), a paper that I believe revealed a lot about how *Ilyanassa* and its larval trematodes function in this world. As studies of snail-trematode systems go, this was exciting stuff. Pointedly, Jeff's work was aimed at determining whether the strangely large size of snails at this site was caused by a faster growth rate, or a longer time growing. The work depended on measuring changes in size (growth) of the sentinel and native snails we released in June of 1996.

I had picked up on the extraordinary longevity of individuals of *Ilyanassa* based on work done on Cape Henlopen (Chapter 9). The general conclusion there was that the snail can live to be about 40 years old. I referred to this in Chapter 9 as my Darwin moment. That was perhaps presumptuous, but it was my moment and I guess I can call it what I want. The reason I called it that was because individual snails in a population and their trematodes have longer histories than I had previously thought. Before the Henlopen work I was wrongly thinking that a snail could live maybe 10 years, which meant that a trematode infection had only that amount of time or less to acquire and exploit a snail host. The host could not be expected to live much longer than that and so the maximum parasite infection longevity was also set at about 10 years. When a host dies any parasites within it (the infrapopulation or infracommunity) must also die. A parasite had better have reproduced successfully before that happened. For a larval trematode infection in an *Ilyanassa* individual, this meant that enough cercariae had to be sent out within that 10-year or shorter window such that enough next hosts in the life cycle got infected. That's what it's all about in biology, maintaining yourself while you do it and perpetuation of your genetic lineage by means of (reproducing) descendants. Of course, being wrong about snail life span meant that I was wrong about the whole system. I originally had no idea of the time span available for all the diversity and complexity in the whole system to come into being. I did understand that any features of the system that had developed by natural selection across generations might have taken a very long time (decades, centuries, millennia) to originate, but I had no idea that many of the more immediate features observed in the system (e.g., snails and the abundance and variety of single- and multiple-species trematode infections in them) had developed over several decades rather than just a few years. You will have the chance to judge whether our reasoning about our data is correct, but one of the fallouts from our project was that potentially the snail could live for 70 or 80 years, about twice as long as I had thought, even based on Cape Henlopen work. Somewhere there may be a so far unstudied snail that can live as long, but it right now appears that *Ilyanassa* has the greatest longevity of any snail on the planet. This puts a whole new cast on its ecological importance.

A host species that is made up of long-lived individuals, by virtue of that fact, can have long-term influences on the environment. Just by living (moving around, feeding, eliminating wastes, reproducing) in a habitat for so long makes the individuals of the species have enduring influences. They imbue their habitat, their environment with a history. If there are parasites living right along with host individuals, there would be even more influences stemming from them. After all parasites have to make their way in the world too. At the least, in the case of these snail parasites, they would keep putting out infective stages (cercariae) to infect a next host. During their great longevity, how many second intermediate and definitive hosts have been infected? How much does that have to do with the ecology of the estuary where this all goes on? Even if unknown in detail, the influence could be immense.

There are a number of marine snails and other organisms in the sea known to have extended longevities. Among snails there are several examples: the turbin snail is known to attain 30, a keyhole limpet lives 20, and other limpets reach 30, one possibly 60 years. Among bivalve mollusks (clams) there are species that reach 35, 50, and even in excess of 100 years. Relatives of sea stars (star fish), there are brittlestar species that live for 20-25 years. Another sea star relative, a sea urchin, probably lives for 100 years. There is a deep sea worm called a vestimentiferan that bears away the palm for longevity. Studies have shown that it lives for 170-250 years. So far as I know, and I keep an eye out for such items, parasites have not been considered for any other examples of long-lived marine animals. They surely have them. About half the species of organisms on earth are parasitic. They all have to be somewhere and a species that supports no parasites is a real rarity.

Let's briefly review what we initially knew about age, growth, and parasitism in *Ilyanassa*. After drifting about in the plankton for two or three weeks, during summer an *Ilyanassa* veliger larva settles out of the plankton and metamorphoses into a bottom-dwelling juvenile snail. At that point it is about 1 mm in shell height. Assuming of course that it lives, by the end of that settlement summer it reaches about 6 mm. With low temperatures it stops growing during winter. By the end of the second summer it would be about 12 mm. In the third summer it becomes a snail capable of reproducing (at about 14 mm) and by the end of that summer it reaches around 18 mm. The chance that a snail will become parasitized by a trematode is only a few percent a year. A small percentage of snails under 18 mm (exposed to infection for just three years or less) are infected, but there is a small chance that a snail in this size range would have been already infected by a trematode. Each year there is always that chance of picking up an infection (being colonized by a miracidium) and as the snail grows and gets larger (more years) there is an ever increasing chance that in one year or another it will have become infected. Another thing has happened in the third summer. The snail has entered that composite size-class made up of snails 18 mm and larger. These guys can't be separated out into year-classes based on size, as could be done during the first three years. The only present way to get even a vague idea of how old a snail in this size-class is requires releasing marked snails and studying their growth rate, as was done in that previous study on Cape Henlopen (Curtis, 1995). If our snail stays uninfected, growth rate slows to around 1 mm a year once it hits the composite size-class, but this is not a regular thing. There's lots of variability: in one year it may grow twice that, in another not at all. If the snail becomes infected, on Cape Henlopen its growth rate is curtailed to something like 0.2 mm per year. That's some five times slower than an uninfected snail might grow. By putting brackets on snail ages it was found on Cape Henlopen that snails might live for perhaps 40 or 50 years. Also, there were strong indications from the Henlopen study that an individual infection can exist in a individual snail for 10 or more years. From other work (Curtis, 1990) it was apparent that snails were not much debilitated (beyond sterilization) by the infections they carried. They moved around at least as much as uninfected snails. Imagine the ground covered over 40 or 50 years (more about this in future chapters).

In essence that's what we knew as Jeff's project began. Henlopen snails over 25 mm in shell height were scarce. However, snails of this size and much larger were common at Savages Ditch. There are a few such large snails at other sites in the Rehoboth Bay area, but they are a rarity. Why should Savages Ditch have so many giant snails? What brought about the difference? Were they growing faster or living longer?

In the previous chapter I described the marked snails that we released at Savages Ditch in June 1996. There were 249 sentinel snails transplanted from low-prevalence sites and 231 native snails collected in the Funnel area. The sentinel snails tested as uninfected while the native snails were all, or virtually all, parasitized. They were released in two batches, sentinels first native snails second. In the published work, size of some of the sentinels were tracked for up to three summers, mainly in 1996, but also in 1997 and 1998. Released natives were found and size tracked over four subsequent summers. Jeff had to operate on data gathered over shorter time, mainly 1996-1997, because he had to deal with the practical matters of getting his thesis written and defended and graduating from the University with his degree in biology.

In terms of growth, what happened with the sentinel snails? These snails, like native snails, were released at the mouth (wide end) of the Funnel where it opens into Rehoboth Bay. We were able to find quite a few sentinels through the summer of 1996, but as indicated in Chapter 11 by the fall of that year they were mostly never to be seen again. Maybe they moved away, or lost their marks as they grew. In any case, almost entirely they could not be found. A total of 86 sentinels, ranging in size at release from 17.4 to 26.2 mm, was tracked over the summer of 1996. Many were found and measured several times. This told us that growth among these uninfected snails was not a regular, or evenly paced, phenomenon. Changes in shell height were highly variable, with both gains and losses being manifested. There appeared to have been a growth spurt for most of these sentinels from 60 and 90 days after being set free. This is reminiscent of what happened on Cape Henlopen, but there the growth spurt happened sooner after release. Anyway, growth clearly takes place in episodes and is not a matter of consistently increasing in size as time proceeds. For our characterization of sentinel growth the data we used were based upon the 86 sighted sentinels. We noted the differences between shell heights when released and found for the last time, i.e., terminal growth measurements.

What did we do with these 86 terminal growth measurements? First, we made a graphic scatter plot. For each of the sentinels, total days free was plotted on the horizontal axis, while shell height change during that time was plotted on the vertical axis. That is, on this plot a single sentinel snail would be represented by a single point, the position of which was determined by the number of days free and by the amount of shell height gain or loss during that time. The whole cluster of 86 data points on the graph showed the relationship between time snails were on the loose in the field and amount of growth. There was a lot of growth variation. Some snails gained size, some lost. To give you an idea of this, in August after being free in the habitat for 60 days there were 14 sentinels on which terminal growth measurements were made. Growth among these snails varied from -0.6 to +1.6 mm. The whole cluster of data points indicates that the longer a snail is out the more it is likely to grow. A mathematical formula can be built (using a statistical procedure called linear regression) allowing prediction of growth if a snail is out for some specific number of days. For completeness, that formula came out $y = -0.4231 + 0.0137x$. In the equation y is the predicted shell height change and x is the number of days out. [I won't go very far into what the other numbers in the equation are. That would take us too far from the main point of this. If you're interested, regression techniques will be covered in a beginning statistics book. You would rapidly discover that -0.4231 is the intercept (where the line intersects the vertical axis) and 0.0137 is the slope (the amount that shell height changes as days out goes up) of the regression line.] As one example, if a snail were loose in the Savages Ditch environment for 140 days (the longest time any of the 86 snails was on the loose) the formula predicts that its growth will be 1.5 mm. Try out the formula, haul out your calculator, and see for yourself that this is indeed what it predicts.

As indicated, 86 sentinels were recovered and measured in 1996. Also, as mentioned in Chapter 11, two sentinels were recovered in the summer of 1997 and one in summer 1998. The two found in 1997 were free during two summers and had grown 2.0 and 2.8 mm. The 1998 find was free during three summers and had grown 4.0 mm. These snails showed about the amount of growth over multiple summers that would have been expected if growth measured in 1996 continued. So a growth of about 1.5 mm a summer season for an uninfected snail at Savages Ditch is a pretty good prediction. It is similar to what was earlier found on Cape Henlopen (average growth there was estimated at 1.2 mm per year). Given the amount of variability in snail growth, it is suggested therefore that growth rate in the two habitats is about the same for uninfected snails. Snails are bigger at Savages Ditch not because they grow appreciably faster; the biggest ones must have been growing for longer.

A second thing we did was push the analysis of uninfected snail growth a bit further. We constructed a general model of growth in size of *Ilyanassa* at Savages Ditch. The model would portray the pattern by which size might be expected to increase over the whole life of a snail. Of course, this portrayal is a model of what actually happens based on the data collected, not what actually happens. In mind you probably have a model of a perfect sunset. A model of whatever sort depends on data (previous sunsets experienced) and assumptions (rules about which sunsets you liked) that go into it in the first place. People who study growth of organisms (patterns according to which body size increases) like to make models because, for one thing, they can be used to predict how long on average it takes for an individual to attain a certain size. The initial step was to take the net terminal shell height change (growth) of each of the 86 sentinel snails and divide it by the number of days the snail was free in the habitat. This gave us average growth per day of each snail. So, for example, if a sentinel snail was out roaming around in the Savages Ditch habitat for 60 days and it had changed its shell height by +0.6 mm during that time, it had grown (0.6 mm divided by 60 days) 0.01 mm per day. That's straightforward enough. After doing the calculations we had 86 measurements of growth per day. A major question is: how many days during the year would a snail grow? It's time for a major assumption (rule) of the model. Based on data going back to work L^2 performed more than 20 years previous, we knew that snails stop growing during the cold part of the year. Our rule would be that snails are able to grow for 180 days each year. This may well be an overestimate. That is, snails might not actually grow during half the days of the year. Recall and that growth is variable through time. Snails have growth spurts. If it is an overestimate, the model will predict more growth in a year than actually occurs. In another sense, this would make the model underestimate, not exaggerate, snail age at a given size, i.e., the model is made conservative. We took the calculated growth per day for each sentinel and multiplied it by 180 to estimate a year's growth. In our example above, if a snail grows at the rate of 0.01 mm per day for 180 days that comes to (0.01 X 180) a growth of 1.8 mm per year. This would be reasonable, if perhaps a bit optimistic. If a snail was 18 mm in June 1996 when released, the prediction is that after one year in the Savages Ditch habitat it would be (18 + 1.8) 19.8 mm. So, one of the 86 uninfected sentinel snails released at an initial shell height in one year in the Savages Ditch habitat could be predicted to attain a certain shell height after one year of growth. With initial size represented on the horizontal axis and predicted size after one year on the vertical axis, the plotted data for the 86 sentinels would be called a Ford-Walford plot. It might be helpful to sketch out on a piece of paper roughly what this graph would look like. On such a plot the 18 mm initial shell height snail, used as an example just above, would be represented by a single point located at 18 mm along the graph's horizontal axis and at 19.8 mm along the vertical. If snails did not grow at all, the 86 points on this plot would fall on a single diagonal line extending from the graph's lower left to upper right. In other words, an 18 mm snail that failed to grow would be represented by a point at 18 mm on the initial size axis (horizontal) and at 18 mm on the size after one year axis (vertical). Likewise a 20 mm snail that did not grow would be represented by a 20 mm point on both axes and so on. Some snails did grow. Points above this zero growth line represent size gains; those below it represent losses.

When you look at the actual Ford-Walford plot of these data (it's in the publication) it can be seen that most snails were above the zero growth line (gained in size) and some were below (lost). Another feature of the graph is the large amount of scatter in the points. Predicted growth per year, based on the growth per day calculation, was not uniform for all snails. Snails were predicted to lose as much as 2 mm and gain as much as 5 mm. The third step in our growth analysis was to assimilate the data in the Ford-Walford plot into a summary growth model for *Ilyanassa*. There are a number of possible models we might have chosen. The von Bertalanffy growth model is applicable to many growth situations and the one that seemed to fit best. This model has a complicated formula, which I won't burden you with, in which the actual growth data from a study are taken into account. What you end up with is another graph, a von Bertalanffy curve. It is presented in all its glory just below. This a modified version of the curve as presented in the published paper.

If for a moment you can contain your wonder at what a thing of beauty this graph is, there are some things to note. When the snail is younger it puts on size relatively rapidly, but size is added more and more slowly as the snail approaches 30 mm in size. Based on the hard-won data from the 86 uninfected snails that went into the construction of this model, the biggest an uninfected snail can get to be is just under 30 mm. Snails get bigger than 30 mm at Savages Ditch, but they are virtually always parasitized. This accords well with observations over many years and many snails. We said in our paper that dissecting (killing) large snails is tantamount to cutting down the oldest trees in an old-growth forest. When you do that you are removing a lot of habitat complexity and history from the whole ecological system that is not easily replaced. When you do a cercarial release test on 30 mm snails of this size and age, you will occasionally run into one that won't produce cercariae, but almost never. Not shedding cercariae does not guarantee that the snail is uninfected and you wouldn't want to kill it to find out for sure. We did as little killing as possible in this work. In the curve it can be seen that an uninfected snail is about three years old at 18 mm, then (based on this standard model's handling of the data) it is predicted to be five years old at 20-21 mm, 10 years at 24-25 mm, about 25 years at 29 mm, and about 40-45 years when it nears 30 mm. By that time (40-45 years) a snail is virtually assured of having been infected at some point along the way. This curve became our summary of uninfected snail growth at Savages Ditch. Snails do not often get as big on Cape Henlopen, but this pretty well matches what was found to be going on there.

The analysis of growth for infected snails was done similarly to that for uninfected snails, but no von Bertalanffy growth model could be constructed. The reason is that once a snail is infected its growth rate diminishes radically. As I have noted previously, on Cape Henlopen it went from around 1 mm per year to around 0.2 mm per year, five times slower. The problem in trying to estimate the age of an infected snail from its size is that you don't know when the snail became infected, i.e., how long it has been growing at the diminished rate. Thus a growth model could not be applied. Nevertheless, it appears that infected snails can be very old. Let me offer an explanation.

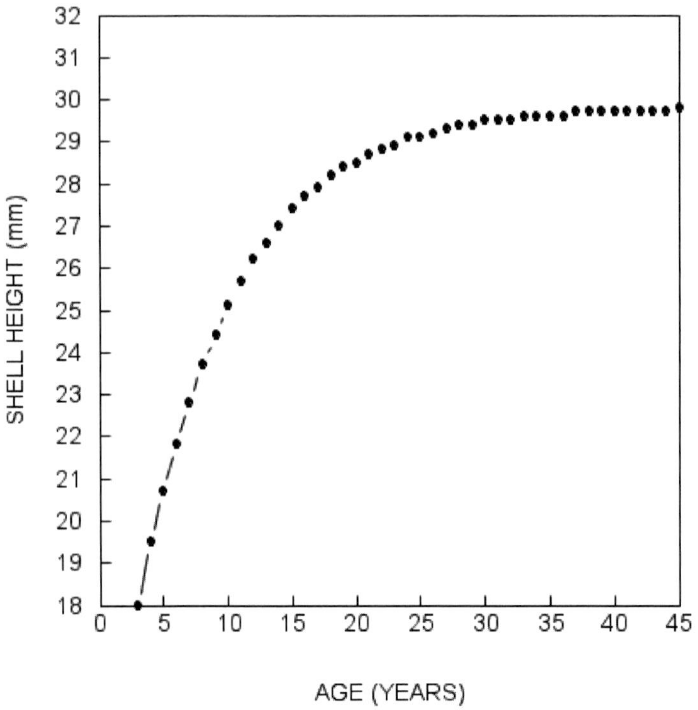

A von Bertalanffy growth curve for uninfected *Ilyanassa obsoleta* showing age at size, based on 86 uninfected snails at Savages Ditch.

Uninfected snails were found and measured mostly in the summer of 1996, the summer when they were released. On the other hand in our paper infected snails were found and measured for up to four summers, through 1999. They just kept showing up in the Funnel and environs every year and this alone, given that uninfected snails quickly disappeared, probably tells us something important about infected versus uninfected snail behavior. We shall get back to that later. In the field we were able to follow 48 *Himasthla quissetensis (Hq)*-infected snails, 11 bearing *Ls*, 26 bearing *Zoogonus rubellus (Zr)*, plus 9 snails doubly infected with *Hq* and *Zr (HZ)*. (For multiple infections, as I have done previously I will use the generic initials of the species involved, e.g., *HZ* for the *Hq-Zr* double and *LZG* for the *Ls-Zr-Ga* triple. The important thing to be aware of as you read is that each of these abbreviations signify a different sort of infection with different species in it.) That gives a total of 94 infected snails for which terminal growth measurements were available. Our plot of number of days out (horizontal axis) versus terminal shell height change (vertical axis) went this way: compared to initial size, snails that were last measured in 1996 (all about 100 days out) ranged between -0.4 and +0.4 mm shell height change; for a different batch of snails last measured in 1997 (about 400 days out) the range was -0.6 to +0.6 mm shell height change; for those last measured in 1998 (about 750 days out) it was -0.2 to +1.2 mm change; and for those last measured in 1999 (about 1200 days out) it was +0.2 to +1.3 mm change. If you think about these data, it can be recognized that infected snails grow slowly and variably, but they continue to add a small amount of shell height each year. If tracked for only a year or two, one might conclude that they don't grow at all. The 12 snails that were free for four summers (terminal growth measurements in 1999) perhaps are the most instructive. Five grew +0.2 to +0.4 mm in that time, three grew +0.6 to +0.8 mm, and four grew +0.9 to +1.3 mm. The clear message is that growth of parasitized snails continues, but at a very slow rate. On Cape Henlopen a very similar thing was observed.

The question arises: do different species of larval trematodes slow the growth rate of snails similarly? As was done with uninfected snails, it can be predicted (using linear regression again) for snails infected with each separate species about how much growth you can expect if a snail has been out for a specific number of days. To refresh you a bit let me present just one formula for prediction, the one for *Hq*-infected snails. It is $y = -0.0606 + 0.0006x$ (again, y is the predicted shell height change and x is the number of days out). If you make a prediction for growth after 1,200 days out (x = 1,200 days, a little more than three years), the expected growth of an *Hq*-infected snail comes out 0.7 mm. Without giving the formulas (they're in the paper), the 1,200 day prediction comes out 0.8 mm for *Ls*- and 1.0 mm for *Zr*-infected snails. The longest an *HZ*-infected snail was out was about 700 days. The growth prediction for that time was 0.4 mm. There are of course brackets around these predictions (growth could have been a little higher or lower), but I will spare you these and simply say: it appears than all these infections impede growth similarly. Again, that's also pretty much what happened on Cape Henlopen.

We did Ford-Walford calculations for infected snails. Suppose that at release in June of 1996 a snail was 28 mm and after 1,200 days (in 1999) had managed to grow 0.7 mm. Growth per day (0.7 mm divided by 1,200 days) works out to 0.0006 mm per day. Assuming that infected snails, as with uninfected snails, can grow for 180 days out of the year, growth in one year amounts to (0.0006 X 180) 0.1 mm. So after one year the snail would be (28 + 0.1) 28.1 mm. We did this for all 94 of the infected snails for which we had terminal growth measurements and plotted data in Ford-Walford format (again, initial size on the horizontal axis and predicted size after one year of growth on the vertical axis). There were a few stray points (possibly measurement errors), but the vast majority of snails landed very near that lower left to upper right diagonal zero growth line. In other words, for parasitized snails size after one year is very nearly equal to what it was last year. It cannot be expected to show much, if any, growth from one year to the next. The average growth per year came out -0.01 mm plus or minus 0.4 mm. What this means is that a parasitized snail might gain (or lose) a few tenths of a mm in a year. As we saw just above, over the long term they do tend to grow, but slowly.

We would like to have an estimate of yearly growth to work with for parasitized snails. Recall that on Cape Henlopen I used 0.2 mm per year as an estimate of infected snail growth. If anything, it appears that Savages Ditch infected snails are growing less rapidly than that. But, where shall we set the estimate? It's a consequential decision. The average yearly growth was essentially zero (-0.01 mm per year), but still snails did tend to increase in size if you followed them over a number of years. So using the average growth per year, essentially zero, won't seem to work. It might be set at 0.1 mm or 0.2 mm per year. In terms of estimating age of a snail, which one you choose makes a big difference. At 0.1 mm per year it would take 10 years to add a millimeter of shell height, at 0.2 mm per year it would take five years. Age estimates would double if 0.1mm per year were chosen and that leads to snail lives that seemed unreasonably long. I shall cover this below, but using 0.1 mm per year snails came out much older than the Savages Ditch habitat in which they lived. We chose to use 0.2 mm per year as an estimate. We did this because most of our infected snails that were out for four years put on a little less than 1 mm. At 0.2 mm per year it would take about five years for 1 mm of growth. This seemed to reasonably justify the choice. Growth actually may be a little faster, but it's probably not much slower. In any case, that's where we came down. Many times in science you don't have the luxury of simply measuring a phenomenon, recording it, and then using that measurement in your reasoning. Sometimes you must make a best estimate based on experience, knowing the system. You must always keep your mind in what you are doing.

This whole effort was directed at determining whether the unusually large snails at Savages Ditch got that way by growing faster or living longer. Large size does not seem to be the result of faster growth. When snails are uninfected they may grow a little faster than Cape Henlopen snails. The average growth estimates were 1.5 versus 1.2 mm per year, but these are quite close to one another and growth is a highly variable phenomenon in this snail. Growth per year could quite possibly actually be equal in the two habitats. Put another way, the two estimates are within the "margin of error," as is often said of results of polls predicting outcomes of political contests. With regard to infected snails, at Savages Ditch they are growing as slowly or possibly slower than on Cape Henlopen. As explained just above, we used the same estimate of growth per year for Savages Ditch infected snails as was used previously for Henlopen snails, two tenths of a millimeter per year.

If growth rate differences are not very important, giant snails at Savages Ditch must be older. To quote Johnny Carson, "How old are they?" I will avoid jokes about Methuselah and just get on with our assessment. As this snail generally goes, snails at the Ditch can be very large. I can recall finding two 30 mm snails on Cape Henlopen among the thousands I have measured. A 27-28 mm snail there is a real rarity. Other than at Savages Ditch, small snails like this, or smaller still, are the rule in habitats in Delaware. In the environs of the Ditch snails greater than 30 mm are routinely encountered. An unmarked one, in Nathan's terminology, is a John Doe. For a long time Nathan had the record by finding a 38.0 mm John Doe. Later (April 2003) I found a 39.1 mm snail (see the frontispiece for this chapter) and that still stands as the record. Snails from 35 mm up can be found routinely if you look for them. In a very large snail, such as the guys I am discussing here, there are at least two things to consider, growth while the snail was uninfected and growth while infected.

Upon infection, growth slows dramatically and the following question becomes of key importance. When (at what size) was a snail infected and therefore started to grow slowly? One way to get at an answer to this question is to figure out how many snails there are in each size-class and what proportion of them are parasitized. Let me re-describe what goes into a size-prevalence curve. Recall that previously I have used 1 mm size-classes. For a couple of examples, all the snails between 20.5 and 21.4 would go in the 21 mm size-class, snails between 21.5 and 22.4 would go in the 22 mm size-class. You would next have to know what proportion were infected in each class. This might be found out by cercarial release tests, or dissection, or some combination. If say 30% of the 21 mm size-class were infected, you could plot this point on a graph with position along the horizontal axis representing size-class (= 21 mm) and position on the vertical axis representing percent infected (= 30%). Likewise, if 35% of the 22 mm size-class were infected, the point would go at 22 mm along the horizontal and at 35% along the vertical. When all of the size-classes have their proportions infected plotted you have a size-prevalence curve. I have dealt with such curves before in Chapters 4 and 9. In this work we made two such curves, one for snails from the Funnel and one for snails from outside the Funnel. The size-prevalence curve for snails from the Funnel was pretty simple. We had 241 snails from the Funnel examined for infection and measured. Virtually all (save 4) were 24 mm or more (up to 33 mm) and all size-classes were between 50 and 100% infected, most closer to 100%. The size -classes that showed the lower percentages infected were actually likely more often infected than had to be plotted. This was the case because many of the included snails had only been tested one time by cercarial release for infections and Karen and I had found cercarial release to be a test with a less than perfect record of performance in detecting infected snails. So the Funnel harbored mainly larger snails, all or most of which were infected. For the other size-prevalence curve, we collected 300 snails from 10 different, well scattered positions outside the Funnel. This curve was a little more complicated and reflected a very different situation. There were 120 snails 24 mm or less and about 10% of these were infected. As size-class increased the proportion infected went up gradually and by 29 mm about 35% were infected. There were not many snails larger than 29 mm, but proportion infected among them went up sharply. In summary, outside the Funnel even quite large snails may be found that have remained uninfected for many years, but it is very unlikely that one will be found that has made it past 30 mm in the uninfected state. Our data for growth of uninfected snails indicates that it takes 40-45 years to hit 30 mm and by that time it is likely that one trematode species or another will have infected a snail. Even if chances of getting an infection in any given year is low, that's a lot of years. On Cape Henlopen, Larry Hurd and I could say that 98% of 23 mm snails were infected, but there is no such simple statement that can be made about Savages Ditch snails. It depends on where you are in the general habitat. In some places (outside the Funnel) many snails, even if quite large, are uninfected, while in other places (e.g., in the Funnel) snails tend to be big and virtually all are infected. There is no one size at Savages Ditch at which a snail may be expected to be infected. All that can be said across the board is that, regardless of where you find it in the habitat, there is only a small chance of a snail getting past 30 mm, and 40-45 years, uninfected. Many questions are raised by this spatial distribution situation of infected and uninfected snails, but we'll leave them alone for now.

With all this in mind what can we say about the age of snails at Savages Ditch? Let's consider snails of fairly modest size that are frequently found in the area. I shall choose 33 mm snails. How old are they likely to be? Well, it is pretty assuredly the case that they are infected at this size. We will give them every break possible (be very conservative) to keep the age estimation down. It's unlikely, but let's say a 33 mm snail made it all the way to 30 mm as uninfected snail, that would be some 40 years of age right there. To put on the 3 mm to get from 30 to 33 mm growing as a parasitized snail would require another 15 years (at five years to the millimeter). Snails with a shell height of 33 mm almost have to be about 55 years old. By the same reasoning a 34 mm snail would have to be 60 years old and a 35 mm snail 65 years old. That largest snail (at 39.1 mm) would be about 85 years old. There is a good probability that a snail will get infected before it hits 30 mm. Considering a 33 mm snail found at the Ditch, let's for the sake of argument say it became infected at 26 mm. There are loads of infected 26 mm snails in the vicinity. According to our von Bertalanffy curve, a 26 mm snail would be about 12 years old. It would have to grow 7 mm to get to 33 mm and that would take it about 35 years. That comes to an age of 47 years. All sorts of infection scenarios are possible. The reasoning makes assumptions that might be violated, e.g., a snail might grow for more or less than 180 days per year, or a snail might consistently grow more rapidly or slowly than estimated. Any such violation would put the estimated age in error. However, the reasoning is based on extensive data and the estimates are not likely to be that far off.

Is the longevity of *Ilyanassa* is extended by being infected with a trematode? This is an attractive idea, but I can't flat out answer yes. It would be a difficult hypothesis to test, given the snail's life span and the fact that virtually all large (very old) snails are infected. The challenge would be to somehow compare the ages of old infected snails with old uninfected snails. In fact, if you look back at the von Bertalanffy curve, both would be growing very slowly. As an experimenter, you probably could not find enough old and uninfected snails on which to assess age and if you tried to keep snails uninfected for life, you would likely not live long enough to get results. Perhaps there's another, smarter way to go about it.

The idea that a trematode might lengthen the life of its snail host makes a lot of sense. A trematode strain that could do that would have an important selective advantage over any strain that couldn't. Lengthening a snail host's life would mean being able to produce and send out cercariae for a longer time. This would increase the number of second intermediate hosts infected and therefore the chance that a parasite's life cycle would be completed. In the final analysis, I don't know if these *Ilyanassa* parasites can do this, but it would make adaptive sense if they could.

A reasonable question to ask is: has the habitat been there long enough to allow snails to reach 70 or 80 years of age. We answered this question by looking into the history of Rehoboth and Indian River Bays. As these Bays currently exist they both exchange with the Atlantic Ocean mainly though Indian River Inlet. Flooding ocean water comes in through the Inlet and directly enters Indian River Bay. Some of it travels west about a mile and swings north around a peninsula known as Burton Island and through a relatively narrow passageway enters into Rehoboth Bay. The water mainly enters Rehoboth Bay that way, but there is an alternate route. A a shortcut is available through some large pipes penetrating a stone bridge, through which some water enters. Approximately a mile and a half (about 2,000 m) north from Burton Island along the east shore of the Rehoboth Bay and north from the Inlet along the outer Atlantic coast is the location of Savages Ditch. Rehoboth Bay is separated from the Atlantic Ocean along its length by a "barrier island," a long strip of land at most maybe half a mile wide. It's a precarious relationship of Bay to Ocean during strong storms. The barrier island might be washed through by wave action. Anyway, that twice daily feeding of Atlantic Ocean water through Indian River Inlet is mainly why the water in the two Bays is salty (up to about 34 ppt) and that's one reason why the snail population can thrive there.

That is essentially the current situation, but has it always been that way? No, over the span of about 200 years things have changed dramatically. Before the Inlet was stabilized in its present location with stone jetties, the ocean beach sand used to drift around under the influence of waves and currents. The Inlet would open and close (exchange between the Bays and the Ocean would be cut off) intermittently and its position would move north and south up and down the coast. In 1800 it was about a mile and a half south of where it is now, by 1843 it was about where it is now, in 1882 it was about three quarters of a mile north of there, and by 1900 it was almost half a mile north of that. Therefore, in the early part of the 1900's the Inlet was positioned about where Savages Ditch is now. It's probably no accident that there is now located about there a store called the "Old Inlet Bait and Tackle." Aerial photographs of the area from as early as 1926 exist. In them you can see the Savages Ditch habitat. It's not quite the same as today, but there it is and it's similar enough. Had I been alive and able then, it's an area I would have liked. In those days it must have been pretty wild out on that long (about five miles) strip of land separating the Bay and the Ocean. There's no discernable road along it in the 1926 photograph. The sand dunes would have grabbed at the tires of early automobiles and I presume that most travel along it was on foot or on horseback. It probably wasn't as idyllic as I would like to think it was, but I'll leave it that way.

In the 1920's an important event happened. Between 1925 and 1928 the Inlet closed and remained closed. In the 1926 photograph the Inlet is closed and simply no where to be found. This was an unusually long closure in comparison to others. It meant that the Bays were isolated from Ocean water input for some three years. They were devastated. Marine life in the Bays, dependent on exchange with the Ocean, died off. There were shellfish and fin fish interests in the Bays that shipped seafood to Philadelphia and other places. These activities required that the Bays be healthy and the Inlet open for navigation. People made their livelihoods from the shellfish and the fish so this was not a small, inconsequential disruption. People made some attempts to dig out and stabilize an Inlet, but storms and waves led to failure. It is not clear, but maybe Savages Ditch was originally one of those attempts. Finally, politicians appropriated some money and the Inlet was officially stabilized with rock jetties in its present location in 1938.

We conclude from our reading of this history that the 1925-1928 closure of the Inlet was a major disruption. It is hard to say exactly what happened to the water in the Bays at that time. The absence of oceanic water input and rainy conditions could have led to nearly freshwater conditions in the Bays. That is often claimed. The snails and other marine life would not have survived that. It is also possible that under dry conditions there would be more evaporation than freshwater input from rain and the water in the Bays could have become hypersaline, too salty. Snails would not be adapted to handle this either and would have died off. Regardless of what exactly happened this was a calamity and it likely killed off all the snails in the Bays. We take it, consequently, that no snail alive and resident there, in say the year 2000, can be older than 72 years and dating back to 1928 when the Inlet reopened. That closure would be, as it were, the back of the stage for snail age. Interestingly, at that time the Inlet would have opened and let Ocean water (and planktonic *Ilyanassa* veligers) directly into Rehoboth Bay just about where Savages Ditch is now, rather than into Indian River Bay. Nobody was there to watch snail recruitment into Rehoboth Bay at that time, but every indication is that snails can be, indeed are, this old and that this is approximately the way it happened. It is amazing to think that snails that came in as veligers around the time of the Great Depression, the Saint Valentines Day Massacre, and the discovery of Pluto could still be there and alive in 2000. Its now 2008-2009 and a few probably still are. In an older habitat the snails might get still older. The community, made up of snails, their parasites, the other hosts in their life cycles, and all the other organisms that are affected, that exists now in Rehoboth Bay does indeed have a long developmental history behind it. If destroyed in the name of coastal "development," it would be impossible to replace it in any human-related time frame.

Another piece of work that built on the work I am discussing here should be included in this chapter. It could have been dealt with in the chapter on Nathan's work (Chapter 11), but it somehow seems more appropriate here because it deals with how long a snail can live with an infection. I'll have to go back a bit to begin. In July of 1996 a doctor recommended to me that I have my left foot amputated. Since my foot debacle at the end of 1995 things had gotten better, but especially my left foot was still in pretty bad shape in 1996. There were open wounds on it that would not heal. I could see there was a serious problem, but I wasn't quite ready to say goodbye to my foot just yet. So I sought out a second medical opinion. The first next doctor I went to had helped me out before. He said that it was unlikely that he, with healthy feet, could heal up those wounds, but he did refer me to another doctor who he thought might be able to do something. This second doctor, Bruce M. Topol, said there is something we could try, but if it doesn't work the foot will have to go. We both agreed to try. After my teaching duties were over in the fall of 1996, he operated in January of 1997. This was a solid month before classes began again in February. He pulled some skin over an ulcer on my heel and took some skin from elsewhere on my body to aid the healing of other wounds on the ball of my foot. There were a lot of stitches. When I eventually got to see it, my foot looked like a confused nexus of railroad tracks going every which way. I had to just stay off that foot. That meant a wheelchair for about a month and teaching on crutches all during the spring semester. During the wheelchair-month I was homebound. A student in one of my fall classes at that time was one Brandy Bishop. We had been talking extensively about the possibility of her doing a research project. We had talked a lot, but as it turned out the project never materialized. Nevertheless, in my confinement she took it upon herself to go shopping for me and come visit with me. That kept me from going utterly stir crazy. I confess to not knowing what has since become of her, but she sure helped to make it possible to survive that month and beyond. Thanks Brandy. I owe you. During my time in the wheelchair and on crutches I was very careful not to put any weight on the foot. Dr. Topol kept a close eye on me and to our mutual amazement the foot healed up. Not quickly, but with physical therapy sessions three times a week it did. A former student and colleague spoken of earlier, Ann Barse, deserves my heartfelt thanks for facilitating my trips to the doctor and to physical therapy. Once again I had lucked out. It wasn't without some serious effort on the part of Dr. Topol, my friends, and me, but things went pretty well. It has been over 10 years since that operation and I still have to be very careful with my feet, even for the shortest trips I never go barefoot, but as of this writing I've not had any more foot wounds. No doubt, if something else doesn't get the job done first, my feet will eventually do me in. However, right now I'm more than 10 years to the good.

By the summer of 1997 I still had to keep my foot dry, but I was able to get a major project done with the help of a new undergraduate, Robert Roy Harman. He's otherwise known as Robb. We'll deal specifically with that project later. I was far from walking normally (never would again), but in 1998 my operation wounds were closed and I could get my foot safely wet again. In the fall of 1999 I was due for another sabbatical. In the summer of that year after spring classes were over I was in good enough shape to begin a rather extensive and complicated field experiment on my own. The main outcome of that will be deferred for now (see Chapter 17), but it is part of the beginning of the extension of Jeff's and Nathan's work that I wish to pursue here. For this experiment, basically in June of 1999 I marked, measured, tested for infections, and released 500 more snails at Savages Ditch. They all came from that habitat. I wasn't walking well and it was a lot of work for one guy, but I was happy to be able to manage. In the process of carrying out this experiment I also kept looking for snails released in 1996. By this time Jeff had graduated and Nathan was about to. They could not be as involved with the ongoing nature of their original project, but from time to time they would join me in searching for snails. They were a big help as they could quickly get to places that I could not. Any 1996 snails found this long after release would be very valuable. It turned out that snails from their project could be located until 2001 and snails released in 1999 could be found until 2002. After that, no amount of looking ever resulted in finding any marked snails. Snails released in 1996 had their tails (shell apexes) painted red with Revlon "Coming up Roses" fingernail varnish. So they wouldn't be confused, I painted 1999 snail tails blueish with Revlon "Ultraviolet". So there were red tails and blue tails out there to be found.

In the current context what was the point of what I was doing? Basically I was searching for any marked snail, whether released in 1996 or 1999. All of these snails had been subjected to cercarial release tests before being released into the wilds of Savages Ditch. This meant that I had an idea (maybe not a perfect idea, but an indication) of their infection status when released. When I happened on a snail that had been out in the habitat for an extended period of time I would collect it and bring it back to the laboratory or to my house. I was making a lot of trips to Savages Ditch and after a while the extra miles to and from the laboratory got burdensome and were unnecessary. I could perform a cercarial release test as well at home as at the laboratory. After getting a snail back I would do a 24-hour release test to see if the infection demonstrated was the same as the one it had when released, or the same one it had when last checked. I would then measure the snail and do any repairs necessary on the tag. Snails were then returned to Savages Ditch and re-released where they were just recollected. Usually they were returned within two days of collection. The point of all this was to keep track of infection statuses of snails through time. Basically, this was to be an extension of the work that Nathan and I had started. We had followed snails and infections from 1996 through 1998. This was a chance to extend some of those observations. Further, the new snails released in 1999 would provide additional information about infection statuses through time. The resulting paper (Curtis, 2003) was relatively short and, like the one Jeff, Nathan, and I published, came out in the *Journal of the Marine Biological Association United Kingdom*. Its title was "Tenure of individual larval trematode infections in an estuarine gastropod," which should give you a clue to the subject of the next few paragraphs.

From 1999 to 2003 I made 117 searches of the Savages Ditch area to try and find marked snails. When ones were found that had not been found recently they were put through a cercarial release test to track infection status. Let me first deal with the 1996 snails that were found during this period. A total of 14 of these were collected and tested in 1999. It doesn't sound like very many, but when you figure that these individually identifiable snails had been loose in the habitat for now four years and hadn't disappeared, it's pretty good. In 2000 I was able to locate six of these 14 and in 2001 five. All but one had been seen previously by Nathan and me. That one was snail RR11. It reappeared first in May 1999 just off the west end of Devil Island after going unobserved since release 1,078 days ago. That finding of a Long Lost John was a momentous occasion. A few beers found their way home in its honor. I didn't see this guy in 2000, but in August 2001 I found him and tested him again. Just to give you a real sense of the kind of thing that goes on in the Savages Ditch habitat, RR11 had a shell height of 29.1 mm and demonstrated a *Zr* infection in June of 1996. It had the same infection in 1999. In 2001 it measured 29.4 mm and still had a *Zr* infection after being free for 1,876 days (over five years). This was the snail observed over the longest time, it had the same infection throughout, and had grown but 0.3 mm.

Of the 14 snails released in 1996 and under discussion here, one had produced no cercariae when tested before it was set free. The other 13 maintained the same infecting species (*Hq*, or *Ls*, or *Zr*) throughout their observed time free in the field (average number of days out = 1,493). The one that appeared not to be infected initially, was later found in June 1998 and it released *Ls* cercariae at that time. It also released *Ls* cercariae when found and tested for the last time in April of 1999. This snail is a good example of how growth history can help inform you about infection history. At release in 1996 the snail was 27.8 mm and when last measured in 1999 (1,060 days out) it was 29.2 mm (growth = 1.4 mm). The question is: was this snail actually infected with *Ls* initially, but failed to produce cercariae? Or, did it pick up the infection after being released? Well, 1.4 mm is a lot of growth for an always infected snail. It turns out that most (71%) of that growth took place between 1996 and 1998, suggesting strongly that the snail was uninfected for part of this time. The conclusion is that this snail got colonized by an enterprising *Ls* miracidium some time between release in 1996 and recapture in 1998. After that it's growth was retarded. Snail infections are numerous at Savages Ditch and they have to form sometime. On the other hand, the remaining 13 snails gave no indication of having changed infection status over years of time.

Of snails released in 1999, 18 were found and retested by cercarial release in 2000, 2001, and 2002. In 2000, I admit that I wasn't too interested in retesting snails that had *only* been released a year ago. It's amazing how one's perceptions of the important things can change. Anyway, in that year I found many, but only tested three. In 2001 I got more ambitious and found and re-tested 12. In 2002 I could only test five. I did a lot of looking in 2003 and 2004, but no marks could be found. I'm sure they were there, but there was a problem. Snails in those years were covered with an unusually heavy coating of mud and algae and the marks were not visible. Oh well, four summers out is hardly a short time as these things go. As opposed to the 1996 snails we have been discussing here, many of the 1999 snails were collected and released external to the Funnel where, even if quite large, there is some chance that a snail is uninfected. Consequently, there is more variety with which to deal. With respect to the 18 snails tested for release of cercariae, they were in the field for an average of 837 days (range from 442 to 1,112 days). (None was dissected to be absolutely sure of infection status because they were part of ongoing experiments and besides I didn't want to kill them.) Twelve of the 18 gave no indication of having changed infection status. One of these remained uninfected for 1,101 days. The other 11 maintained *Hq*, *Zr*, or *LZ* infections throughout their observed time in the field. The other six snails demonstrated some change: four failed to release cercariae initially, but later did; yellow blue 39 (YB39) demonstrated only an *Hq* infection initially, but in 2001 and 2002 revealed *Zr* as well, and YB49 demonstrated *Hq* in 1999, but failed to do so in 2000. The four that released no cercariae initially, but later did, included one that was 28.5 mm in 1999. It was tested and measured for the last time in June 2001 (739 days out). Then it demonstrated an *Hq* infection and was still 28.5 mm. I concluded that it had been infected all along. I came to the same judgement on two others of these snails that produced no cercariae in 1999, but produced *Hq* larvae when tested in 2001. Both had grown very little and were most likely infected all along. The fourth snail that initially appeared parasite-free in 1999 and was from outside the Funnel. It showed up in 2002 (1,096 days out) with a *Ls* infection. It had grown from 29.7 to 30.7 mm in that time, substantial growth for such a large snail. I tentatively conclude that another enterprising *Ls* miracidium colonized this snail sometime between 1999 and 2002.

I introduced back in Chapter 11 the ideas of interactive and isolationist parasite infracommunities. An interactive parasite community is an assemblage of species infecting a host that is in effect bombarded by parasite infective stages. New parasite species try to infect often, interact with any resident species, and evictions of parasites may often occur. The community is in constant tumult with species composition changing often. Conversely, an isolationist community is one in which new parasites don't often try to colonize a host. Consequently, there is little turnover of species and species composition of whatever community is present seldom changes. With that small review let me consider as a single group the two groups of snails that I have been discussing and see what these snails have to say about whether larval trematodes infecting *Ilyanassa* exist in interactive or isolationist communities. Of the 32 total snails followed in this particular work, 25 did not change infection status during times free ranging from 2-6 summer seasons. I used a more appropriate measure of time than that. As an example, if two men do a job that takes eight hours, the job took 16 man-hours. Similarly, add up the days that these 25 snails were out available to be infected and it comes to 29,660 snail-days without being assaulted by a successful miracidium. That's a long time with no new infections colonizing these snails. Among the other seven, above I discounted all but two as probably falsely indicative of change. Those two in all likelihood picked up new single *Ls* infections. Anyway, if I am correct about those five snails, and there is no way to be absolutely sure, that would be more snail-days free of successful miracidia. The 32 snails were loose in the habitat for an average of 1,095 days (nearly 3 years). Assuming that I am correct in my assessment that only two of the 32 experienced real changes, it comes out that 2.1% of snails changed infection status per year. If we make the extremely unlikely assumption that all seven apparent changes were in fact real, it still comes out at only 7.3% per year. Recalling Nathan's rule (Chapter 11), even at the unlikely higher rate, the average snail would be around in the habitat for in excess of 10 years before it would experience a new infection. I'll go out on a limb here and say that is not interactive. This work confirms Nathan's and my conclusion that in this snail host infracommunities of infecting trematodes are isolationist in nature. An infecting species most likely exists alone in a host snail with no other infecting species trying to colonize that host. The colonization rate seems way too low for the communities to be interactive.

In Chapters 8 and 10 I made the point that *Ilyanassa* are frequently infected by more than one species of larval trematode. It is true that you are more likely to find a multiply infected *Ilyanassa* than is the case for most other snails. Nevertheless, far and away most infections are made up of a single species. The main reason so many multiples exist is that this snail has a great longevity. There is lots of time for parasites to accumulate and, once established in a snail, for the most part an infection is not likely to be removed. Infections, even multiple, generally seem to persist right along with the host. What evidence is there to support the last statement? In a future chapter (14) I shall present some detailed laboratory evidence for this. Here I present evidence from the field. Instances of multi-species infections observed persisting in the field will of necessity be few. Multiply infected snails are scarce to begin with and it is very unlikely to collect, detect the infection initially, mark and release the snail, find that snail again in the field after considerable time has elapsed, and then detect the infection again. Nevertheless, it has happened at Savages Ditch. Snail YB39 was observed to carry an *HZ* infection for at least 355 days. Nathan and I had a snail (#267) that carried another *HZ* infection for at least 728 days. A third snail (BB23) bore an *LZ* infection for at least 674 days. If the parasites are "fighting" for possession of snails, it must be going on at a very slow, indeed an unobservable, rate. I keep saying "at least" because, and this goes for the many long-persisting single infections as well, there is almost no likelihood that an observed infected snail got its infection immediately before it was collected, tested, and marked and then lost it immediately when I stopped finding it in the field. Infections were surely resident in snails long before and long after the short period when they were tracked. This is why I think that a snail infection can probably persist for decades even though I have only actually observed them doing so for up to 6 years.

This period of time from about 1996 to 2004 was my last experience with doing field biology. I miss it and I hopefully will say so far. During the latter part of this time, basically from 1999 on, I was working at Savages Ditch for the most part alone. Every once in while Karen or Jeff or Nathan would come out to the field and help to do whatever had to be done, but their undergraduate careers were over and they were all on to other things by this time. In 1999 I bought a Mad River Kevlar Explorer. This canoe weighs only about 50 pounds and I was stable enough on my feet to load it on my truck, get it off and into the water, paddle it around, and then get it back on again. This gave me great freedom. I had a place, Haven Road south of Savages Ditch, where I could get up right next to the water and get my canoe launched without having to carry or drag it more than a few yards. That access point was very important to me and I felt bereaved of a meaningful place when they ultimately closed down that road. Well, it was there for a few of my Kevlar Explorer years and I like to think I made good use of it. For one thing, I later got an account of one day on the water in 2001 published. It was appropriately titled “A Small Canoe Adventure.” That constituted my first venture into non-scientific prose. This is my second. Well, both ventures contain some science, but I hope always explained in a way available to a non-specialist.

My days of solitude at and around Savages Ditch in Rehoboth Bay gave me some cherished memories that will give you some taste for the place and the circumstances under which the science was done. Some are bright with reward and cheer, but there are risks and dangers and others a bit darker. For example, getting around in the mud can lead to mishaps. When searching for snails in the intertidal zone I would use two seven foot poles to help me get around and stay upright on my undependable legs. Many is the time that I stood in one place too long, maybe fixing the mark on a found snail, or reading vernier calipers, or a compass, or a distance with a rangefinder. One foot or the other, or both, would sink in. I would have to perform some move or other, my weight would have to shift some, and I found I couldn’t move my stuck foot, or feet, to adjust for the new balance requirements. Down I would go. Most of the time no real harm was done. I would just have to get myself up to a knee in the mud and then use a pole to pry myself upright again. There would be some cleaning up to do and it might cost me a soaked and destroyed tape recorder, but that was all there was to it. But, it wasn’t always so. One time, in July of 1999, I was going along the very muddy south shore of Devil Island and fell into the mussels that line the shore there. A mussel’s sharp shell put a hell of a gash in one of my lower legs. It took a while to get the bleeding stopped.

Another time I was doing something around my field equipment at the release stake, my operation base point. I had a small, collapsible metal stool there. When you are in the field sometimes for four to five hours you need a place to sit down once in a while. A mud-stuck foot cost me my balance and I crashed down on the stool legs. They were not sharp, but the force of the fall against one of them was sufficient to rip a deep wound in my left shin. How it bled, and it wasn’t slowing up! For a moment I didn’t know if I was even going to get out of there. The very shallow, quiet water around the release stake was red with spilled blood. For quite a while I applied direct pressure to the wound with a towel I had with me and eventually the bleeding slowed down. I had an elastic bandage in my pack and some material that could be used like gauze to cover the wound. I wrapped it tightly. A little blood leaked through, but I was not losing the blood I was before. I let things settle down for a while sitting on my resurrected stool. Then I judged that I could go on searching for snails. I did so, and with pretty good success, for an hour or so. Then I packed up and left. The hike back to the truck on that leg with all my gear piled on my back was a bit dicey. When I got home and into the shower to clean up I took the wrap off. I got a stern lesson in how easy it would be to bleed yourself out in warm water. I had to quickly use a towel to wrap the wound back up. I had to keep it tightly bound for days. That wound was with me for quite a while. In fact, I still sport the scar now nearly 10 years later.

Storms. When working at Savages Ditch in hot, humid weather there is always a chance that a thunderstorm will ruin your day. Late one June afternoon, it was about 5 PM, I was down there, had my gear set up, and was searching for snails. I was moving along minding my own business when I heard a little rumbling and saw a few dark clouds off in the southwest sky. Well, that happens quite often and nothing much comes of it. A little rain doesn't cancel a snail search. Getting from the truck down to the release stake is a bit of a project. It's about a third of a mile walk through the mud and it would take me 20 or 30 minutes to do it. I would carry all my equipment in or attached to a frame pack. There was the stool, my distance reckoning flag pole, the range finder, a 60 meter measuring tape, a tape recorder (and a backup), some food and water, a thermometer, a salinity measuring device (refractometer), my two seven foot props, and a few other items. For the trip I was pretty loaded down. Once I got there the pack had to be gotten off, a balance threatening act, and all the stuff deployed. The reverse, packing up to leave, was a challenge under the friendliest of conditions. Everything had to be loaded and the pack had to be remounted on my back. From the west a few sprinkles started, the sky got darker, and thunder boomed louder. I got the clue that I was about to be driven off site, but I was too slow. All of a sudden a gusty wind started to blow maybe 30 or 40 mph. Some stinging hail came with it. I had to get myself packed up and vacate the site quickly. There could be no orderly packing up. The tape was a particular problem. Normally I would roll it up on its winder and stow it, but the wind had about 30 or 40 meters of it flapping around and tangled. All I could do was gather it in as best I could and stuff it disorganized in the pack, along with the mud that came with it. I would untangle it later. With my equipment sort of all inside and the flag pole and stool onboard outside, the pack was hard to get on my back in the wind. Once I was unceremoniously blown over into the mud. Finally I got the pack mounted and my prop poles in the right places. I slowly and with great misery made my hail-stung, muddy, and soggy way back to the truck. In short, I got pretty well shit on. Maybe I should have taken up a laboratory science.

Humans are not the only ones at risk. In November of 2001 I was out in my canoe and made a disturbing discovery. In my first piece of published prose (mentioned just above) I related a 1996 circumstance in which Nathan and I were in a canoe exploring in southeast Rehoboth Bay. We came upon a common loon floundering around in shallow water and all tangled up in discarded nylon fishing line. Now there, carelessly discarded line, is a real danger to wildlife. Thanks mostly to Nathan's grit the bird got relieved of the entanglement and swam away, somewhat disheveled and miffed, but away and O.K.. Later I had this experience, which did not have a similar happy ending. The place where I used to put the canoe in is in a sort of embayment in the very southeast of Rehoboth Bay. These days it's a dead end in there. There's no way out for the water except for the way it came in from the north end near where the Haven Road canoe launch site was. Unlike the present situation, in the early 1900's water used to enter and exit Rehoboth Bay right though there. The Bay was connected to Indian River Bay right near the Inlet there and much water passed through. The embayment is called Bottom Hills Drain in honor I suppose of its former geographical role. Near the western shore of this dead-end body of water, about two thirds of the way south into it, there are some pier ruins. There are a number of pilings protruding some eight feet out of the water when the tide is down. The tide was pretty low when I was there this time and I was nosing my canoe around among these pilings. Near the top of a piling there was a dead cormorant hanging by its feet in some discarded fishing line. I had seen this sort of spectacle before, but on those occasions it had been a skeleton hanging there. This was a more recent tragedy. I can only imagine what that bird must have gone through in its dying. Cormorants land and sit on the tops of those pilings all the time. This unlucky guy just happened to run into some fishing line that had snagged there near the top of a piling on a very high tide. Maybe the wind blew it there. The cormorant landed, got tangled, couldn't get free, and slowly starved to death. When finally too weak to stay on top of the piling, the bird fell off, and too far gone to regain the piling top it just hung there by its feet, struggling upside down, and completing its slow starvation. For how many days did that go on? When death came it was surely a relief. Call me a softy, but it doesn't seem that any animal should have to die like that, certainly not because some idiot was too lazy to retrieve and safely discard his tangled fishing line. I don't want to be anthropomorphic and attribute human emotions to a cormorant, but when I found this bird's body it still had quite bright intact eyes. There was some sort of expression in them that I'm still trying, but failing, to capture in words. All I can say is that this slow, agonizing death must have been endured in agitated "hopelessness." For me it was a very risky thing to do, but being the heroic guy I am while using the piling for stability I managed to stand in the canoe, and cut the fishing line. I collected and gently placed the bird in the water and he drifted away. I don't imagine it made any difference to the cormorant, but it made me feel a little better about the situation. Too late for this unlucky specimen of wildlife, but whenever I can I collect errant fishing line and throw it away in a trash bin on land. Nylon's half-life is very long and lots of land trash is ultimately dumped in the ocean. Undoubtedly, despite my small efforts, that won't end it.

There were of course lots of brighter experiences during this time. At this point, based on the foregoing in this chapter you, or I, could fairly easily summarize the happenings. But, when the work is in progress you (I) can't know what is going to happen next, what, if anything, is going to be found out. Cape Henlopen gave me a few experiences with finding marks (Long Lost Johns) that had been loose in the habitat and not seen for a long time. Savages Ditch is a smaller place, the same snails frequent the same areas each summer, and such encounters are much more likely. I recounted the case of RR11 above, but there were many such finds. Each in its own way was downright thrilling. A lot of the fascination with doing this kind of work is in such moments. There you are trudging along, walking back and forth, searching let's say in the Funnel. Let's further say, to keep things wonderful, that it's a bright sunny summer day with just enough wind to keep insects at bay. Your eye catches on something a little out of the ordinary. Was that a tiny flash of red? You investigate. You use your poles to safely bend down to the mud and pick up the suspect snail and scrape him off. Ah ha, there's a trace of red on his shell apex, not much, but a trace. Does he still have the numbered tag? You do a little more washing and scraping. Yes, there it is! Now, can you read the number? You get out the magnifying glass you carry for just such occasions. Yippee, it's a green tag and you can make out a 77 on it. Here it is July 2001 and from the red on his tail you know this guy was originally released back in 1996. He's been out here for six summers. You haven't seen him yet this year. How about that! Here he is again. You can't remember and will have to contain yourself until you can get back to your computer to find out what the deal is with this guy. How big was he and what infection status did he demonstrate six years ago? How many time has he been found since? Later, records tell you he was 26.6 mm in June of 1996 and you last found him in April of 2000. You measure him. So, he's grown just 0.5 mm since release in 1996. He must be parasitized to grow so little. According to your record for GR77 he released *Zr* cercariae when tested prior to release and it was *Zr* when again tested in 1998, 1999, and 2000. Does he still have *Zr*? Maybe the infecting species has changed since last year. To find out you pour some filtered Savages Ditch water into a clean jar and plunk the snail down into it. You put a loose top on the jar (snails can be troublesome with their escapes) and situate it out on the porch for the night. You expectantly awake at sunrise the next morning and you set up the microscope and examine the jar. Sure enough, there are hundreds of *Zr* cercariae in there and you look carefully, but there are no other species of cercariae present.

This guy has the same infection today he had six years ago! Now you have to recheck the size measurement, make any repairs to the tag (more clear nail varnish), repaint his tail red, let the nail varnish dry thoroughly, and then start think about taking him back to Savages Ditch for re-release where you found him yesterday. When you do all that and he's back in the mud and motoring away, it may be a one-sided relationship, but it's almost like saying goodbye to a friend. Will you ever see him again? Probably not, but there is always that chance. Well, I never did see that snail again. I feel sure he is still alive down there and motoring around in the mud. But, finding a snail with such a long known history is quite literally like finding a diamond in the rough. We could argue about which is more valuable, the snail or the gem. For my part, I come down on the snail side of that argument.

Birds. Aside from being of interest as definitive hosts for some of the trematodes that infect *Ilyanassa*, birds, live ones, are loads of fun just to watch. Around Savages Ditch birds of many types are likely to be in the area feeding, nesting, or just passing through. Because I was there alone most of the time, they seemed to sense there was no threat and it was usual for my surroundings to be populated by plumed co-habitants. To give you a flavor there would routinely be herring gulls, laughing gulls, black-backed gulls, ruddy turnstones, sanderlings, willets, clapper rails, ospreys, egrets, oyster-catchers, common terns, and least terns in attendance. Such species as the black skimmer, hudsonian curlew, Louisiana heron, or glossy ibis might briefly be there. Black skimmers are really cool. They fly along just above the water with their lower bill skimming the water's surface, apparently to attract fish. To do my work I always had a small tape recorder with me to save data and observations. Listening to some of those tapes now it's fun to hear again the bird cacophony most always going on in the background. In the summers of 2000 and 2001 a pair (I'm assuming they were the same pair) of oyster-catchers nested on the western end of Angel Island. I was down in the general area quite often, but I made it a point to avoid their nesting site. Generally, one would stay around the nest while the mate foraged. They were extremely busy and, as you might expect, very protective of their area. They didn't get too excited when I came by, but black-backed gulls were another matter. An oyster-catcher is about 18 inches head to tail; these gulls are about a foot larger in the same dimension. On several occasions one or two gulls would attempt to spoil the nest. They probably wanted to eat the eggs or, if hatched, to kill and eat the young. Oyster-catchers make a sharp, piercing *wheep* sound and when a gull would encroach both parents would take to the air. They both aggressively flew at and harassed the larger bird, filling the air with their whistled screams. This would go on for maybe half an hour and eventually the gulls would be driven off and peace and serenity would be restored, for a little while anyway. Not wanting to encroach myself, I never got close enough to actually see any chicks. The only oyster-catchers I ever saw were flying, so I guess the nesting was a biological and resounding success.

As a final event to convey something of the environment and experiences of my last years of field work here's a solo trip of exploration that I took in early August 2004. From Savages Ditch, when you look west onto Rehoboth Bay and across to the other side there is an isolated strip of beach that just shines out at you, especially on sunny days. It beckons you. It seems one is always drawn to what's on the horizon. The beach is on the east face of a long, pine-forested island, called Lynch Thicket, and almost always there is nobody there. On the Thicket's western side is a waterway called Massey's Ditch and across that on the mainland there no shortage of human activity. There is no bridge to the island and it's therefore mostly isolated from humankind. That's what attracted me to that beach. From the other side of the Bay the idea of it was, you might say, irresistible. In its more northern parts Rehoboth Bay is about three miles across. In its southern reaches it narrows down before it merges with Indian River Bay. At the level of Savages Ditch and this beckoning beach it's just over a full mile across. In the east it is small sky country and the horizon is pretty close at hand. Finally I could resist no longer. One morning I loaded my canoe on my truck along with sustenance and other equipment and headed out. I drove the 14 miles down to Haven Road at Bottom Hills Drain and launched my canoe. It was 8 AM as I pushed off, still a bit chilly and raining a little. To get to the Bay proper and across from there you have to get past a collection of saltmarsh islands about a 1000 meters wide. There are tried and true ways of travel through these islands, but naturally I had to take a route not taken before. The waterways among the islands are narrow and unpredictable. Many of them are dead ends for even a canoe. If the tide is down, any of them can get prohibitively shallow. The tide was low and falling that morning. I was heading westerly and could see the open Bay at the end of my chosen waterway. About half way through my canoe began dragging its bottom on very soft mud. In the shallow water I could see a huge population of large (old) *Ilyanassa*. That was comforting, but almost immediately the water became so shallow that I couldn't make forward or backward progress in the canoe. I was stuck. There wasn't anything for it but to get out of the canoe and get intimate with the mud and the population of snails. I have written enough of my leg infirmities and you will quickly apprehend that my abilities moving about in the mud were minimal. In that rainy, slimey mud they were appalling. It was not a graceful affair, but I got enough purchase in the knee-deep mud to crawl slowly along and haul the canoe past the no-water obstacle. Here it was only about 9 AM and I was covered in muck and already worn out. The silver lining is that, covered as I was, the greenhead flies weren't much interested in me. In that inter-island waterway, they would have had a bloody field day at my expense had I simply sat there and waited for rising water to make the escape possible. Anyway, I came to appreciate fully just how many snails there can be in these waterways among the islands. I had earlier done some sampling around these islands and in these waterways and that led to a future paper, a tale for next chapter.

Once I got my muddy self out into Rehoboth Bay I had about a mile to go to get across to my target beach. It had gratuitously stopped raining. In the middle of the Bay there are sandbars that come out of the water on many low tides. They are not a risk to a plodding canoe, but to a speeding powerboat or a sail boat they can be a serious surprise. The wind was blowing out of the northwest and there was some chop to deal with. Heading west, the wind and chop were a bit of a challenge, but nothing insurmountable. I must admit though that when you are alone in a canoe in a mile-wide expanse of water a feeling of vulnerability does come over you. I just got into my long-distance paddle mode and before too long I had gotten across. The beach was just ahead. Nathan, in one of his pieces, wrote of his canoe "sighing" onto the shore. That always struck me as a great way to express the sound of that moment and I will unabashedly steal it. My canoe sighed onto the beach. I was there. And, it was everything I expected. There was no one there, no footprints anywhere, and to top it off by now the sun was shining through brilliantly. A little inspection told me that there were *Ilyanassa* in residence on the beach face. This really is home. My first desire was to be rid of the mud that I had worn as I paddled across the Bay. There was a strong current and a five foot deep cut right in next to the beach and that seemed like a logical place to get bathed. I managed to get out of the canoe and leaning on it got myself into the water. It felt super good. Lots of birds hang out around this isolated beach in what I later found out was called Raccoon Cove. Good as the water felt I was hoping that the birds had not brought in *Austrobilharzia variglandis (Av)* to infect snails. I didn't need a body-wide case of swimmer's itch as a consequence of my curiosity about this beach. Thankfully, that didn't happen. After lolling about in the water for a while, I damn near upset the canoe and all its contents (paddle, food, tape recorder, camera, wallet, truck keys, etc.) into the drink while steadying myself on it to regain the beach. With the current there that would have been a very troublesome disaster. Well almost, but that didn't finally happen either. This is some great day I was having. I got a prop pole out of the canoe, got up, and began walking the beach. I had sighed in at about the midpoint, but the beach is about 150 meters long north to south. The sand was soft going. Covering its length I could see many *Ilyanassa* scattered about on the beach face. I took a few as a sample. The next day I dissected them and they were indeed infected with *Gynaecotyla adunca (Ga)*. Snails wouldn't be up on the beach face unless infected with that species. It occurred to me that this beach would be a great place to sample and do experiments on the crustacean next host in the parasite's life cycle. A thing necessary to make the closing point about the *Ga* phenomenon, and a thing as yet undone, is to show conclusively that where there are parasitized snails with altered behavior (Chapter 6) the crustacean next hosts in the area are more likely to be infected with *Ga* metacercariae. Some 30 years ago on Cape Henlopen, when you would walk along the high beach at dusk or after, there used to be an abundance of those amphipod beach-hoppers bouncing around. That was before I got onto the parasite connection between the snail and the beach-hopper. Now, Cape Henlopen doesn't have the beach-hoppers it once did, but on this essentially peopleless beach it's likely that they're still here. It would be a problem getting to and from this site, especially at night, but an energetic young researcher could handle it. Maybe I'll take it up in my next life.

When I looked east, offshore a little from what I came to call Raccoon Beach, I could see that there was something else of interest. The tide was yet pretty low and it was evident that there was an intertidal sand island just on the other side of that deep cut right along the beach, my swimming hole. I managed to get across the cut and up onto the island. It was very roughly circular, about 50-75 meters in diameter, with a slight depression in the center. Walking around on it, the water at the time was about ankle deep and there were snails, *Ilyanassa*, in great abundance all over it. No way I could get an idea of parasites or anything else by collecting a few snails from here and dissecting them. I left them all alone and just observed for a while. There were many birds of many types clustered on this isolated, intertidal island. I could see fish in the shallow water. Definitive hosts, it should be a hotbed of infected snails. Who knows what as yet unappreciated phenomena occur here? Oh, to be looking for a research project at this time.

That day I guess I spent two hours on and around that beach. I found a nice old fallen cedar trunk to sit on while I ate some lunch. I did see a few boats go by out in the Bay, but they and the people in them came no closer than maybe a quarter of a mile. It was a great two hours. Raccoon Beach and Lynch Thicket were essentially all mine and so they remain in my consciousness. After eating and getting my stuff ready for departure, I made my usual awkward entry into the canoe, got myself together, and shoved off. I went north a little bit and looked into a cove up there, but it was now well after noon and I was feeling a bit taxed from all my exertions getting here. I figured I had better get back across the Bay. The wind was still blowing northwest. When I had come across the Bay a few hours earlier the tide was ebbing and wind and tide were more or less moving in the same direction. Now it was flood tide and they were at cross purposes. I would have the wind at my back, but waves would make it a rougher ride going the other way. It was a fun, wet and fairly adventurous ride. I had to pay close attention to canoeing all the way and got to my launch site at Haven Road pretty well spent. In not too spry fashion I hauled the canoe back to and loaded it on the truck and up the road I went. I got back home at 3:30 or 4 PM. Sleep, not thinking, was the order for much of the rest of the day.

Recalling this trip, and others during this time, make me realize how fortunate I was to have pursued field biology as my professional interest. The actual doing of research and the peripheral stuff around it, such as Racoon Beach, were and are inspiring to me. Circumstances would soon change, again. Before long I would retire from teaching and enter a period wherein research amounted to writing papers on data gathered maybe five or 10 years earlier that I never had time to write up and publish. Well, to some degree all papers are that, just not based so deeply in the past. That's where I am now, so you will be happy to learn there is more to come. Then too this book project, which of course just rolls out with no effort, came to mind. For me, my eight or nine years, 1995-2004, working at Savages Ditch in Rehoboth Bay have been productive, enjoyable, and challenging. My students can claim responsibility for much of that characterization of those years. They and the years certainly have added to the worlds's lore and understanding about the hero of this tale. In the little bit of them that I have here related I hope you have found these years likewise enjoyable. There is more to tell.

As for Jeff, he defended his undergraduate thesis with aplomb. It was a difficult case to make that *Ilyanassa* could live 70 or 80 years, but he handled it very nicely and learned a lot while doing it. After graduating from the University of Delaware with success he went on for a Masters Degree at Wollongong University down under in Australia. In connection with that exploit he got to spend several months in Antarctica doing research on the effects of ultraviolet radiation on marine diatoms. I see him now and then, but I have now pretty much lost contact with him. He has given me some have good memories. Jeff was great to spend time with. I trust his memories are similar and that he feels he got a lot out of his experiences with snail research. At last rumor he was located and being productive somewhere in California. Keep on top of it Jeff.

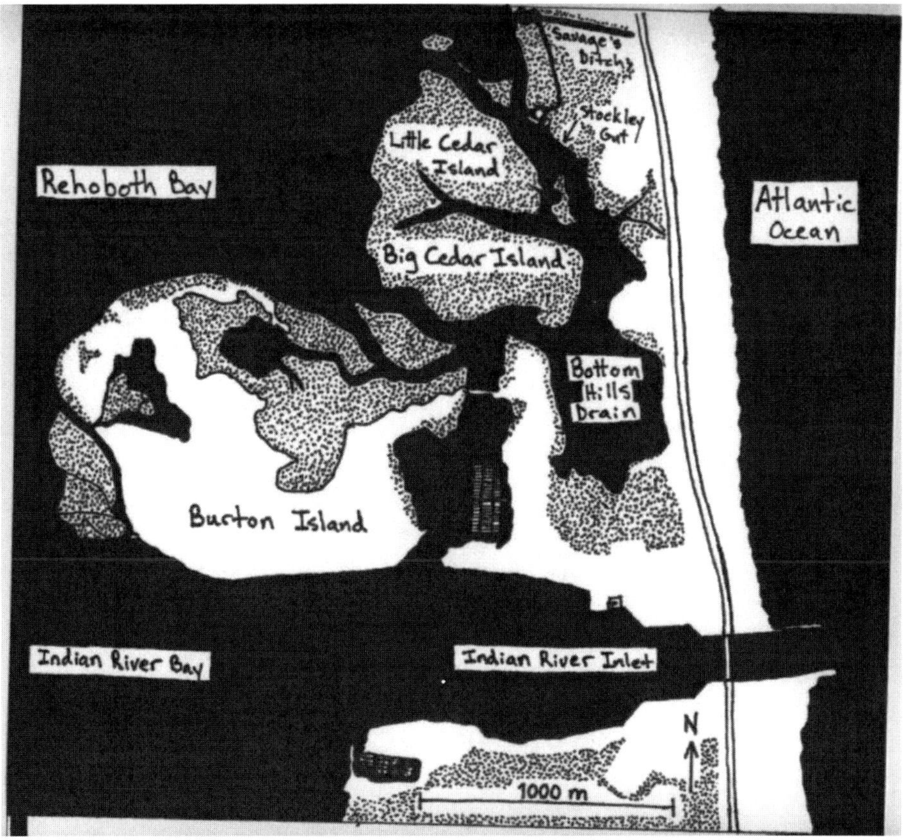

Southeastern Rehoboth Bay, the area of the study described in this chapter.

Chapter 13

PARASITES IN THE WILDS OF SOUTHEASTERN REHOBOTH BAY

I mentioned a couple of times in the preceding chapter a 1997 study done among those saltmarsh islands south of Savages Ditch and with the aid of Robert Roy Harman. Data were collected in the summer after the operation to "fix" my left foot. For most of that summer I still had to basically stay off that foot and avoid getting it wet. Not withstanding this limitation Robb wanted to involve himself in a research project and I had something I wanted to find out. With Robb's assistance it was possible for me to get out in a canoe to collect the necessary samples. This was pre-Kevlar Explorer days. I had a Grumman aluminum canoe that weighed about 75 pounds and was about 17 and half feet long. At the time it was too heavy for me to manage alone, but Robb was a capable guy and once we got it in the water it was a superb sampling canoe. The doing of this project is front and center in my memory. As I write this, the paper that resulted is in press. It will probably become Curtis (2009) when it comes out next year.

In the previous chapter I described this southeastern area of Rehoboth Bay. It is adjacent to and south of Savages Ditch. If you wish you could use Google Earth on the internet to visualize the area. Here I will stick to words, but see the frontispiece for this chapter. The area that we sampled is about 1,500 meters north to south by 1,000 m east to west, i.e., about 1.5 km^2 in area. [A kilometer (km), 1,000 meters, is 0.62 miles.] There are a number of differently sized islands in the area all of them covered with saltmarsh grass. I took some photographs from the southern edge of the flat expanse once and the area looks like a giant green meadow. Devil and Angel Islands (of Savages Ditch fame) are right on the northern edge. The islands are separated by variously narrow waterways. That's why the canoe was necessary. Hell's Gate and Heaven's Gate are examples of these waterways, but there are many others. Heaven's Gate opens to the south into a major and quite wide (about 50 meters) thoroughfare that carries much tidal water essentially north and south through the area. This waterway is officially called Stockley Gut. At its southern end is what I have written of as Bottom Hills Drain. There are other major waterways (similarly wide) that feed into Stockley Gut from the west or northwest where the body of Rehoboth Bay is. Still other waterways (ditches) are much narrower. The tides takes water depth up and down usually about one meter. At high tide there are places in the area where the water is naturally about two meters deep, but that's about it. There is twice daily input of Atlantic Ocean water so the salinity throughout the area is usually quite high, reaching nearly ocean water strength. So, as far as salinity goes the area is conducive to the occurrence of *Ilyanassa.* Snails occur pretty much throughout the area along the intertidal shorelines of waterways. They also occur, sometimes in great abundance, on the mostly subtidal areas between the islands and just offshore of the islands in Rehoboth Bay proper. In short this is a fairly large area, very different from previously studied sites, with a large population of snails about which I knew almost nothing. We couldn't have that.

The properties of a host population are a major determinant of the (component) community of parasites that infect it. Parasite species are more important contributors to the functioning of an ecosystem than most people realize, especially when they parasitize an important ecosystem component such as *Ilyanassa.* If you would understand how an ecological system works, you had better understand the important hosts and their parasites. With regard to the nature of a host population: how are individuals distributed in space, what about reproduction and recruitment of new individuals, how abundant is it, and how big are the individuals that make it up? Of most of these host properties I already had a pretty good notion, but not in the habitat in question. Though there is high salinity throughout, this is essentially a saltmarsh and I had not extensively studied a snail population in such a habitat. When I started this study I did not have a particular hypothesis in mind to test. I had observed this population of snails a little and I simply had questions to which I wanted answers. They fell into two main categories: (1) what are the properties of the host snail population in this habitat; and (2) how does the set of trematodes that infect snails in this habitat (the component parasite community) make use of the host resource?

To begin to get a handle on these general questions, a method different from what had been used previously had to be brought into play. It had to do with sampling. First, why do we take samples of the objects of our investigations? It's because we can't or don't want to process every last individual in the population under study. Rather, the idea is to take random and representative examples (samples) from the population that will allow us to make accurate inferences about the whole population. This is largely the business of statistics. To make this possible with this snail population and its trematode parasites, sampling had to be quantitative. In much of my previous work when I needed to sample snails I could just collect some "by eye" from the population of interest. For this purpose the problem with doing that was that it would not be possible to say how many snails of certain kinds [e.g., females, males, females with imposex, or infected with *Himasthla quissetensis (Hq)*] were present *per unit area*. This is not a critical ability in some work, but to characterize the use of a snail population by a set of trematodes I needed to be able to do this. Why? I will attempt to answer clearly.

Here's how Robb and I went about sampling. We looked at an aerial photograph of the 1.5 km^2 area and selected 21 shoreline locations within it that would serve to canvas the whole area. There is a photograph with the locations indicated in the published paper. In the summer of 1997 we started going to these locations, one at a time, and taking a set of samples at each. We wanted to sample where snails were present and we used sites at locations where at least a few snails were evident. We recognized that the random-sample assumption in our sampling program was violated by doing this, but not in a major way. Snails occur almost everywhere along shorelines in the area. The equipment was not too fancy. My foot kept me basically directing from the canoe or dry land. Robb did almost all the hard work. This is the way it would go. We had a square frame made out of PVC pipe (Jeff Kinley's creation) that measured one meter on a side. This would be the area from which a sample of snails was collected, 1 m^2. It was quite a large area as samples usually go, a good bit of work. At low water we would go to a location, find a place where snails were at least in evidence, and haphazardly cast the 1 m^2 frame into that place. Wherever it landed that would be our first sample. The frame was fastened to the mud or sand surface with stakes. Next we had a basket-like clam rake fitted with a perforated metal lining that Robb would use to systematically rake (like mowing a lawn) and collect the upper layers of sediment (where the snails are) inside the frame. As the sediment was collected it would be transferred to a perforated bucket for further sieving out of mud or sand. The largest hole in the clam rake or the bucket was 4.4 millimeters and any snail that was too big to pass through such a hole would be retained. This meant that even very small snails, in their first summer, would be collected. After checking for and collecting any snails in the frame missed by the rake, we would have all snails from a 1 m^2 area of bottom in the bucket. All retained materials (sediment, snails, clams, worms, debris) were then transferred to a labeled plastic bag for return to the laboratory. After taking that first sample, the frame would be flipped over twice, generally along shore, to define the next sample. There were a couple of exceptions that I won't belabor, but this meant that adjacent samples within a location were taken mostly one meter apart. Sample number two at a location was collected in the same way as the first and then a third. Three 1 m^2 samples of snails were taken at each of 21 locations. This gave a total area sampled of 63 m^2. This alone, exclusive of laboratory procedures, represents a lot of work. The effort took us most of the summer, lots of man-hours.

So, over the summer we gradually brought our 63 samples back to the laboratory. Once in the laboratory a sample would be emptied into a white rectangular pan. With the aid of some bay water, a light, and a magnifying glass we would search through a sample and pick out all snails. If there were 20 or fewer snails, all would be measured for shell height and dissected to determine sex, imposex, and parasitism. If there were more than 20 snails in a sample, we did something else. I had some pieces of plastic grating that once covered and protected florescent lights. The many squares in the grating were very useful for lining up and separating snails. All snails were lined up in the grating, one to a compartment, and assigned numbers based on which compartment they happened to land in. Let's suppose there were 60 or fewer snails in a sample and they were in the grid. A computer program generated 20 random numbers and 20 snails would be chosen based on the numbers that came up. The 20 snails randomly selected would be unlucky enough to be dissected. The rest were measured, but survived. If more than 60 snails were in a sample a random 60 were chosen. The first 20 random snails were dissected and all 60 measured. Why did we only dissect 20 snails? Well, by this time, with knowledge of how old they might be, I did not want to kill any more than that and 20 random snails out of all in the sample was adequate to characterize the sample. All snails that avoided being dissected were taken back alive to their habitat and repatriated.

How did I handle data from a sample? As an example, let's consider snails infected with *Hq*. Suppose there are only 10 snails in a m^2 sample. Because there are 20 or fewer snails they all would be dissected. If there are two *Hq*-infected snails among them, then straightforwardly there were two *Hq*-infected snails in that m^2 patch of ground ($2Hq/m^2$). If rather there are 20 snails in a sample, they would still all would be dissected. Suppose five are infected with *Hq*. Then, the density of *Hq*-infected snails in that m^2 patch of ground was five ($5Hq/m^2$). That is, 25% (5/20) of the snails in the 1 m^2 sample plot were *Hq*-infected. Now suppose there are more than 20 snails, say 200, in a sample. We would not dissect them all. Rather, we would choose a random 20 to dissect. Suppose that five among the 20 randomly chosen happen to be *Hq*-infected. It is reasonable to extrapolate infection percentage among the random 20 dissected to the 200 snails in the sample: that is, if 25% of the random 20 dissected are *Hq*-infected, then 25% of the 200 snails in the plot (0.25 X 200 = 50) are likely to have been *Hq*-infected. The estimate for that sample is, had all 200 snails in that m^2 of habitat bottom been dissected, there would have been 50 snails infected with *Hq*, a density of 50 *Hq*-infected snails per m^2 (50*Hq*/m^2). Of course, similar estimates could be made for uninfected snails or any other infection category that might be encountered in the sample. I take you through all this procedure because you can't really judge or understand a study unless you understand how the data for the study were gathered and handled. It matters that all snails in samples are taken into account.

Robb and I had some interesting times in the process of getting and processing these samples. In 1997 Nathan's and Jeff's projects were also in full flight. The three guys all knew each other. The two "old timers" got a rise out of introducing the rookie to the techniques and conventions of the snail-parasite system. Robb was a little older than the typical undergraduate. He had been out working for a while assembling bicycles for a large store chain and was still doing it. We scheduled our work around his employment duties. The idea was that he would take part in the project I had in mind, use the data to write a paper on the project, and hand it in for a grade on research credits. If it turned out that a published paper came from it all, he would have the option to be in on that too. What Robb seemed to want more than anything else was to find out what biological research was like and to be involved. Thank goodness for his labors. I would not have been able to do the project without him. Aside from being an agile, fit guy and a willing accomplice, Robb had a devilish sense of humor. His perceptions about what we were doing and events around us made him a lot of fun to be around.

Robb had never done any canoeing before this. At the time I had a Chevy pickup and we used it to carry the Grumman canoe down to the launch point in Bottom Hills Drain. That canoe was rugged. I don't believe you could hurt it even if you wanted to. It was like the surviving canoe in the movie "Deliverance." With my one-footed help we would get it not too gently down off the truck. He would drag it to the water with all the gear aboard. After I got myself and a dry foot in the rear seat, Robb would muscle the canoe the rest of the way into the water and afloat. He would get in the front seat and off we would go. Pretty quickly he got adept at handling the paddle. In water such as we had to deal with you can usually steer a canoe well enough from the rear. One day I showed Robb how to do a cross-bow rudder. When you are making headway you can rest the paddle shaft on the gunnel next to the bow. By sticking the blade forward into the water at an angle the bow of the canoe is sharply drawn toward the paddle. It can be a very useful maneuver. Robb got utterly enamored with the move. Canoeing among and around these islands requires dealing with the wind and quite strong currents, but there are no rocks. White water canoeing it is not. If you hit something, it's likely mud or maybe grass. There were almost no occasions where a cross-bow rudder was actually needed, but much to my amusement there was Robb up front with a little devilish glee trying it out and complicating whatever steerage I was attempting to perform from the stern. Experimentation is good.

On a given low tide we could maybe accomplish the sampling of two locations. One day in early August we went to take samples at our westernmost locations in a place called Sawpit Cove. Once we got on the water it took something like half an hour to paddle there from the launch point. We would travel along the Old Ship Channel. It was sunny, hot, and humid in the late afternoon, great conditions for a thunderstorm. The sunny weather held as we got the six samples from two locations. At about 5:30 PM, as we took the last sample, there was ominous thunder and black clouds in the western sky. Ten minutes later the clouds were upon us and there was lightning visible in them, snapping around and momentarily lighting up the blackened sky. All of a sudden there was a sense of urgency to our situation. It wasn't raining yet, but that would certainly come. There was no shelter to be had anywhere, we had gear to load, and about 30 minutes of canoeing in front of us before we could get back to the truck, luckily to the east away from the dark clouds. We had better get going, and fast. The lightning was the main worry. It's not good to have that crackling about when you are on the water in a metal boat. So, with the wind, black clouds, thunder, and lightning at our back we began the paddle back to our launch site. I can well remember that trip. Not because it was a disaster, but because it was exciting and a success. The darkness of the sky and lightning bolts finding their way to the ground were threatening. There was no talking, just intense, purposeful paddling. As usual I had a tape recorder with me and I turned it on for part of the trip. Listening to it now it brings back memories to hear no sound other than thunder among the calls of gulls and the Grumman gurgling and hissing through the water. The sounds associated with the passage of an aluminum canoe through the water have a more strident character than those of my later acquired, more silent, Kevlar Explorer. Despite the threatening sky we made it back and got things into and on the truck just about as it started to rain. As we went north up the road, rain splattering against the windshield, we could look back and see that all hell was breaking loose where we had just been. We had definitely lucked out.

All the circumstances that had allowed this to happen make you wonder about the tortuous connection between cause and effect. Let's stay close to the immediate moment. Given the timing of the storm, had low tide been a little later that day we would have been later getting our samples taken. Had there been a need to stay longer, we no doubt would have. By that time of the summer my foot was pretty well healed and I did not have to be quite so careful with it. That made me quicker and more useful. By then, Robb was fully aware of what had to be done to obtain samples. He was efficient and skilled. Besides he had gotten very good at paddling. It might be added that we were very practiced by this time at loading the canoe on the truck. In short, we were a well-oiled, fast machine. All these factors made it possible for us to finish our work sooner and get out of there in a timely fashion. It was a conspiracy to make it possible for us to "luck out." If any of these factors had not been in place, it might well have turned out otherwise. The effect is pretty clear. We lucked out. However, "the" cause of that is pretty muddled. Anyway, Robb and I had some interesting times that summer doing this project.

With that little interlude we now go on to some of the project's product. The first issue is, roughly how large is the snail resource available to trematodes in the area? Are we talking tens of thousands, hundreds of thousands, or millions of snails? This is one of the questions to which I wanted and needed an answer. Robb and I had collected 63 samples, one square meter each, from along shorelines in the study area. There was considerable variation in the number of snails in these 63 samples. It ranged from zero to 760/m^2. Using a whole number, there was an average density of 109 snails per sample, i.e., 109/m^2. To make a rough estimate of total snails in the area, I made an assumption based on what I had observed. I knew where snails occurred in the area. Along most of these shores snails existed in an intertidal band about two meters wide. If snails existed exclusively in such a band, how many would there be? I used a digital map measuring device and a topographical map of the area to determine that there are about 19,743 m of shoreline around the islands. If snails exist in a two meter band along this shoreline, that's (2 X 19,743 =) 39,486 square meters of snail territory available. If there is an average of 109 snails/m^2, that comes to (109 X 39,486 =) about 4,303,974 total snails. Of course, there are some stretches of shoreline where snails are absent or occupy a narrower band. If these dominated, that estimate would be too high. On the other hand, there are many inter- and subtidal areas where snails are very dense and occupy much more than a two-meter band. Recall the snail-laden mud I crawled along in for a while trying to get to my idyllic beach across the Bay in the last chapter. Such areas would make the estimate too low. In the final analysis I can't say exactly how big the snail resource is, but it is likely that the estimate of about 4.3 million is too low by a factor of two, three, or possibly four. Whatever the true number of snails, it is safe to say that there are millions upon millions of snails in the study area. Trematodes in the study area have a vast snail resource available to them.

I also wanted to know something about how this large resource was distributed in space. There are statistical tests (e.g., the Kruskal-Wallis test again) that can be used to justify the patterns I'll mention below, but I won't mire you down in them. If you compare the number of snails per square meter (snail density) at the 21 different locations, it turns out that snail density varied significantly. As we might suspect from the previous paragraph, snails were more abundant at some locations than at others. There is no constant, usual snail density. In fact, you don't even have to go from one location to another to get a major density change. At some locations samples, only separated by a meter, had very different snail densities. As an extreme example, at one location density ranged from a single snail per m^2 in one sample to $584/m^2$ in another only a few meters away. There was an overall pattern to this variation. Where snails were on average larger (over about 25 mm), they tended to be less dense (under about $50/m^2$). Where snails were on average smaller, density varied greatly (up to $760/m^2$). The lesson here is that while the snail resource in the area is vast it is not evenly distributed in space.

What about reproduction and recruitment of the snail resource? From the point of view of the parasites this would be an important consideration, having to do with the stability of their resource. First, it is not a given that all *Ilyanassa* populations regularly receive (recruit) new individuals. As we have covered before (Chapter 5) this snail reproduces by males and females producing fertilized eggs that go into an egg case produced by the female. From the fertilized eggs swimming veliger larvae develop. These emerge from the egg case and exist in the plankton for two or three weeks. If while drifting about a veliger is lucky, it will come upon a suitable place to settle out of the plankton onto the bottom where it can successfully change into a tiny juvenile snail. Happy metamorphosis, that's the origin of new snail individuals. I have found populations of this snail made up entirely of large (old) individuals, no juvenile snails having been delivered from the plankton in many years. So, is the population in southeast Rehoboth Bay a regularly recruiting one? Yes. How do I know that? Larry Hurd and I figured out that (uninfected) snails in the 12 mm size-class or smaller had settled out of the plankton within the last two years. In this study area, locations that had such small snails present must have recruited them recently. Where were they found? When the tides ebb from Rehoboth Bay water moves from north to south, eventually entering Indian River Bay and maybe exiting from there into the Atlantic Ocean by way of Indian River Inlet. In this process water just flows into the study area from the north. If you stand in Heaven's Gate on the northern edge of this study area, the north to south ebbing current can sometimes nearly sweep you off your feet. At times I certainly have had to depend on my staffs to stay upright. Without going into the details of the distribution of less than or equal to 12 mm snails, suffice it to say that they were found only at places in the direct path of these ebbing currents. Apparently the source of settling *Ilyanassa* larvae in the area is Rehoboth Bay. Does this mean that the veliger larvae were spawned there? No. Veligers exist in the plankton for long enough that they could have originated outside the Bay. Wherever they originated, maybe Cape Henlopen, or Canary Creek Saltmarsh, or Indian River Bay, or even Rehoboth Bay, they are mostly delivered to the study site by Rehoboth Bay ebbing currents.

If young snails have a limited distribution, only where ebbing currents flow directly in from the Bay, how can the widespread distribution of snails in the study area be explained? For one factor the habitat has a long history. In past years swatches of larvae could have been carried by currents to other parts of the study area. Veligers might have settled there and the population of adult snails they produced is still resident today. Put more briefly, from this study it's pretty clear where juveniles settled in the previous two years, but there's no reason to assume that settlement always occurs only in those places. In the past juvenile snails could have settled elsewhere contributing to the present widespread occurrence of adult snails in the study area. For another factor snails live for decades and there are snail movements as adults to consider. It would be possible for snails to move into locations where larvae have never settled. No doubt some of this goes on. *Ilyanassa* movement in the field will be addressed in some detail later. For now just let me make this point. Snails can move around quite a bit, possibly 50 or a 100 meters in a matter of days. However, obstacles, such as the islands and convoluted shorelines in the study area, impede or stop such extensive movement. While some movement of adult snails into new areas undoubtedly occurs, there is no way that a general snail intermixing among the 21 locations used here could take place. They are generally too far apart and separated by islands and/or convoluted shorelines.

So that's some the properties of the host resource in the study area. It is vast, widespread, locally variable in snail size and density makeup, and it is renewable. Now, how do the parasites make use of this resource? The first thing is prevalence. What proportion of snails in the area are infected with trematodes and with what species? In this study we collected 6,862 snails. Again, if in a sample there were 20 or fewer, we dissected them all, otherwise we dissected a random selection of 20. In all 864 were dissected. If only actually dissected snails are considered, then 250 out of 874 were infected, a prevalence of 29%. If the proportion of infected snails among all snails collected are considered, then an estimated 956 out of 6,862 were infected, a prevalence of 14%. Why should there be such a difference in prevalence depending on which snails you consider? In short, it's because 20 snails per square meter is an artificial, imposed cap on the number of snails that might be there. If there are more than 20 snails present in a sample (most samples) and you consider only the 20 dissected, the many younger, probably uninfected snails likely present get ignored. Using only dissected snails to estimate prevalence inflates the estimate above what it should be. To get an accurate estimate, reflective of prevalence in the field, you need to consider all snails per unit area (i.e., in this case number per m^2). So 14% is the better estimate of prevalence in this study area.

Even though prevalence was quite low, compared for example to Savages Ditch, or Cape Henlopen or Gray's Point in Indian River Bay, there was a good diversity of infections. In Delaware there are nine species of trematodes infecting *Ilyanassa* and six of them were found here. There were of course many uninfected snails. In quick summary, as suggested above 86% of 6,862 snails collected were uninfected. Obviously, the other 14% of snails were infected. Single infections observed were: *Hq* (5.5% of all 6,862 snails were infected with *Hq*), *Lepocreadium setiferoides (Ls)* (0.8%), *Zoogonus rubellus (Zr)* (3.6%), *Austrobilharzia variglandis (Av)* (0.5%), *Gynaecotyla adunca (Ga)* (1.3%), and *Stephanostomum tenue (St)* (0.5%). Using the method of abbreviation for multiple infections given before, five different double infections were encountered, *HZ* (*Hq* with *Zr*) (0.5%), *LZ* (0.1%), *LG* (0.2%), *ZG* (0.4%), and *ZA* (0.01%). It can be seen from this that *Hq*, *Zr*, and *Ga* were the most common infections, each occurring in more than 1% of all snails. The rest of the observed infections were scarce or outright rare.

I constructed a size-prevalence curve for snails collected in this study to assess whether the size of a snail can be used to predict whether or not it is likely to be infected. Let's just consider the overall curve, whether a snail is infected with any trematode(s). This curve is like an old friend from a previous life (Chapters 4 and 9): proportion of infected snails goes up gradually with snail size-class. All snails under 14 mm were uninfected, when you get to 24 mm snails about 30% are infected, and when you get to snails 30 mm and larger almost all are infected. There's nothing new about that. What is new is the connection between size and age made possible by the von Bertalanffy growth model for uninfected snails presented in Chapter 12. It is not an exact connection, but it gives an idea of how long snails are likely to remain uninfected in this habitat. Snails 18 mm or less are three years old or younger. The size-prevalence curve tells me that 92% of snails 18 mm and under have lived that long and remained free of parasites. By the time an uninfected snail hits 24 mm it is about eight years old and about 70% of 24 mm snails are still uninfected after that many years in the field. If we jump ahead to 29 mm snails, they are about 28 years old and about 60% are still uninfected. Snails in this habitat are likely to go unmolested by trematode parasites for many, many years, even though there are several species around that might infect them. Samples were taken differently for example on Cape Henlopen or at Savages Ditch, but snails in those habitats are not likely to exist for so long free of parasites.

A major question to be answered is: in this study area how are the infections distributed in space among the available snail (resource) population? First, overall prevalence was significantly higher at some locations, lower in others. Mostly this was caused by whether there were large snails present. Larger (older) snails are more likely to be infected. Where there are mostly only those present prevalence will tend to be higher.

That's true enough, but what about individual infection categories (e.g, uninfected snails, or those infected by *Hq, Ls, HZ,* or *LG* ...)? How are they distributed in space (i.e., among the 63 samples) in the available snail resource? We use the term dispersion when we talk about distribution of something in space. I brought this up back in Chapter 5. An organism can be dispersed in space in three basic ways, in a random, or clumped, or uniform pattern. In a clumped dispersion, the pattern I mostly want to think about here, an organism is likely to be found where other organisms of the same type are found. They tend to cluster together in a few patches of real estate. For example, in the present context you might expect *Ls*-infected snails to be found mostly clustered in certain samples. You can test for dispersion patterns using a variety of statistical procedures. I, and quite probably you, don't want to get drawn off into those, but a little bit will be useful. If you are interested, a little digging in a statistics book would take you further than I will go here. For starters imagine a 100 m^2 plot of ground. Further imagine that this plot of ground is divided into 100 square meter areas (sample areas) in which you will search for and count the organism of interest. A representation is presented below. The dots in the boxes (the 100 m^2 plots) represent individual organisms of some type (e.g. some species) and I illustrate what the three patterns of dispersion in space might look like. In terms of a parasite that infects a host, it naturally can't be where a host is not. So dispersion of parasites in space is at least partly dependent on how hosts are dispersed in space. For example, you might wonder (test) if snails infected with *Hq* are clustered in samples as illustrated in the center box. A way to do this is to divide the whole patch of ground up into areas (in the context of this work, into 63 square meter plots) and figure how many snails in each are infected with an infection category (e.g., uninfected, *Hq, Ls, HZ,* or *LG* ...) in each plot. You would determine the average (mean) number of snails infected with *Hq* in samples. You would also determine a measure of the variation in the number of *Hq*-infected snails in samples, something in statistics called the variance. With these two statistics in hand you divide the mean into the variance and thereby calculate a variance to mean ratio. In brief, if the variance in the number of *Hq*-infected snails in plots equals the mean number of *Hq*-infected in plots (variance/mean ratio = 1), then the indication is that the dispersion is random; if the variance is larger than the mean (variance/mean ratio greater than 1), then the indication is that the dispersion is clumped; and if the variance is less than the mean (ratio less than 1), then the indication is that the dispersion is uniform. In short, the variance to mean ratio (divide the variance by the mean) gives an indication of whether your organism is dispersed in space according to a random, clumped, or uniform pattern. As one example, the variance in the number of *Hq*-infected snails in the 63 samples was 224.5 and the mean number of *Hq*-infected snails in samples was 6.5. The variance to mean ratio (224.5/6.5) comes out 34.5. Obviously, this is much greater than one and the dispersion of *Hq*-infected snails is pretty definitely clumped. When I did this (calculated the variance to mean ratio) for all infection categories (including uninfected) present in the collective 63 samples in this study area the dispersions always came out clumped (variance/mean greater than 1). I was not surprised by this. Random and uniform spatial distributions of organisms are only occasionally found in nature. It is more usual to encounter clumped dispersions. That's because members of the same species interact, associate with each other and for various reasons (e.g., feeding, breeding, protection from predators) and tend to occur in the same places.

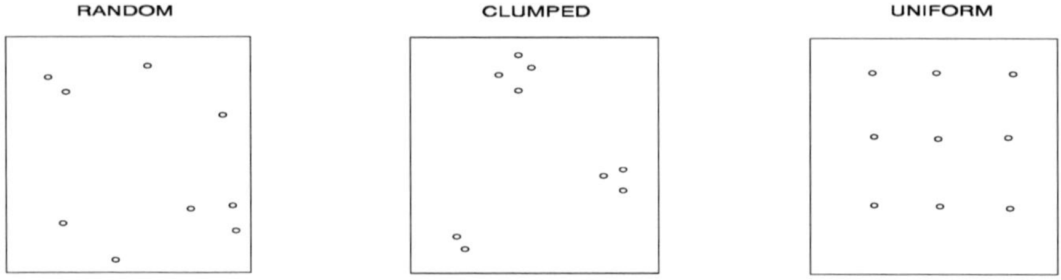

An illustration of what the distribution of organisms (dots) in an environment might look like if dispersed at randomly, in a clumped fashion, and uniformly

So, snails infected with the same category of infection (including uninfected) tend to be distributed in space in a clumped or aggregated fashion. There are lots of them in some places, not so many in others. That's not too surprising, but its not really the point of interest. Snails have a clumped distribution and the distributions of the parasites depend (at least in part) upon the distribution of snail hosts. So you would expect the parasites in snails to be spatially clumped too. The question of interest boils down to one of relative clumpedness. Clumping is not something you have or don't have. There are degrees of it. Because snails are the resource for the parasites, what we want to know is the degree of clumping of snails infected with various trematode species relative to the degree of clumping by snails in general. There are a number of ways to measure degrees of clumping. The best one I could come up with for this purpose is something called Green's Index. [To not leave this a total mystery I will give the formula, but you need not grapple with it unless you wish to. It is based on the variance to mean ratio: Green's Index = (variance/mean-1)/N-1, where variance and mean are as described above and N is the total number of snails in an infection category. To apply the formula, you would probably have to look up how to calculate variance in a statistics book. Your calculator will probably do it for you. It is usually symbolized on a calculator's keys by s^2.] The key thing to understand is that Green's Index can take values anywhere from zero to one. If Green's Index = 0, degree of clumping is minimal. If Green's Index = 1, degree of clumping is maximal.

Green's Index for uninfected snails come out at 0.04. That's not zero, but pretty close to it. While it is evident from the variance to mean ratio that uninfected snails are clumped in their distribution, degree of clumping is close to minimal. Most snails in the habitat are uninfected and this therefore is a good measure of degree of clumping for snails in general. Let me take an extreme example to illustrate what I did to assess relative degree of clumping: Green's Index for *HZ*-infected snails came out 0.75. If you divide 0.75 by 0.04, the ratio of clumping by *HZ*-infected snails to uninfected snails (the resource), it comes out at 17.77. Another way to say the same thing is that both categories have clumped spatial distributions, but *HZ*-infected snails are 17 (almost 18) times as clumped as uninfected snails. With one exception, calculations showed that all infection categories were at least twice as clumped as uninfected snails (smallest ratio = 2.12). The exception was *Zr*-infected snails. They had about the same degree of clumping as uninfected snails (*Zr* Green's Index divided by uninfected snail Green's Index = about 1). What did this broad result tell me? It told me that, with the exception of *Zr*, there's more to the distribution of parasites in the snail population than simply the distribution of host snails. Apparently trematode species infect snails as a result of factors over and above whether or not there is a snail there to infect. What might those factors be? We shall have to look into that. It has much to do with the way the parasites make use of the snail resource.

Another thing I wanted to know was whether snails infected with particular trematodes tended to be distributed in the same locations with other infecting species. Put another way, were trematode species distributed within the snail population (in space) independently of one another? I employed a species association test to find out. There were six trematode species and 21 locations. The test requires a listing of which species were present, and which were absent, at each of the 21 locations. The assembled data is a table made up of six rows (one for each parasite species) and 21 columns (one for each location). If a parasite species was present in at least one snail at a location there's a one in the table, if not a zero is entered. This table of ones and zeros is then statistically analyzed with the species association test. I will, as usual, spare you the workings of the test and just say that species were not associated in the presence/absence table (the table of 1's and 0's). Parasite species did not co-occur or avoid one another at the 21 locations. The hypothesis that the six species of parasites use the snail resource independently of one another was accepted. Acceptance of this hypothesis could arise simply from there being no associations, or to a cancelling of positive and negative associations. A positive association would be one in which two species tended to co-occur at locations; a negative one would be one in which two species tended to avoid occurring together at locations. One pair of species, *Av* and *Ga*, did show signs of a positive association in space, but that was it. (Definitive host for both parasites are shorebirds. I suspect that where these definitive hosts cluster, both infections will tend to show up in snails.) In any case, there were no positive and negative associations to cancel one another. The conclusion is that parasites, almost entirely, use the resource snail population without regard to where other parasite species occur. There's no predicting which parasite species will occur with which others at the 21 locations used in this study area. They colonize snails independently of one another.

Bear with me a moment. There is one more analysis of data in the table described in the previous paragraph that is instructive as to how parasites make use the available snail resource. At the 21 locations some assemblage of parasite species was present. For example, at location 21 *Hq, Ls, Zr, Av,* and *Ga* were all present. At most locations some smaller subset of the six species was present. The question is: are the assemblages of parasite species nested by location? When you look for nestedness in the assemblages of species at locations within a community of species, you are searching for some sort of structure in that community. Here I am looking for structure in the trematode component community infecting *Ilyanassa* in southeast Rehoboth Bay. In brief, if species are nested, the set of species at a location will not be a random draw from the available pool of six species. If not nested, the set of species at locations will be a random draw from the available pool. This can perhaps be better understood by thinking of cards (the species) dealt from a deck (the species pool). If the deck is well shuffled and the cards are dealt randomly, you don't expect to see any particular cards, for example a king, a queen, and a deuce, routinely in poker hands. However, you might expect this if the deck were habitually rigged or stacked in a particular way. Put in terms of a deck of cards, the question is: is the "deck" (parasite species pool) somehow rigged so that certain "cards" (parasite species) are dealt to locations in the habitat. The concepts of nestedness and species associations are closely linked and easily confused, but they are not the same. We shall come back to the distinction when we finish with whether or not assemblages of parasite species are nested by location in this habitat.

As you probably suspect by now, there is a calculation to determine whether there is nestedness or not. This calculation is not something you routinely do by hand. There are programs available in the literature (on the internet) that do the job. The data that one of these programs uses is the same presence/absence table as used for the species association test. The presence or absence of the six species at each of the 21 locations is "the system." What the program does is calculate the "temperature" of the system. Colder temperatures indicate more ordered (nonrandom) systems, more nested. Hotter temperatures indicate less ordered (more randomized) systems, less nested. The temperature of the system can be thought of in terms of water. The molecules (species) in cold water are more ordered than in hot water. The "temperature" of this system calculated out at 16.1 Celsius. Given possible temperatures of between zero and 100 degrees, that's a "cold" system and it is definitely nested. So parasite species assemblages in this study area, in this host, were nested by location. Locations in this study were all within about 1 km of one another. In another paper (Esch, Curtis, and Barger, 2001) we analyzed parasite assemblages in *Ilyanassa* for nestedness at nine more widely separated sites (several kilometers apart). That system was nested too, and apparently for the same reason. In both studies the most widespread species, the ones occurring at the most sites, were *Hq, Ls,* and *Zr*. In this study snails infected with *Zr* occurred at all locations, something that is generally characteristic of this parasite in this host. If there are snails present and they are infected with any trematode, it is likely to be *Zr*. Trematode species assemblages are nested because all or some of these species (*Hq, Ls, Zr*) are the ones most likely to be "dealt" to snails at a site from the pool of available species. You may well be asking yourself, why is this not a positive species association? Why didn't the species association test detect these three species as associated? This is a perfectly valid question and it certainly caused me some pondering. The answer is that these are the species most likely to be present at a location, but they are not "dealt" as a set to locations. At some locations maybe only *Zr* is present. At others maybe *Hq* is there too, at others maybe *Ls*, and at still others all three are present. There is no predicting which of these three species will infect snails at a location, but *Zr* will and maybe *Hq* and/or *Ls* as well. As a consequence the assemblage, the species present at a location, is not a random draw from the system's "deck." The deck is somehow rigged so that assemblages are nested, likely to contain *Zr* and *Hq*, or *Ls*, but these species are not tightly associated in space because there's no predicting which ones of the three will be present. It would be of interest to understand what goes into the "rigging" of the "deck" to most likely deliver these parasite species (in one combination or another) to locations. We probably don't yet know enough about infection of *Ilyanassa* by trematode miracidia to truly understand what is going on here, but it is certainly worth pursuing. It has much to do with how the parasites use the host resource.

One of the biggest bugaboos in figuring out any of this is our low-level understanding of the life cycles of these trematodes. We have gone through this before, but the basic life cycle involves a first intermediate host (the snail) in which sporocysts or rediae produce cercariae that find their way out of the snail and swim to a second intermediate host (an invertebrate or fish) where they encyst as metacercariae. When an infected 2nd host is eaten by a definitive host (usually a fish or bird) the metacercariae develop into adult worms in the body of the definitive host. The adult worms produce and trade sperm, which result in fertilized parasite eggs. The eggs are packaged in capsules, which get into the definitive host's digestive tract and are voided from that host's body with fecal material. At this point the capsules, with their contained fertilized parasite egg, are in the outside world. So far things are pretty well understood, but now into the unknown. The next step is for fertilized eggs in capsules to develop into miracidia and somehow find their way to a snail that it can infect. If successful, the life cycle would be continued and the genes of the involved trematode genetic strain preserved. Those genes get to function for at least part of one more generation. About this to-the-snail life cycle step we know very little. To begin with we can note that miracidia of only two of the nine trematode species, *Zr* and *Av*, involved in Delaware *Ilyanassa* have ever even been observed in the laboratory. Back in the 1930's a researcher that I have written of before (Chapter 4), one H. W. Stunkard, was able to obtain *Zr* adult worms from eels (the definitive host for *Zr*). The *Zr* adults contained capsules with live miracidia in them. In the laboratory, he was able to infect two *Ilyanassa* with miracidia he had obtained. This is the only case I know of where experimental infection of *Ilyanassa* with any of its trematodes has been accomplished. He watched for a long time (months) and in the end dissection revealed *Zr* sporocysts in the snails, but neither of the experimentally infected snails ever produced any *Zr* cercariae. In Hawaii another snail is infected by *Av*. Miracidia were observed in that system, but I'll not further pursue that case here. I have attempted to experimentally infect snails. I maybe had some luck, but it was never conclusively successful. The two species of miracidia that have been observed in the laboratory (*Zr, Av*) emerged from capsules and swam to snails. The other seven miracidia have never been observed. It's not because people haven't tried to see them. Primarily it's because miracidia fail to hatch out of capsules in the laboratory, even when capsules have been watched for months. Based on their close relatives in the trematode world, which have been observed, the other seven miracidia most likely swim to snails, but miracidia in general constitute an essentially unfathomed step in life cycles. Whether seen in the laboratory or not, the actions of miracidia in the field that lead to *Ilyanassa* infections are an utter unknown. There is no question that we have to operate from ignorance. We can reason about what must go on in the field, but so far these miracidia are unobservable and unavailable to true scientific investigation.

Allow me to take a couple of paragraphs to summarize what we have found out concerning the host resource in southeast corner of Rehoboth Bay and the use of it by the resident trematodes. There is a vast resource (millions upon millions of snails) and it is widespread. Snails occur in inter- and subtidal zones almost everywhere throughout the area. As you go from place to place (location to location, and even from sample to sample within locations) snails are likely to vary in density (number/m^2) and average size. The rule seems to be that local populations of larger (older) snails are sparser than those of smaller (younger) snails. The resource is renewable. It can be shown that juvenile snails have recruited from the plankton at least during the two years previous to this study. It is a virtual certainty that recruitment has occurred in other years as well. The resource snail population does not freely and generally mix. Snails (whether infected or not) have the potential to move around quite a bit (maybe as much as 100 meters in a matter of days), but other work has shown that they do not readily move around even small obstacles such as islands. The snails at locations sampled in this study were mostly separated by hundreds of meters, convoluted shorelines, and even whole large islands. In other words, the local populations of snails sampled were effectively isolated from one another. There would be no chance of a general intermingling of snails from different locations. If this could happen, my analyses and conclusions would have to be very different.

If that's the basic nature of the resource, how may the use of it by trematodes be summed up? There were a total of six species using the snails in the study area. All six showed up as single infections in snails, plus there were five different double combinations of them in individual snails. Even though diversity of infections was fairly high, relatively few snails were infected. In some samples percent infected was high, but the overall prevalence of infection among snails was only about 14%. All 12 considered infection categories (e.g., uninfected, *Hq*-infected, *Ls*-infected, *LZ*-infected, and etc.) gave indication of having clumped dispersion patterns in the 63 square meter samples. That is, for each category the numbers in square meter plots of ground were not always the same (uniform), nor did they vary randomly. Rather, snails in all infection categories (including uninfected) tended to cluster (aggregate) in particular samples. If snails (the resource) are clumped, it would follow that infections in snails would also have clumped dispersions. That they are clumped is not the surprise. The important question is: how strongly are infection categories clumped relative to the clumping of the snail resource? I used something call Green's Index to measure the degree of clumping of the 12 infection categories. It turned out that all infected categories, except *Zr*, were two or more times (sometimes much more) as clumped as uninfected snails. It is a given that a host snail must be there for a parasite to be there. Comparatively elevated clumping of parasitized snails means that there is more to the spatial distribution of parasites in the resource snail population than simply the availability of hosts to infect. When I looked to see if species of parasites were associated in their spatial distributions, i.e., tended to occur in snails at the same or different locations as other parasite species, the answer was no. Trematode species occupied snails at locations uninfluenced by whether other species were there or not. That is, conditions (e.g., maybe currents) that result in one species of miracidium infecting snails at a location do not guarantee that any other kind of miracidium will (or will not) also get there to infect snails. Miracidia of different species appear to operate as "lone rangers." At the same time, there was some inherent community structure. Assemblages of parasite species at locations were found to be nested. Nestedness is an elusive concept. Using an analogy, what this means is that the "deck" from which parasite species are "dealt" to the 21 locations is somehow "rigged" so that a location's "hand" is likely to contain *Zr, Hq,* and/or *Ls*. The basis for that deck rigging is the real question that needs answering. For some reason, snails infected with *Zr* were dealt to all locations and *Hq* and/or *Ls* were the next most likely. These three species had an enhanced chance of getting "dealt" to snails at locations. Consequently, species present at locations show nestedness. They are not a random draw from the pool of six species.

These results suggest that there is not much order to the way the parasites are spread through the snail population in this study area. What little order there is comes from nestedness caused by *Zr* being present at every location and *Hq* and *Ls* being next most widespread. Other than that, except for possibly one species pair, just because one parasite species was present or absent at a location gave no prediction that any other species would be present or absent at that location. As mentioned parenthetically above, that one possible exception of *Av* and *Ga*, may involve both species being carried by bird definitive hosts. Where birds gather the chance that both species will be found in snails at that location increases. However, other pairs of bird parasites (e.g., *Hq* and *Av*, *Hq* and *Ga*) did not have associated spatial distributions. Neither were associations in evidence for pairs of fish parasites. If having similar definitive hosts in common were an influential factor, more positive associations should have been indicated and this makes me unsure of what to make of the *Av* and *Ga* case. Anyway, the distribution of parasites in the study area's snails appears to be basically a hodgepodge. The haphazard parasite distribution suggests that the parasite larval stages that infect snails get haphazardly distributed in the study area. Only *Zr* miracidia, from eels, seem to get everywhere. The infections they produce, though not abundant (only 3.6% of snails), have the same degree of clumping as uninfected snails. I would like to know what brings this about, but I suspect not in this lifetime.

It is apparent that snails at different locations are exposed to infection by different parasite species. Why aren't snails everywhere infected with the same trematodes and why is prevalence so low? Given the state of our knowledge about how snails get infected in the field, this will be a tough question to answer. It is nevertheless worth thinking about because it is, once again, at the root of how the parasites make use of the host resource. At least two factors must be involved. One is the delivery of infective stages to locations by infected definitive hosts. Essentially, birds bring *Hq, Av,* and *Ga* to habitats while fish bring *Zr, Ls,* and *St.* If definitive hosts were seldom to visit a site, then seldom would snails there get infected and diversity and prevalence of infections there would be routinely low. If this were the case, prevalence would be routinely low at a site because infected snails can't easily move from site to site in this study area. So patterns of definitive host visitation to sites are one thing to consider. This has been studied in other snail-trematode systems and found to be a good explanation of why snails in some places are frequently infected while in others they are not. This factor is therefore not to be disregarded.

Another factor to consider is that once capsules (with fertilized parasite eggs in them) are deposited in a place by a definitive host, what's the likelihood that the capsules themselves, or miracidia from them, will be dispersed from that place? For example, intact capsules might drift away in water currents. Also, if a bird lands and walks through the mud, it might pick up capsules on its feet and fly them to another site. Beyond the transport of intact capsules we must also think about swimming miracidia that emerge from them. It is true that some snail species get infected by ingesting (eating) the eggs of trematodes, but that does not seem to be the way it works with *Ilyanassa*. Though most of the miracidia that infect *Ilyanassa* have never been observed, the two that have been came out of capsules and swam to snails. It is likely that the others do so as well. Maybe future studies will tell us differently, but for now it looks as though we are dealing with swimming miracidia. Miracidia are microscopic, ciliated larvae. If you had some suspended in a jar of seawater and lighted them just right, you would maybe be able to discern them as very tiny specks with the naked eye. Being ciliated means they swim by means of small undulating “hairs” on their bodies. Strong swimmers they would not be and they are not a feeding stage. Most miracidia run out of gas (stored nutrition) and sink to oblivion within a day. If up in the water, they would be dispersed by all but the very weakest of currents. During their short lives, miracidia would be spread by tidal currents once emerged from capsules.

If there are infected definitive hosts around, they would deposit capsules in their feces. Is it likely that there are infected bird and fish hosts around and in the study area? In the immediate area there are several places with many infected snails that answer yes to this question. We can point to Cape Henlopen, only 15 km distant, Gray’s Point in Indian River Bay, only eight km distant, and Savages Ditch, right on the northern edge of this study area. These sites had sufficient infected definitive hosts around to over the years generate abundant infected snails. This study area is close to those sites and birds and fish infected with trematodes would have frequented it as well. I’m at a loss trying to postulate why they would avoid it. I have spent much time in the area and, while I can’t speak as certainly about fish, I can say that birds are diverse, active, and abundant in the area. A good supply of trematode capsules in the study area is a well-founded, reasonable expectation.

If a lack of infected definitive hosts is not likely the main determinant of low prevalence in the snail population, what about the dispersal of capsules, or miracidia from them, after they are deposited? The places mentioned in the previous paragraph, where there are abundant infected snails, all have at least one thing in common: they all consist of, flat intertidal areas where birds congregate. Birds can and do congregate in this study area too. There are shallow areas all over where this happens (the Savages Ditch site is one of them), but they are smaller in size compared to say Cape Henlopen. Maybe this reduces the delivery of parasite capsules to the area and partly explains why there are fewer infected snails present. Again, we should not dismiss this as a factor. Another factor that would come into play is this: when tides pile into the study area strong currents develop in the narrow waterways among the islands. This is so on flood tides, but especially so when the water ebbs through the area from Rehoboth Bay. Capsules, and more importantly miracidia, would be swept along. To infect a snail a miracidium must be able to hit on a target snail, attach to it, hang on, and penetrate the target's skin. The whole operation takes something like 10, 20, or 30 minutes. In a strong current a miracidium would have at least a difficult time accomplishing all the necessary steps. An explanation of why only something like 14% of millions of snails in the area are infected, and that's the product of years or decades of parasite accumulation, may well partly or wholly reside in this factor: miracidia can not easily stop long enough to infect snails in the environment's strong currents. On a broad intertidal flat currents would be present, but less channeled and intense. Ciliated miracidia would not be so likely to be swept utterly away from snails, possibly even out of the snail environment. In a more conducive infection environment, such as an intertidal flat, miracidia still have trouble finding and infecting snails. In such places only a few percent of snails get infected each year (Chapters 9 and 11). In an environment with strong, directional tidal currents it is likely that success of miracidia at infecting snails would be severely cut back. The yearly rate of snail colonizations by trematodes in southeast Rehoboth Bay must be even lower than it is say on Cape Henlopen, where it was measured as 1.6% per year (Chapter 9).

From a scientific point of view I recognize that all this reasoning is just that. To actually make the point that definitive hosts shed capsules in the study area one would have to collect those hosts and demonstrate that they are infected. This would require killing some of them and identifying the adult worms in their intestines. That would be a wonderful job! Further, it would be good to document where in the study area these infected hosts hang out, collecting their fecal material and demonstrating the presence of capsules therein. That would be another joyous pursuit! Some of these things have been done with these parasites in other areas. For example, eels and shorebirds are often found to be infected. To say the least this would be a gigantic undertaking, not impossible, but requiring lots of man-hours and resources. Another necessity would be a laboratory study of the behavior and abilities of miracidia in variously strong currents. At present I have no clue how this could be done in the field. And, if you recall that most of these species of miracidia have never been observed even in the laboratory and thus far it has not been possible to experimentally infect this snail with miracidia (except for H. W. Stunkard's limited success with *Zr* back in the 1930's), a moment of reflection will tell you that such a study would face major difficulties.

There's no way that anyone could do the necessary research without abundant people, time, money, and equipment. While it is not the best scientific approach, these difficulties give me some license to approach the problem by using what I have found out and reasoning my way through it. It is a likely assessment of what goes on, but hard data would better make the case. Nevertheless I shall continue on for a little bit with my reasoning. Please come with me.

In my work following infected snails at Savages Ditch for a number of years (Chapter 12) I noted, as Karen and I had discovered (Chapter 8), that infected snails can put out hundreds or thousands of cercariae each day. Because snails live with infections for many years, lifetime output of cercariae by an infected snail stands to be immense. Now think of the many chances for failure in the completion of a trematode life cycle. Most of these parasites must get to and infect three different hosts [a 1^{rst} (the snail), a 2nd (an invertebrate or fish), and a definitive host (fish or bird)] and then get back to another snail to complete a life cycle. Getting from a snail to a 2nd host involves risky time outside any host as a short-lived cercaria. To transmit from a 2nd host to a definitive host involves the chancy business of predation, the latter host must eat the former. The trip from a definitive host to a 1^{rst} (snail) host involves risky time in the outside world as a short-lived miracidium. The likelihood that a trematode in a snail will manage to send its various stages through all these improbable steps successfully is vanishingly small. In previous chapters we have been covering some of the parasite adaptations that enhance the chances of completing a life cycle. Perhaps the most striking discovered so far is the alteration of snail behavior by *Ga* putting host snails high on the beach so cercariae can get to a beach-dwelling 2nd host. A thing these parasites all do is put out immense numbers of cercariae while they are resident in a snail. The value in putting out lots of cercariae is that at least one will likely make it to and infect a second intermediate host. Success means the life cycle gets to continue. If the parasite can alter the behavior of the second intermediate host so it is more likely to be preyed upon by a definitive host, that too would increase the chances of the life cycle continuing. Adult worms in the definitive host produce many miracidia, which makes it likely that at least one will get to and infect a snail. Such adaptations not withstanding, life cycle completion for a trematode is still very unsure business. A snail infection, a genetic strain of a species of trematode infecting *Ilyanassa*, can only increase the probability of life cycle completion by producing an abundance of cercariae. That's all that stage in the life cycle can do to that end. This requires not killing or impeding the snail host. It must be kept alive and feeding year after year while abundant cercariae are produced and released. The genes of any genetic strain of parasite incapable of doing this would be quickly be eliminated by natural selection from the parasite population, weeded out, lost.

The question arises: what specifically is being compensated for by adaptations to produce so many cercariae? It appears that the most unlikely step in the series of steps necessary to complete a life cycle is the step from definitive host to snail by miracidia. All the other life cycle steps must be made more probable mainly to absorb (compensate for) this unlikely event. After all, if abundant cercariae did not result in sufficient infected 2nd hosts, there would be fewer infected definitive hosts. But, these two sorts of hosts are in fact often infected. As a consequence, many snails wind up infected, though it takes many years to develop high prevalence in a snail population.

As so far presented, my idea is that miracidia are poorly suited (adapted) to accomplish their role in the life cycle, infecting snails. One would be wise to wonder how this could be. Why didn't natural selection act across the generations to produce better adapted miracidia? Would not better success at infecting snails benefit the parasites? I argue not because the life cycles of these parasites quite possibly would be damaged or destroyed if miracidia were much better at their role. Here's a possible reason why. I credit and agree with another worker, N. Dronen, Jr., for this way of looking at the problem. Back in the 1970's he studied all the steps in the life cycle of a trematode in a pond. This would be very hard to do with an *Ilyanassa* trematode in the marine environment. His trematode infected snails, damsel flies, and frogs. He was able to measured the chance of success at each step. When you consider collectively all the stages of a parasite species in all host species in its life cycle, plus transmission stages (e.g. miracidia) between hosts, you are talking about the parasite's suprapopulation. Sorry for one more term, but it's important. The chance that a miracidium would be successful at infecting a snail was found by Dronen to be the lowest of all steps in the life cycle, something like one or two miracidia in ten thousand would be successful. He reasoned that if miracidia had a greater success rate, many more snails would be infected, and therefore many more second intermediate hosts would be infected, and then many more infected definitive hosts would follow. The system of hosts for the parasite's suprapopulation stood to be overburdened with parasite infections. If even one critical host population, such as the involved snail, succumbed under the burden, the parasite's life cycle (the suprapopulation) would succumb with it. Obviously, this would not be of survival value for the parasite. Any genetic strains that promoted much better success of miracidia (perhaps with stronger swimming ability) would long ago have seen their suprapopulation eliminated by natural selection. Such "improved" miracidia could not be long tolerated, as any suprapopulation in which they became run-of-the-mill would be lost. The effect would be that miracidia would remain inefficient as a means to infect snails. If they didn't remain inefficient the whole parasite suprapopulation would be placed in jeopardy. Actually, miracidia are not badly adapted at all. To restate the case in terms of the hero of this tale, if miracidia were supremely effective at infecting snails, most snails would become infected and sterilized. Snail reproduction might be severely curtailed or halted and *Ilyanassa* driven to rarity or extinction. There wouldn't be enough 1^{rst} intermediate hosts to infect to carry on life cycles. In point of fact, in their inefficiency, miracidia are well adapted to promote the ultimate successful reproduction of a parasite's suprapopulation.

This project has told me a lot about how a set of parasite species makes use of a host population, but we surely don't know everything we need to. There is some structure in the parasite component community. It is brought about by the greater likelihood of *Zr, Hq,* and *Ls* infections getting delivered to snails at locations, whatever the cause of that may be. That and the essentially disordered, haphazard way snails get infected in this habitat are likely the key things I was able to find out. There appear to be no consistent forces in the environment that dictate where snails are most likely to get infected. Examples of such forces might be predictable current patterns that deliver miracidia (or capsules) of several trematodes to certain sites so snails there are likely to be infected, or possibly behavior patterns of definitive hosts that dictate where they cluster, thus making it more likely that snails there will be infected. Such forces do not seem to rule in this habitat and the parasites apparently make hodgepodge use of the resource. If I am right about this, it tells us much about how this system works. You may be thinking that of course it's a hodgepodge affair. Isn't that the way things work in nature? I reply, well sometimes, but cases are rare where you find a disordered system and have a probable hypothesis explaining why it's that way. That's about as far as I can go with it. I'm thinking that it would be useful to work up a comparison study on a place like Cape Henlopen where currents would not be as dominant a factor and moreover snails would more easily move among sampling locations. Such a study would be informative, but it would have it's own major difficulties. A long time ago Karen and I did some work on Cape Henlopen that might serve such a purpose. It would be a lot of work, but maybe I can interest her to the point of action in pursuing that project. Maybe?

Back in 1997 after Robb and I had collected the data for this study, Robb wasn't able (time-wise) to go through all the data processing and analysis required to produce a publishable paper. I have not kept up with him, nor he with me, since about that time. He was more than a great help. He made the data collection possible and I can't thank him enough for that. We had many interesting times, even a little adventurous, that were filled with good humor during our work together. In addition to his good grade, I hope he got what he wanted out of this project. I think he did. As I mentioned earlier, I just recently finished putting the manuscript for this project together. Had I done that 10 years ago I wonder how it would have been different. I was very different then and my product too surely would have been different, maybe only in emphasis. It's one of those contingencies about which we'll never know.

Two parasite species (A and B) relaxing and ignoring one another.

Chapter 14

WITHIN-SNAIL BATTLES?

It is time to treat my experience with, to now, my last student co-author. In 1999 Mike Hendrickson called southern Delaware home, but he was living and taking classes on the main campus of the University of Delaware in the northern part of the state. Working toward a degree in biological sciences in the home department there, he had nothing to do with the academic unit where I taught. Mike was about half way through his degree work and decided he wanted to meaningfully involve himself in a research project. He understood that if you want to get accepted (with funding) into a graduate school, it is beneficial to get some research experience while you are an undergraduate. There would have been opportunities to work with faculty located on the main campus, but Mike could not be there during the summer months. He needed a project that could be done nearer home in southern Delaware. In my view, and as I have said previously, one of the better things the University does is to support an undergraduate research office. Over the years I contributed to the efforts of this office and support had been awarded to my students. The works of Karen Hubbard, Jeff Kinley, and Nathan Tanner and others had all been supported, at least in part, through the undergraduate research office. As a result my and my student's work was included in that office's website and through that Mike became aware of me and my work on snails and parasites in southern Delaware. No doubt Mike's initial attraction was to my location rather than to me, but there is no harm in that. During the winter of 1999-2000 we traded a couple of emails and ultimately made an arrangement to meet in my laboratory in Lewes. Undergraduate students that I had worked with in the past were generally at an earlier stage in their undergraduate careers than Mike was at this point. He was in his junior year and pretty sure about what he wanted to do. The fact that several of my students had been co-authors with me on published papers was an important thing to him. All too often students get to help out on faculty research projects by doing menial tasks such as washing dishes or entering data. They don't get to contribute to the idea of the research, to help ask the question, to help in how it will be answered, and participate in the execution. As I have communicated before, a student doesn't get a complete research experience unless he or she takes part in the whole process and that was what I had in mind with all my research students. That was what Mike wanted to do.

I have advised graduate students in the past. Linda Collins and Kathy Corbett, whom I mentioned way back in Chapter 3, and one Diane Woodard all completed graduate degrees under my guidance. A student that has gone through their undergraduate career and gone on to graduate school in pursuit of a Masters or Ph.D. degree can be expected to come up with a research problem on their own. Such a student has enough experience and read enough literature to be able to do this, although some guidance will be necessary. An undergraduate student however, as smart and capable as he or she may be, will not have had much experience with research and very little or no exposure to the scientific literature. Usually at least, one cannot expect an undergraduate to propose a project that stands a good chance of ending up as a published paper. Mike wanted to do something along these lines and so in our several meetings that winter we began the process.

Ultimately, the project we came up with was based on work that was in Karen's thesis and in our later publication (Curtis and Hubbard, 1993). It is all covered in better detail in Chapter 8, but let me briefly re-summarize here. This is the topic about which my discussions with Mike began. Results from Karen's and my work indicated that when pairs (or threes, even fours) of trematode species infected the same snail they did not much affect one another. There was no indication that any species displaced any other from regions of the snail's body, no indication that any species shut off production of cercariae by any other, and we could not demonstrate that any species significantly curtailed production of cercariae by any other. With regard to adaptations to other species in this snail-trematode system, we concluded that the parasites would be adapted to make use of the snail host. After all, in every iteration of one of these parasite's life cycles, i.e., every time one is completed, an *Ilyanassa* host must be encountered and used by the parasite. It makes sense that these parasites would be adapted to use this host. Any parasite genetic strain within a species that was not so adapted would have been unable to complete its life cycle and very quickly it, and genes possessed, would have been eliminated from the parasite's gene pool. On the other hand, our conclusion was that host snails could not be adapted to resist or deal with trematode parasites. This was based on the fact that the snail is spread by planktonic larvae and there are just too many chances for snails to live free of parasites. Parasites are not abundant in all, or even most, snail populations to which snail larvae might be dispersed. Consider from the last chapter the number of snails that live parasite free in southeastern Rehoboth Bay. A snail genetic lineage would not routinely encounter parasites generation after generation and thus could not undergo adaptation to deal with them. A consistent selection pressure, parasite on snail, across snail generations would in all probability be lacking. For a similar reason we said that parasite species cannot be adapted to deal with one another. They don't occur together in the same snail often enough for there to be a consistent selection pressure from one parasite species on another. You, as a careful reader of these adventures, have an advantage in understanding all of this. Getting this basis of information across to an undergraduate student who had never had occasion to think about snail or trematode life cycles, or indeed the process of adaptation in detail, took much time. Among other courses, Mike had taken a course in ecology and understood the principle behind how natural selection led to adaptations. He was prepared to understand what I was talking about, but there was a lot to cover and there is a difference between knowing something as a matter of book-learning and knowing it "in your gut," as you have to when you are about to do some original research on it. This was some six or seven years after Karen's and my paper came out. After our initial lengthy discussion was over I gave him a copy of the paper to take home with him. Now I had to wait for another session to find out how well all this material had been assimilated. When that session came I was very pleased. Mike had grasped everything I had said, raised some good questions, and was starting to think effectively about how the knowledge might be profitably extended. He was an unknown at the start, but he turned out to be a very smart and capable guy. Good! This would likely work and we were on our way. Now we had to decide on specifically what we would do. We had some observations (previous work by Karen and me). What would be the question pursued? What hypothesis might be tested? How would we test it? How would we analyze results? What would various outcomes mean? We may have been on our way, but the journey would be long. There was much to think about and do before this bit of science would be able to flower.

This is all about infracommunities and how the species that make them up get along as they infect the same snail. Recall that an infracommunity is the assemblage of parasite species infecting a single host individual. In that host each species has an infrapopulation. One way that this has been approached is to look at component community patterns, that is, to look at which trematode species are present in a snail population (the component parasite community) and how often they co-occur in single snails. From that information it might be possible to infer what must go on in infracommunities. This in my opinion is very unsure business involving all manner of sampling problems. Anyway, whatever happens between species when they must coexist it happens within infracommunities and that is where we should look if we wish to find out what goes on there. As indicated above, Karen and I did an extensive piece of work on this back in the early 1990's and the general result, with some notable exceptions, was that the species in an infracommunity seem not to affect one another very much. A problem with that work was that we were limited to viewing the state of affairs in infracommunities as quick "snapshots" in time. We collected snails that had one or more infecting species and ran our procedures on them at that time (cercarial release and determining where sporocysts or rediae and cercariae were distributed in the snail body). Since that time Nathan and I had followed individual snails with various infecting species in the field for long periods and I had learned that multiple-species infections can persist for years on end. What we needed was some means to gage the effects of longer-term coexistence of trematode species in a snail. There was no way to maintain infected snails in the laboratory over the required time, at least under anything like natural conditions. If we could do that, we could take some sort of measurements on the state of the infracommunity as time proceeded. As that was not practicable, snails would still have to be collected from the field at a time and examined. The current state of whatever infracommunity existed at that time would be the final product of however long a history that infracommunity had in that snail. The strategy of collecting an infected snail from the field and measuring the state of the infection would not change, but the thing measured would. Whatever we measured should give us an idea of what had been going on over the history of whatever infection was present in the snail.

What measurement would reflect long-term history? When ecologists take an interest in species interactions they usually try to determine the effect of one species on the population size of another. For one example, when two species interact (the members deal with each other) they both may have a smaller population size compared to when they are not interacting, at which time both would have larger population sizes. This species interaction can be symbolized as follows, a - - situation when interacting and + + when not. The - - means that both populations have reduced sizes when members of the species interact; the ++ means the populations of both are larger when species members do not interact. If an interaction can be correctly symbolized in this particular way, it is a signal that when interacting the two species are competitors for some resource that is in short supply. The interaction is in the mode called competition. A situation in which neither species affects the other would be symbolized as 0,0 when interacting and 0,0 when not (see frontispiece for this chapter). If individuals of one species prey on (kill, totally or partly consume) individuals of another, that would be a mode of interaction termed predation. In terms of effects on population sizes, when the species are interacting, the predator's population increases while the prey's population declines. When not interacting it's the reverse, the predator's population gets smaller and the prey's gets larger. This can be symbolized as a + - situation when predator and prey interact and a - + situation when they do not.

That's just two of several modes of species interaction. As a relevant aside, a third is parasitism. That's the relation between host (snail) and parasite (trematode). Parasitism is like predation in terms of the effects on populations, + - when parasite and host interact, - + when they don't, but the kind of interaction is different. When they are interacting, a parasite uses a host individual for a time as a source of nutrition, without totally consuming it (outright killing it). As a result the host's population size is made smaller. In the case of trematodes in *Ilyanassa*, the parasites reduce the host population by rendering infected snails sterile. Infected snails don't contribute to reproduction of the host population and it is thus smaller than it would be if the parasites were not present (no interaction situation). As the parasites use a host (i.e., interact with it) they get to reproduce and thereby increase their own population.

In the context of species interactions I mention especially competition and predation because when you think of two (or more) trematode species drawing on the same individual host snail for nutrition you might expect these modes of interaction would be involved. Indeed, these two modes (especially predation) have often been invoked to describe the relationship between species of trematodes infecting the same snail host. Sporocysts have no mouth and make their living by absorbing nutrients from the host. One can imagine two sporocyst species inhabiting the same snail and absorbing the same nutrient, which is in short supply, i.e., competition. Rediae, on the other hand, are equipped with a mouth and a sucking pharynx (throat). They can ingest bits of host tissue and might be capable of preying upon sporocysts (or rediae) of another species inhabiting the same snail. There are in fact a few striking photographs in the literature of rediae in the process of ingesting a sporocyst of another species. If there is competition or predation going on between species in parasite infracommunities, it should be reflected in the relative population sizes of the parasite species. Consider two trematode species, A and B. If competition were occurring between them, that is, if there were some host nutrient they both needed, but not enough to go around, we should see a - - situation when A and B are together in snails and a + + situation when they occur alone in snails. Put another way, populations of A and B should both be larger when singly infected snails are examined and smaller when doubly infected snails are examined. If predation were occurring between them (A the predator and B the prey) we should see a + - situation when they occurred together in a snail and a - + situation when they occur singly. These were the kinds of things we needed to somehow measure. How might we measure them?

We decided that if we could collect some snails from the field that were doubly infected with some pair of species and count the number of stages (sporocysts or rediae) present and also some snails singly infected with each of these species, we should be should be able to get some results indicating what is going on between the species. The problem was that multiply infected snails of any kind are scarce. If we could do it at all, it would be difficult to find enough to analyze. I had a clue what we might do. Over the years (Curtis, 1997) I had found that snails doubly infected with *Himasthla quissetensis* (*Hq*) and *Zoogonus rubellus* (*Zr*) were relatively abundant. (A reminder, the symbol for a double infection with these species is *HZ*.) On Cape Henlopen 275 out of 7500 snails (3.7%) were *HZ*-infected. In 1995 when for the first time I collected snails at Savages Ditch, 20 out of 80 (25%) were *HZ*-infected. We might have chosen another, but this would be a good a species pair to go after. It was of widespread occurrence and relatively easy to find compared to other species pairs. It would be at least possible to collect enough *HZ*-infected snails to analyze. Singly infected snails were quite abundant and would be fairly easy to come by.

With that choice made there were still many things to overcome. For one major thing I had to show Mike how to measure and dissect snails. In the spring of 2000 we made many trips out onto Cape Henlopen, returning with snails which served to instruct on the nature and procedures associated with a variety of infections, including of course *Hq* and *Zr*. Mike learned quickly and soon he became expert at examining snails and their infections. A major problem was how to count the number of parasite stages in a snail. This is far from a simple matter. The number of rediae or sporocysts in an infected snail routinely number into the hundreds or thousands. They occur throughout a snail and are intimately bound up with and stuck to snail tissue. Infections with *Hq* required the counting of rediae. The rediae of this species are not especially tightly enmeshed with snail tissues and other rediae. It is not an easy matter to get them separated so they can be counted, but it can be done. The sporocysts of *Zr* are more of a problem. They stubbornly cling together and to snail tissue. How can a singly or doubly infected snail body be processed once it is removed from the shell and the desire is to count the rediae and/or sporocysts present? Here's what we came up with. When a snail was brought back to the laboratory it was first tested by cercarial release to see if it released *Hq*, or *Zr* cercariae, or both. During 2000 Mike collected a total of 734 snails, mostly from Savages Ditch, but some also from Cape Henlopen. Most of them released no cercariae, or were infected with something else. If a snail did not have *Hq* and/or *Zr*, it was of no use to us and such snails were returned alive to their habitat. A snail that released the appropriate cercariae was treated in the following way. It was first measured for shell height and then the shell was removed. Once the snail was out of its shell its head and foot were cut off and discarded. Karen's work had determined that neither rediae nor sporocysts inhabited this region of a snail and it would just be extra tissue to separate from parasite stages later on. The remainder of the snail's body was then cut longitudinally (long-ways) into two halves. Lastly, the two halves were cut cross-ways in into a total of 12 to 24 smaller pieces, depending on the size of the snail. All the pieces of the snail were then put into a single vial containing Clorox bleach and stored there. We tried a number of digesting agents, but found that bleach was readily available and served to digest away most snail tissue and separated most *Zr* sporocysts from one another. This made counting rediae and sporocysts possible. We found that approximately one week in the vial was sufficient for this to happen. When the contents of a vial was to be further processed the following was done. We had some culture dishes in the laboratory that each had 96 small wells in them, rather than a single large chamber, to hold fluid. The vial was shaken to loosen any sporocysts and/or rediae therein. A small pipette was used to remove the resulting vial fluid and suspended material. The contents of the vial was divided up, in very small quantities, among the 96 wells in a culture dish. The next task, and a tedious one indeed, was to use a dissecting microscope to count all the sporocysts and/or rediae in each well. The very small quantity of material in each well facilitated the identification and counting of parasite stages. The hand tally counter clicked away. The pipette and vial were microscopically examined to detect any sporocysts or rediae that might have adhered to them and been missed. In the end we knew accurately just how many sporocysts and/or rediae, were present in a snail. This is just what we needed to know.

Mike is a very independent guy. Once he knew what the job was and how to do it he was happy to work pretty much alone. All through the summer and into the winter of 2000 he collected and processed snails. During the summer he would make trips to the field and was in and around my laboratory in Lewes. When he had to be back at school I arranged for some space where he could work in Buck Cornell's (remember him?) laboratory in the biology department on the main campus. Thanks Buck. We had agreed on the number of snails needed and the end result was that rediae and/or sporocysts were counted in 15 *Hq*-infected, 15 *Zr*-

infected, and 15 *HZ*-infected snails. The main problem was coming up with the 15 *HZ* snails, but it was ultimately done. For each of these 45 infected snails we knew with great accuracy how many sporocysts, or rediae, or both were resident in the snail's body. We could compare the numbers of *Zr* sporocysts and the numbers of *Hq* rediae in singly and doubly infected snails. The goal, once again, was to find out if anything had been going on in these infracommunities that resembled competition or predation (or some other species interaction) between these two parasite species during the time they had coexisted in the same snails as double infections.

So what happened? I don't want to deluge you with a lot of graphs and tables and the like, which would be the usual way to present such information to an audience of scientists. Graphs are available in the published paper (Hendrickson and Curtis, 2002), but I'll try to get along without them here. Let me first just give you the basic information. The 15 snails infected singly with *Hq* had an average of 2,216 rediae in them (maximum number = 4,569, minimum number = 597). The 15 snails singly infected with *Zr* had an average of 4,099 sporocysts in them (max = 6,225, min = 1,428). Of course, the 15 doubly infected snails (*HZ*) had both *Hq* rediae and *Zr* sporocysts in them. These snails had an average of 2,963 rediae (max = 5,130, min = 1874) and an average of 2,865 sporocysts (max = 4,160, min = 1,809). There will be an interesting complication, but you can get a pretty good idea of what happened by taking a minute to digest these numbers. Based just on the average numbers, the number of *Hq* rediae in doubly infected snails was approximately the same as in singly infected snails. Put another way, *Hq* infrapopulation sizes in snails seem to be about the same regardless of whether *Zr* was present in the same snail or not. On the other hand, numbers of *Zr* sporocysts seem to be reduced when a *Zr* infection had to coexist in a host snail with an *Hq* infection. It is apparent, then, that there is some negative effect of coexistence with the other parasite in the case of *Zr* infrapopulations, but *Hq* infrapopulations seem to be unaffected when coexisting.

The interesting complication was that snail size mattered. The 15 snails infected singly with *Hq* ranged in shell height from about 19 to 32 mm. The 15 singly infected with *Zr* ranged from about 22 to 33 mm. Doubly infected (*HZ*) snails ranged from about 25 to 33 mm in shell height. We did a statistical test (something called and analysis of variance) that compared shell heights in the three groups of snails. Results of this test told us that the two groups of singly infected snails were of equal size. In the language of statistics, any differences between shell heights in these two groups of snails were not significant. We could think of the two groups as essentially equal in shell height. Doubly infected (*HZ*) snails, on the other hand, were significantly larger than the other two groups.

[Why should doubly infected snails be larger than singly infected snails? That was not the point of our investigation, nor a question that could be easily answered. It would be interesting to pursue. Perhaps starting with and evaluating a hypothesis that it took longer, and therefore snails got bigger, for two parasites to accumulate in a snail than just one would help. Alternatively, we might pose that smaller snails cannot support a double infection. My experience finding multiple infections in some smaller snails makes me favor the first hypothesis.]

We used another statistical technique (linear regression) to see if the number of parasite stages in a snail was related to snail size. It turned out that for snails singly infected with *Hq*, snail size was significantly related to the number of rediae in a snail. Larger snails had more rediae in them. For *Zr*-infected snails, the relationship between size of the snail and the number of sporocysts present was not clear-cut (not statistically significant). There was lots of variation in the number of sporocysts present, but it was not always clearly related to snail size. The variation was surely caused by something, but the cause(s) of the variation was and is unknown to us. All we could say was that larger *Zr*-infected snails *tended* to have more sporocysts.

If we are going to compare the number of parasite stages in singly versus doubly infected snails, especially with respect to *Hq* rediae, we shall have to take snail size into account. To do this required a statistical test neither of us had made use of before. It is called an analysis of covariance. When you have to use a statistical test with which you are unfamiliar it takes a while to get comfortable with it and use it correctly. As usual I won't go into it all that deeply, but it did require both of us to work the test through and get the mechanics of it straight. What the test will do, for example, is compare the numbers of *Hq* rediae in singly infected snails with the numbers in doubly infected (*HZ*) snails while taking into account that larger snails should have larger infrapopulations of rediae. Another analysis of covariance compared the numbers of *Zr* sporocysts in singly infected snails with the number in doubly infected (*HZ*) snails, again while considering the sizes of snails. When the two tests were done it turned out that numbers of *Hq* rediae in singly versus doubly infected snails, taking size into consideration, did not differ, but numbers of *Zr* sporocysts did. There were more *Zr* sporocysts in singly infected snails and fewer in doubly infected snails. This is not different from what was apparent from the average numbers of these stages given above, but now we have a proper statistical test to back up the following claim: population sizes of *Zr* are negatively affected (smaller) when they have to coexist with *Hq*, but *Hq* population sizes are the same size whether they coexist with *Zr* or not.

The result was pretty clear, but what does it mean for the relationship between the two species when they have to coexist in a snail, however infrequently that might be? What should that relationship be called? For the *Hq* and *Zr* pair the situation seems to be 0 - when interacting and 0 + when not. The “0” signifies a zero effect on *Hq* infrapopulation sizes whether coexisting with *Zr* or not. The presence of a “0” effect on population sizes of *Hq* emphasizes that an asymmetrical situation exists, *Zr* is affected, but *Hq* is not. Recall that, in terms of population sizes, competition required a - - situation when a pair of species was interacting and a + + situation when not. So competition does not seem to capture what goes on between these two parasites when they coexist. From another angle, rediae and sporocysts feed differently and it is unlikely that they would have some resource they both need and was in short supply. Does predation work any better? With predator symbolized first and prey second, this interaction requires a + - situation when interacting and a - + situation when not. This doesn’t fit what goes on between this pair of parasites any better.

I can add some other observations about the possibility of predation to this. In all the *HZ* infections I have looked at, which is quite a few, I have never seen an *Hq* rediae with a *Zr* sporocyst in its mouth. That doesn’t guarantee that predation never happens, but at least it appears to be of uncommon occurrence. Further, if *Hq* rediae were eating *Zr* sporocysts, one would expect *Hq* to utilize the extra nutrition and produce more rediae. There was no increase in *Hq* infrapopulations when they coexisted with *Zr*, which indicates that this did not happen. So predation doesn’t seem very promising either.

Maybe some other species interaction will fit. One that works, at least on the surface of it, is a species interaction called amensalism. The usual way this works is that populations of one member of a species pair (the amensal) are unaffected, but they produce some toxic substance that impedes the growth of the other’s populations. A sort of classical example of amensalism is the creosote bush. This bush releases creosote into the immediately surrounding soil, which prevents other species of plants from growing there. Another standard example is an antibiotic produced by a fungus called *Penicillium*. The fungus produces and disperses penicillin into its environment and it inhibits the growth of bacteria. This situation can be symbolized (amensal first) with 0 - when interacting and 0 + when not. This certainly fits the *Hq-Zr* situation better. The problem is that no substance toxic to *Zr* is known to be produced by *Hq*. Maybe one exists, but it is yet to be demonstrated. Further, it is unlikely that *Hq* could be adapted to produce an anti-*Zr* chemical because it co-occurs in the same snails with *Zr* so seldom. The selection pressure to produce such a chemical just wouldn’t be consistent or strong enough. It was the subject of extensive discussions between Mike and myself, but in the end we were not satisfied that competition, predation, or amensalism adequately captured what went on when these two species co-occurred in the same snail. Still, they are likely to influence one another somehow because in a double infection infrapopulations of both must coexist in a confined, necessary space (the snail) that they both ultimately draw on for nutrition.

What did we end up thinking must be going on? The key observation that led us to our assessment of the situation was one stemming from Karen's and my work. In a nutshell it is that multiply infected snails are scarce. This is true for multiple infections in general, but particularly true for any specific pair of parasite species. In certain habitats a co-infecting pair may be relatively common, such as *HZ*-infected snails a Savages Ditch. If it were not for such localities, this work, or Karen's and my work, would not have been possible. We certainly took advantage of it in this work. But, this snail occurs all up and down the Atlantic seaboard of North America. There is a huge global population. The proportion of snails co-infected with any pair of trematodes in the global population is vanishingly small, even the relatively common pair *Hq* and *Zr*. Again, just think of the large snail population and infrequent double infections in the southeastern part of Rehoboth Bay discussed in the previous chapter. The point is that members of a parasite species pair very seldom encounter one another in *Ilyanassa* hosts.

If two species rarely encounter one another, they cannot become (or be) adapted to one another. Why not? Let me briefly restate something I have covered previously, as early as Chapter 5. The process of becoming adapted requires that natural selection be going on. Natural selection is a mechanism and like any mechanism it requires that certain elements be in place and functioning for it to work. Try to imagine a chain-driven bicycle with no sprockets to engage the chain. For one species to become adapted to deal with another requires that members of the two species frequently encounter one another. That way it is possible that genetic strains of one parasite will encounter the other parasite species generation after generation. Under these conditions any genes that produce features in parasite A that help in dealing with parasite B can be preserved (by natural selection) in the A parasite's gene pool. A similar natural selection scenario can occur in the B parasite's gene pool over time (parasite generations) and the two species can become co-adapted to deal with one another. Pretty clearly, the trematodes found in *Ilyanassa* (including *Hq* and *Zr*) do not encounter one another often enough for the process of adaptation, one species to another, to work. They just don't "see" each other often enough.

If one parasite species cannot become adapted to another in this collection of species, what can adaptively be going on? Karen and I took this as a given, but the claim is that these parasites must be adapted to make use of the snail host. What's the basis for this belief? All these parasites must inhabit *Ilyanassa* in each parasite generation, even if in the process they won't likely be exposed to another parasite species. In each generation any genes possessed by a parasite that promote excessive ill-use of the snail host (e.g., drawing on it too much, debilitating it) will be weeded out of the parasite's gene pool by natural selection. Put slightly differently, what this means is that any parasite genetic strains that use the snail host inefficiently are either outright eliminated or produce fewer cercariae than more efficient strains. In short, efficient host-use genes would be preserved in the gene pool, while inefficient host-use genes would be filtered (weeded) out by natural selection. This is the process of adaptation and why the parasites would be adapted to make efficient use of the snail host. This certainly is the case.

Far and away the most common situation observed is a single parasite species infecting a single snail. In all probability (see above) an infecting parasite is adapted to deal with the snail host. That's what these parasites have become, adapted to deal with the host, but not other parasites. It follows that adaptively whatever trematodes do when they infect an *Ilyanassa*, in some way it serves the need to deal with the host and not to deal with whatever other parasite species might be present.

Now, when we speak of ecological relationships between species (e.g., competition, predation, amensalism) there is a chance that they arose by accident. For example, *Hq* might produce a chemical for some adaptive reason having to do with the host that accidentally happens to inhibit *Zr*. Or, *Hq* rediae are equipped with a mouth that is an adaptation to consume bits of host tissue, but with it they accidentally ingest an occasional *Zr* sporocyst. The former example when analyzed would appear to be amensalism, while the latter would appear to be predation. Interactions arising in a manner such as in these examples would be ecologically operational as cases of amensalism or predation, but they would not be cases of one species being adapted to interact that way with another. In the case of parasites infecting *Ilyanassa*, any ecological relationships that appear to be between parasites would actually result from the way the parasites are each adapted to use the host snail. Adaptation to the host is non-accidental and whatever else goes on follows from that. If with respect to parasites we want to explain anything about what goes on in a snail, it likely stems from the relationship of parasite to host, not parasite to parasite. In my view we should emphasize adaptive relationships in ecological systems, which would be non-accidental, as this will give us the best idea of what consistently goes on in them.

One rule of adaptation to utilize *Ilyanassa* exhibited by all these parasite species seems to be "do no harm (beyond sterilization) to thy snail host." This is illustrated by at least two things: (1) infected snails move around in the field at least as much as uninfected snails do (Chapter 6, more on this later) and (2) individual infections persist in individual snails for years, probably decades (Chapter 12). Single *Hq* and *Zr* infections have been observed persisting in individual snails for up to six years. It is likely they persisted for even longer than that. Even multiple infections long persist. Nathan and I observed six individual *HZ*-infected snails that persisted for at least 106-134 days and another snail was observed wherein an *HZ* infection persisted for three years. The survival value to the parasites in keeping snail hosts alive and "farming" them for many years seems pretty clearly to be extensive cercarial output. This increases the chances of parasites completing their life cycles. The strategy is: when you have a "cash cow" and it's not likely you'll get another you had better hang on to it. Apparently, there was strong enough selection pressure that all the trematodes infecting *Ilyanassa* are adapted to exercise this strategy.

Our conclusions from this work suggested that the ecological interactions competition and predation did not fit what went on between *Hq* and *Zr* when they infected the same snail. Amensalism fit better, but no *Zr* inhibitory chemical produced by *Hq* is known. There's always the chance that one exists, but if it does it will have to do with adaptation to the host snail by *Hq*. That parasite (*Hq*) does not co-occur with *Zr* often enough for it to be adapted to produce it as a defense against *Zr*. At the least, production of such a chemical for that purpose would be a waste of energy virtually all of the time and the ability to produce a useless and energy-costly chemical would be eliminated by natural selection. If these modes of species interaction are not applicable to these parasites in multiple infections, just what is going on?

We sought to provide a new way of thinking about how two (or more) parasite species get along when infecting the same snail. The prime "concern" of a trematode infection in *Ilyanassa* is to not overly damage the host. This would be true whether there was more than one species infecting the snail or not. Somehow, the parasites manage to limit their infrapopulations below numbers where they would become too much of a burden on snail hosts. If they didn't, they would threaten the host's and their own survival. Infrapopulation sizes of *Hq* rediae and *Zr* sporocysts were always in the few thousands. How does an infection "know" not to produce infrapopulations of say tens of thousands? Now that would debilitate a snail.

Well, Mike and I had to admit that we didn't know in any detail, but we could reasonably speculate. These parasite infections are closely associated with the snail and communicate chemically with it. For example, they all sterilize the host and one of them (*Zr*) is known to do this by means of a chemical it produces. We reasoned that parasites read some chemical signal produced by the host when it becomes nutritionally stressed. This chemical might be a snail hormone released under stress conditions, or possibly some chemical product released when tissues begin to breakdown, anyway a chemical signal. When a singly infecting parasite reads this signal, it "knows" that it would be of survival value, for the host and ultimately itself, to curtail within-snail reproduction of new rediae or sporocysts. The infrapopulation is thereby kept within bounds. This ability would serve nicely when an infection exists alone and also when it has to co-exist with another infection in an infracommunity. The other infection(s) would draw down snail resources and make the snail go into nutritional stress. A parasite would respond to its chemical signal by curtailing its reproduction under these circumstances. This could apply to *Zr* when it coexists with *Hq*. After all, its infrapopulation sizes were smaller when it co-occurred in snails with *Hq*. Note that the direct response is to conditions in the host and not directly to the presence of *Hq*. The other parasite (*Hq*) requires additional explanation. It can regulate infrapopulation size. Larger snails, which presumably provided more room and resources for the infection, clearly harbored more rediae than smaller snails. It must be adapted to pick up on a host-produced signal to do this. But, *Hq* within-snail populations of rediae seem insensitive to the presence of *Zr*. We postulated that *Zr* is more sensitive to its chemical signal (reads it at a lower level of stress) than *Hq* is to its. This might explain why *Zr* is sensitive to the presence of *Hq*, but *Hq* is not sensitive to the presence of *Zr*. In any case, the parasites are probably unaware that another species of parasite is present in the snail. What they would be aware of is the condition of the host. It did not escape us that this would be useful for parasites with regard to anything that might lead to snail stress. Not only another parasite infrapopulation might do this, but also a food shortage in a poor foraging environment. A parasite living in a snail in such an environment, and able to read a host signal, would be able to reduce its demand on the host by curtailing its own reproduction, reducing its infrapopulation size, to an even lower level than usual. This would help the snail, and the parasite, to survive. [As an aside, though I have not done it, this is something that could be experimented with: compare the infrapopulation sizes in snails maintained on good diets versus snails maintained on poor diets.] Through their adaptation to the host, parasites would have a pliable strategy to maintain a host in the living condition for years on end and guarantee the production of hoards of cercariae.

Since Mike graduated from the University I have not seen anything of him. In 2003 he contacted me to write a letter of recommendation for him to a Masters Degree program in a department of statistics at a branch of the University of California. I was happy to do that for him. Mike has more facility with mathematics than I do. To boot we had employed some fairly complicated statistics in our work. In my view Mike would do very well in such a program and I could recommend him without reservation. How this application came out I don't know. Last I knew he was in California. He was living and working in San Francisco trying to achieve state residency to reduce tuition costs at a California university.

There can be no doubt that as a research student Mike was a different sort of experience for me. During the time that I knew him he was as capable as any of the students with whom I had worked. The main difference was that he was happiest working alone. Once I had shown him what to do and how to do it, off he went. The work he did was exacting and very tedious. Counting and keeping track of all those sporocysts and rediae in those little wells while peering through a microscope made for a daunting 45-snail results quest. There are some, but very few undergraduate students that I have known that would have had the patience and dedication to see this project through to completion. I am hopeful that I have brought you along in these pages so you don't hold this view, but I wouldn't hold it against you if you thought: "So what, who cares how many parasite stages there are in a drab little marine snail?" All I can say is that such concerns are the stuff, the raw materials, of science. By taking account of such seeming trivia as the relative numbers of rediae and sporocysts in snails, it is possible to get an insight into the way things work in nature, the relationships, the co-operations, the antagonisms. It does not matter whether there are 2,508 or 3,459 sporocysts in a snail. The exact numbers don't matter. What matters are the relationships between and among different kinds of organisms and above all our understanding of them, why they are what they are. We are supposedly the "smart ones" on this planet and we have the responsibility to understand these things as best we can.

You have to give someone their head, especially when they are being scientific. Mike preferred to work alone and that's O.K. with me. Regardless, I think he fully understood what I am about to say. He had strongly held opinions about how to interpret the data, as did I, and our views were not always initially the same. We had many long discussions (moderately heated) about what was going on in parasite infracommunities in snails and other organisms. I learned from his points of view, which brought about changes in my positions on issues. He no doubt had the same experience. Out of our discussions we arrived at our best appreciation of what must be going on between the snail and the parasite or parasites that inhabit it. Of course, all of these discussions were grounded in truth by being grounded in data, which came with care and unbiased from the natural world, nature at work. This was, after all, an exercise in natural science and Mike and I were exhilarated, rewarded, and informed by doing it.

The Cape Henlopen sandflat in 1991 where movements of individual snails were tracked.

Chapter 15

DISTANCES TRAVELED AT A SNAIL'S PACE

Back in Chapter 6 I covered some work that, with the aid of Harry Hudson, I did in 1985 (Curtis, 1990). We followed the movements of individually marked snails in the vicinity of a sandbar over 16 low tides (about one week). The object in that work was mainly to keep track of the movements of snails infected with *Gynaecotyla adunca (Ga)* and compare them with the movements of snails not so infected. Individual snails infected with *Ga* made repeated trips to the top of the same sandbar. As I later found out (Curtis, 1993), the purpose of these trips was to release *Ga* cercariae into the mucus trails of host snails as they were being stranded in the intertidal zone by receding tides. It would then be possible for cercariae to get to and infect the next host in the parasite's life cycle, a crustacean that lives out of the water high up in the intertidal zone. A lot was learned about how much snails move around in their habitat, and the effects of trematode parasites on that movement. Because *Ga*-infected snails were making more or less regular trips to the sandbar peak, they were moving greater distances than other snails. Importantly, snails infected with other species of trematodes moved as much as uninfected snails did. So, in terms of movement, infected snails did not appear to be debilitated by their infections. This finding was counter to what other workers had found to be the case with other species of snails and their trematode infections. There were several cases in the literature where movement of infected snails was found to be inhibited, but something different seemed to be going on in the case of *Ilyanassa*. My observations in 1985 were all made in a fairly restricted area, around a single sandbar. What movement do snails, infected and not, show on the broader sandflat? That was still an open question and one worth pursuing. How much snails move around on a sandflat is an important aspect of their ecology.

You probably recall (Chapter 9) that in 1991 Karen and I released 1,200 individually marked snails onto Cape Henlopen that had tested as uninfected by cercarial release. I previously covered something about the wondrous and thrilling days on the sandflat. They contributed importantly to the work, but I'll not delve much into that aspect of the operation again. Get ready for the facts, just the facts. Movements of these snails, and others released in 1993, were ultimately reported upon in Curtis (2005). It was a long delay between getting the data and writing and publishing the paper. My only defense of the long delay is that I had to put the analysis of data and writing off because other papers had to be published while my co-authors were still in, or nearly in, their undergraduate careers. All 1991 snails came from the Canary Creek saltmarsh ditch site (Chapter 3). They were all released at the "release stake," which was located about 10 meters out on the sandflat and about 1,200 - 1,300 meters along shore northeast of the base of the Henlopen Fishing Pier. The release area is depicted in this chapter's frontispiece. One of the things we did with these snails was to note their locations (distance and direction) relative to the release stake when any were subsequently found. As we have seen, infected snails can behave differently. Among the 153 snails recovered and dissected, five were found to be infected (plus one still uninfected not found until two years later in August 1993) and I excluded these six from data I report on here. The rest of released snails (altogether 1,194 of them) were presumed (on good grounds) to have been uninfected when they were released in June and followed throughout the summer and fall of 1991. Let me provide some basic information. We were able to find (sight) 554 individual snails (of the 1,194) after they were released. Collectively they were sighted 971 times: 301 were sighted only once, 152 were sighted twice, 59 three times, 28 four times, and 14 from five to seven times. All sightings were between 22 June and 26 October. At the time, Karen and I could not be sure we would ever see any of these snails again, but as it turned out right after release the area immediately around the release stake was "swamped" with marked snails. We could not measure distance and direction to all of these snails. At most, tides allowed us about an hour and a half on both sides of low-water to work. We couldn't handle that many snails in that kind of time and still have time to search a wider area for marked snails. We took to doing the following. If a marked snail was found within a 15 meter radius of the release stake its identity (mark number) was noted and then later it was assigned a random distance (within 15 meters) and direction. Thus the "random radius" for 1991 snails was set at 15 meters. These snails would all be reasonably "located" by doing this and it saved us a lot of time. Of 971 total snail sightings, 634 were within 15 meters of the stake and 337 were outside that. These snail sightings were the basis for characterizing snail movement during 1991.

In June 1993 more marked snails were released at the same release stake. Among these were 200 from the Canary Creek site. All of them were tested for infection by cercarial release and determined to be uninfected. The other 300 snails were collected from a tide pool just southwest of the release stake and right next to the beach. I wanted to include snails native to Cape Henlopen to see if they behaved differently from snails transplanted from Canary Creek. All were tested by cercarial release to see if they carried infections. Many snails from this area of the sandflat would be expected to be infected. Of the 300, 157 tested as uninfected and 143 released cercariae. For these native snails, I did not assume that those testing as uninfected were actually uninfected. Rather, with all these guys I determined whether or not they were infected and with what species on the basis of dissection and/or cercariae released. Snail sightings in 1993 were handled a bit differently than in 1991. The "random radius" was 10 meters rather than 15. Also, marked snails found between 10 and 15 meters out were assigned a distance of 15 meters, but the actual direction, degrees from north, was recorded for each snail.

In 1993 searches were conducted on 38 daytime low tides between June 7 and November 14. Movement of uninfected and infected snails will be described separately. Some basic information about uninfected snails follows. Uninfected snails that were located consisted mostly of snails transplanted from the saltmarsh. Only 11 of the 170 uninfected snails seen were originally native to Cape Henlopen. There were 184 sightings of uninfected snails within 10 meters of the release stake, while 262 were sighted outside 10 meters. Of the 170 individual snails sighted, 61 were seen only once, 36 were seen twice, 32 three times, 16 four times, and 25 from five to 11 times. Here's basic information on infected snails. Of snails infected with *Himasthla quissetensis (Hq)* there were 47 sightings within 10 meters radius and 166 outside. Thirty-eight individuals were sighted one to five times, while 15 others were sighted six to 14 times. One of these snails, still infected with *Hq*, was found in August 1994. Sightings of *Lepocreadium setiferoides (Ls)*-infected snails included 16 within and 36 outside 10 meters; 14 individuals were sighted 1-5 times, while three were sighted 6-8 times. Sightings of *Zoogonus rubellus (Zr)*- infected snails included 28 within and 63 outside 10 meters; 26 individuals were sighted 1-5 times, while four were sighted 6-13 times. A single *Austrobilharzia variglandis (Av)*-infected individual was sighted twice about 10 meters distant from release. Snails infected with both *Hq* and *Zr (HZ)* were sighted 12 times within and 14 times outside 10 meters; 11 individuals were sighted 1-5 times. More sightings of fewer snails were made in 1993 (949 sightings of 350 out of 500 snails) compared to in 1991 (971 sightings of 554 out of 1,194 snails). In 1991 we had learned, by way of another pesky statistical test, that snails marked with red nail varnish on the pointy end of the shell (the apex) were more likely to be seen. Therefore, all snails in 1993 were marked with red.

With that not insubstantial amount of information out of the way, what was learned about snail movement on the broad sandflat? Snail movement was very different in the two years. My approach will be to describe what happened first in 1991 and then in 1993. Below is a drawing showing the positions of major sandbars, the beach, the position of the release stake, and all sightings of released snails June through October in 1991. It depicts an area 200 by 150 meters, a big piece of sandflat. Released snails spread out over an area of about three football fields.

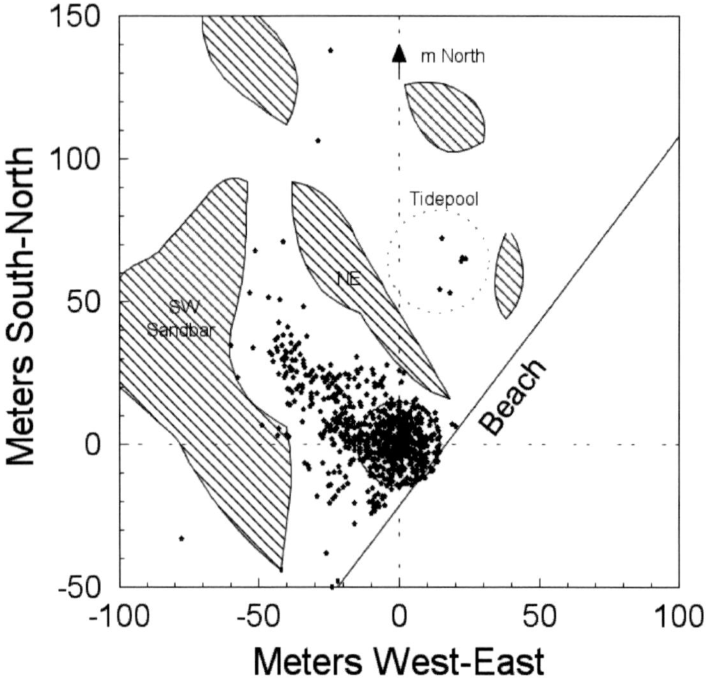

A plot of how snails dispersed on Cape Henlopen in 1991 (Used with kind permission from Springer+Business Media: Marine Biology, Movements of *Ilyanassa obsoleta* (Gastropoda) on an intertidal sandflat, 148, 2005, 307-317, Lawrence A. Curtis, Figure 1.)

In the drawing the release stake is located where the axes cross at the 0,0 point. Note that the beach runs southwest to northeast and the offshore direction is to the northwest. It can be seen that most snails were sighted either right around the release stake (i.e., within the 15 meter random radius) or offshore in a gully between the southwest (SW) and northeast (NE) sandbars. The drawing gives a good idea of what happened with these snails from June through October in 1991, but of course I did not let the analysis stop there. More can be gleaned from the data. One of the things I wanted to know was how rapidly snails dispersed from the release point. Snails were tracked on trips at least a day apart so observations on a scale of minutes, or even hours, are not available. One way to get at rate of dispersal is to simply measure the distance snails traveled versus time since release. We measured the number of meters from release routinely whenever a snail was sighted and we also knew how long the sighted snail had been free. So, ignoring direction moved for the moment, I simply plotted distances moved versus days out. The plot (presented in the published paper) had 971 points on it and it is a bit confused, but the message is very clear. In about five days snails had moved as much as about 25 m from release. [We will have to state a lot of distances. A meter is a touch (~3 in.) longer than a yard and m is the standard abbreviation for it.] By 10 days they had moved as much as about 50 m away from release. Basically, snails tended to remain at most about 50 m away from the release stake throughout the summer and into the fall. In late October, around 130 days after release, some snails were found even farther away from the release stake. The snail farthest from release was found 180 m out 130 days after release. That location is off the above drawing to the northwest. The snail had a pretty long time to do it, but nevertheless at a snail's pace it's a long way for a snail to travel. That was the only time that snail was sighted. So, where it had been all summer will never be known. The annual migration of snails offshore to the lower intertidal zone for winter likely explains the more distantly located snails at about this time. That migration takes place in October and November.

Another way to get at dispersal rate is to use a dispersal index and see how it increases over time. The advantage of doing this, rather than just using distances individual snails traveled after release as above, is that you get an overall estimate of average dispersal of the whole released population as time proceeds. Back in the early 1960's a worker named D. P. Clarke who worked on grasshoppers came up with an index for their dispersal. Clarke figured that you should measure the distance from release of each organism sighted during a given time period after release. Each distance would be squared and the squared values added up. Then whatever that total is would be divided by the number of sightings during the time period. Finally, you take the square root of that value. This would give you an index of dispersal (D), sort of an average distance from release during a given time period. As a formula the index looks like this $D = \sqrt{((\sum d^2)/N)}$. It's a shorter representation that way. The $\sum$ symbol means sum up the d^2 (distance squared) values and N is the total number of sightings in a given time period. For my purposes I divided times after release into five-day periods. There was one exception, days one to four were categorized as day two. A graph with dispersal indexes (D) on the vertical axis and five-day periods on the horizontal showed that as a group snails reached their maximum dispersal (index roughly 20) in about 10 days and remained at about the same degree of dispersal until the October offshore migration began. What this means is that all summer most snails were satisfied to stay in that gully northwest of the release stake at an average of about 25 m away.

Taking account of direction of movement requires considering something called net moves. Suppose you know where a snail was on one day, say at release. Then, say 15 days later, the snail is found 30 m northwest of the release stake. The minimum distance that the snail had to have traveled in those 15 days was 30 m to the northwest. That would be one snail net move. Of course, the snail may have traveled much farther than that and in a variety of directions, but with any additions or retraces of movement, that is the snail's net move over that 15-day period and it was to the northwest. A particular snail has to be found twice to obtain one net move, three times gives two net moves, four times gives three, and so forth. I noted above that many snails were sighted more than once and (forgetting about distance moved for the moment) directions of many net moves could be determined. All pairs of sightings could not be used. For example, if a snail was sighted twice within 15 m of the release stake its net move would be worthless because the directions were assigned randomly within the random radius. In 1991, in all 378 net moves executed by 252 snails could be assigned directions. [I just now found an error in the published paper (Curtis, 2005). There's always at least one no matter how careful you try to be. It is sort of inconsequential because it can be figured out and corrected, but nevertheless it's embarrassing. I reversed the entries in Table I, putting the 1991 entries where the 1993 entries should have gone and vice versa. Oh well, there is no help for it now.] Anyway, of the 378 net moves in 1991, 221 of them went northwest, 30 went northeast, 59 went southeast, and 68 went southwest. I did some statistical analyses to back up the points I will make below, but I won't belabor that aspect of the work. Especially during the first 10 days after release there was a decided preponderance of northwest net moves. Snails mostly moved away from the release stake out into the gully between the two major sandbars. There were few moves to the northeast (a sandbar in the way there), a goodly number to the southeast (once they had moved northwest, they could move back), and quite a few southwest net moves were observed (there a sandbar was farther away). As a characterization then, in 1991 snails moved at most 25-50 m within about 10 days after release in June. They averaged out at about 20-30 m away from release.

Most did not disperse farther until the offshore migration in late October. Movement was mostly to the northwest and most snails wound up in an area between a pair of sandbars where they remained all summer.

Given this general characterization, there are some other matters to take up. One is that sandbars inhibit snail dispersal. On many low tides we would get to the release site at about the time the water was receding off the peaks of sandbars. While we waited for the water to fall about half a meter lower to expose the sandflat we could search the sandbar peaks for marked snails. We never found any there then, nor did we ever find any there at any other time. This is evident in the drawing. Based on Chapter 6, if any marked snails were infected with *Ga* we might have expected to find them on sandbars. Of course, all 1991 snails tested as uninfected before release. We were hoping that some would get infected with *Ga* after release and begin to show up on sandbar peaks or the beach, but no such luck. Only a very few 1991 marked snails got infected with any species during their time in the field. The fact that no marked snail ever showed up on a sandbar was a strong indication that higher sandbars were a barrier to snail movement. That surprised me. Before this I had thought that snails not infected with *Ga* could at least briefly visit the beach and sandbars, but only *Ga*-infected snails remained. Apparently, uninfected snails almost never mount sandbars, meaning that snail dispersal (except for *Ga* snails) depends on sandbar positions. That sandbars block snail movement is supported by other evidence. In the foregoing drawing a few marked snail sightings can be noted in the tide pool on the east side of the northeast sandbar. We searched there often from the start, but it took nearly two months from release for any of the snails to wander over there and show up. They most likely moved around the northeast sandbar by exiting the northwest end of the gully where most of the snails remained and eventually came toward shore to inhabit the tide pool. There is a chance that they went around the southeast end of that sandbar near the beach, but no marked snails were ever seen transiting through there. There is one snail in the drawing (Green 46, released on 9 June) that was seen on the west side of the southwest sandbar. We searched that area routinely. It was only sighted once on 26 October, more than four months after release. I can't be absolutely sure it was not there unseen before that, but it appears to have taken a long time (months) getting around the southwest sandbar. These two sandbars seem to have confined most snails between them for the whole summer.

This would be important to know if you wanted to monitor the same group of snails in the field for a summer experiment of some sort. Perhaps you need to keep track of the average physiological condition of snails in a specific population over the summer. Whatever you want to do, you had better pick a group of snails well hemmed in by sandbars. Otherwise your snails might disperse away and other snails might disperse in. Your experiment would fail because you think you have been monitoring one group of snails, but actually several have been assessed.

Another matter is how far did snails move in one day's time? If you take net distance moved by a snail and divide it by the number of days between the sightings, you get the average net distance moved per day. In 1991 it turned out that (considering all sightings) the average number of days elapsed between sightings of a snail was 22 days and the average (mean) distance moved in a single day came out at 1.7 m. The range of values that go into that average is from 0 to 46 m/day. Sightings of the snail that moved 46 m in one day were only one day apart and therefore, in 1991, this was a well-founded maximum movement per day. It was found in one place on one day and 46 m away on the next. Now the question is: did it move that distance under its own power or was it somehow carried over part or all of that distance. I have, for example, seen snails tumbled along in a water current. Also, small snails sometimes attach to bits of floating algae. We can't know how this snail traversed the 46 m. All we know is that it somehow did. So, the question becomes: could it have moved this distance under its own power? Back in the 1980's a worker by the name of R. V. Dimock, Jr. studied locomotion by *Ilyanassa* in the laboratory. The snail moves by the action of cilia (like tiny mobile "hairs") on its foot, so-called pedal cilia, and Dimock measured that snails cover about 2 millimeters (mm) per second when they are motoring along. Divide a meter into 1,000 equal pieces and each piece would be a millimeter long. What this speed (2 mm/sec) means is that under constant movement in a straight line a snail can cover 10 m in 1.4 hours. It would take 6.4 hours to move 46 m. On a single tide many parts of the sandflat are submerged for this long so it is at least possible that a snail could have covered 46 m during one high tide. In the field snails often stay about in the same place for days, but they frequently move 10-20 meters per day. While we do not know the means by which this particular snail executed the maximum snail movement in this study, we can be sure that the average of 1.7 m per day drastically underestimates the typical amount snails move around. Most studies of snail movement in the field have been done with snails living on steeply sloping rocky shores. There snails move at most a meter or two over very long times. Only a few studies have been done with individually marked snails living on broad, gently sloping sand- and mudflats, but they clearly move around decidedly more.

From 1991, through 1992, and into 1993 the intertidal landscape of the release area underwent major changes. It had to do with the drifting around of sandbars. Back in Chapter 9 I described what happened. I was unaware of all this as it was happening, but I was able to later reconstruct what happened from photographs I had taken. Briefly, the 1991 northeast sandbar moved southwest. It moved across the 1991 gully, which, as in the foregoing drawing, was essentially perpendicular to shore extending out to the northwest. The old gully was obliterated. As the sandbar moved, another similar gully developed on its northeast side. It was evident as early as late 1992. This gully slowly moved southwest (along the shore) and by summer 1993 it looked very similar to the old 1991 gully, but it was positioned a bit to the northeast. These changes resulted in me not being able to find 1991 snails after 1991. In the 1993 release area the release stake was in the same place and there was a central gully bordered by southwest and northeast sandbars. The situation looked very similar in 1993, but was in fact very different. Dispersal of marked snails in 1993 would reflect this fact.

In June 1993 Karen and I, with the aid of another student, Chris Morrisey, released 500 individually marked *Ilyanassa*. As described earlier in this chapter, from the saltmarsh site came 200 that had tested as uninfected and from a tidal pool next to the beach and southwest of the release stake came the other 300. A drawing, similar to the previous one, of what happened in 1993 is included below. This drawing is presented at a larger scale than the preceding one because it is necessary to distinguish different symbols for snails with different infections. Recall that in 1993 these infection categories were assigned based on (1) the fact that they came from the saltmarsh (all considered uninfected), or (2) dissection or cercariae released. As indicated in the drawing, uninfected snails (not infected, NI) are represented by solid triangles, *Hq*-infected snails by open circles, *Ls*-infected snails by solid squares, *Zr*-infected snails by solid lower half circles, and *HZ*-infected snails by solid circles. The positions of the new northeast and southwest (the old 1991 northeast sandbar) sandbars are also indicated. The area covered by the drawing (250 X 150 m) is even larger than that in the 1991 drawing.

In 1993 there were 350 snails sighted 948 times and each point in the drawing represents a single snail sighting. From a quick comparison of the 1991 and 1993 drawings it can readily be seen that the range of dispersal was much wider in 1993. Maybe it's already obvious to my discerning readers, but it will be of interest to figure out why. Let's first deal exclusively with uninfected snails. As stated above, the bulk of saltmarsh snails were seen after being released (159/200). Among native Henlopen snails another 11 were seen and ultimately known (by dissection) to be uninfected. A total of 184 sightings of these uninfected snails were made within the 10 m random radius of the release stake. The other 262 sightings were outside that radius and this alone suggests something different compared to two years ago, more sightings farther away from the stake. Inspection of the drawing shows that these sightings were well scattered in the release area. Within about ten days after release uninfected snails were found as much as about 100 m away. Snails were not often found more distantly located throughout the summer and into the fall. That's about twice as far as snails dispersed in 1991. There was no apparent up-tick in association with the annual late-fall offshore migration. Using all snails sighted more than once, an average of 20 days passed between sightings of the same snail. Average net move distance traveled by snails was 2.2 m/day. Range of net move distances in that 2.2 m average was 0-21 m/day. So the greatest observed daily net distance moved in 1993 was 21 m. Not only did snails get farther offshore, they also got farther along-shore. Within 10 days of release snails got east of the northeast sandbar and west of the southwest sandbar. In a short sentence, uninfected snails pretty clearly dispersed about twice as far from the same release point in the same amount of time in 1993 compared to in 1991.

Average dispersal with time [based on Dispersal Indexes (D)] for the whole population of marked, uninfected snails in 1993 reflected the same thing. In the four-day time interval after release the D value was about 15 for the 127 snails sighted during that time. In the next time interval (5-9 days) it reached almost 30 for the 81 snails sighted then. In the next interval (10-14 days) it reached nearly 40. After that it leveled off in the range of 50-60 for the rest of the summer. If we keep in mind that a D value is approximately the average distance in meters from the release stake of sightings in a time interval, we can see that Dispersal Indexes too suggest about twice the dispersal in 1993 compared to 1991. I did a statistical analysis that compared D values for uninfected snails in the two years. The analysis considered two factors, year and snail size. Snails sizes were put into three size categories, small, medium, and large. [The analysis was a two factor analysis of variance (an ANOVA). Whether dispersal was in 1991 or 1993 was obviously important to consider. Size categories were divided up based on the following. The mean size of snails was 19.1 mm and variation around that mean (plus or minus two standard deviations) gave a bracket of 18.3 to 20.4 mm for medium sized snails. Small snails were below this size range and large snails were above it.] Results of the analysis indicated that year was an important factor, with D values greater in 1993. They were not just greater, they were statistically significantly greater in 1993. Whether snails were small, medium, or large did not affect D values. So snail size did not matter, but statistical analysis backs up the way things appear in terms of dispersal distances in the two years. This is always comforting.

With regard to directions of net moves, there were a total of 281 net moves by 125 uninfected snails in 1993. Net moves were most frequent in the NW direction category (about 36% of the total net moves) and least frequent (about 18%) in the SE category. The rest of net moves were along-shore (SW and NE direction categories). So there was a statistical preponderance of net moves to the NW, but it was less pronounced than in 1991. As in 1991, snails were only more likely to move in the northwest direction (offshore) in the first 10 days after release. After that snails were more likely to move in other directions.

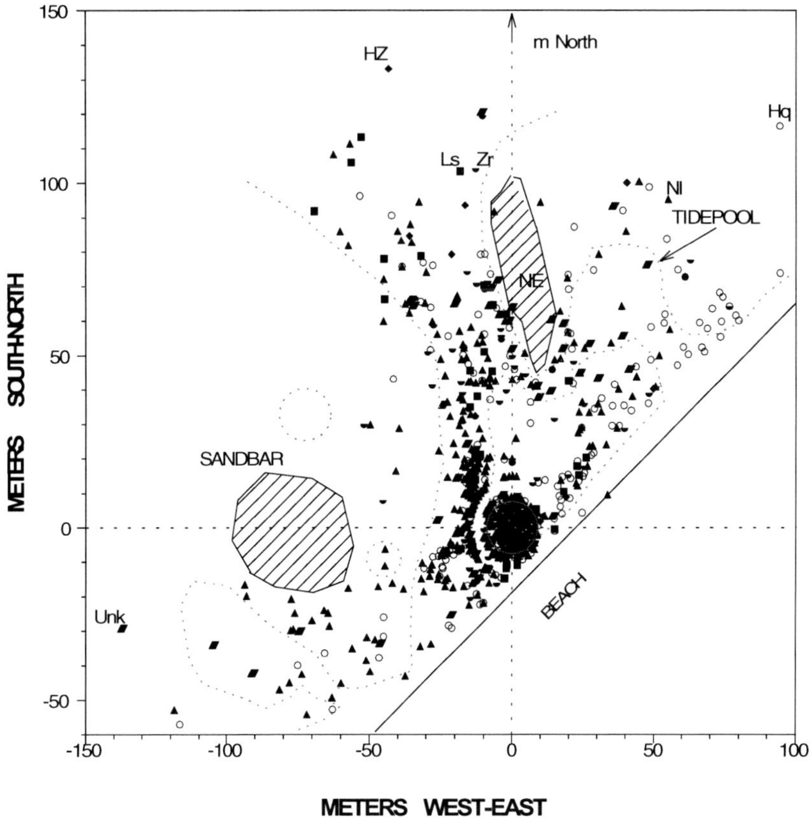

A plot of how snails dispersed on Cape Henlopen in 1993. (Used with kind permission from Springer+Business Media: Marine Biology, Movements of *Ilyanassa obsoleta* (Gastropoda) on an intertidal sandflat, 148, 2005, 307-317, Lawrence A. Curtis, Figure 5.)

Why was dispersal of uninfected snails wider in 1993 than 1991 and more likely to be in off- and along-shore directions? The fact that snails dispersed toward the beach less than in other directions may be explained by them having less option to move that way. The beach was close at hand and in the way in that direction. If you compare the topography in the two foregoing drawings, it can be seen that snails were less confined by sandbars in 1993. The sandbars were less massive than in 1991 and had changed positions. In 1993 snails were not found on sandbars anymore than in 1991, which is not at all. So sandbars again determined the dispersal routes taken by snails. The northeast sandbar was, for one thing, farther offshore. Additionally, there was a shallow channel on the shoreward end that usually had water in it for most or all of low-water. It connected the 1993 gully with a tidal pool on the northeast sandbar's east side. During low tide snails were seen moving through in the shallow water in this connector and snails appeared in that tidepool on the other side of the sandbar within 10 days of release. In 1991, when there was no such connector, it took nearly two months for snails to go around a sandbar and reach a similarly placed pool. This connector and the generally more wide open topography must be the reason for the faster and wider dispersal in 1993.

What happened with regard to infected snails in 1993? First, in terms of distance dispersed, they all showed the same basic pattern as uninfected snails. Plots of D values for *Hq-, Ls-, Zr-*, and *HZ*-infected snails versus time, as had been done with uninfected snails, showed that all demonstrated a burst of dispersion in the first 10-20 days after release followed by a leveling off. For the whole experiment in 1993 I compared the mean values of D for uninfected snails (mean D = 46.6) and snails infected with *Hq* (mean D = 52.0), *Ls* (= 44.3), *Zr* (= 42.9), and *HZ* (= 50.2). A statistical test (an ANOVA) told me that these average D's were not significantly different from one another. Infected snails in all categories dispersed equally as far as uninfected snails. That result corroborates what I found around a sandbar in 1985. As far as movement in the natural habitat goes, surprisingly, infected snails are not inhibited by their infections.

In terms of directions moved there were differences from uninfected snails, which showed a statistical preference for northwest net moves, and also differences among snails infected with different trematodes. Snails infected with *Hq* uniquely showed a preference for along-shore net moves. They mostly dispersed to the northeast, but also to the southwest. Originally, these snails all came from a near-shore pool southwest of the release stake. They mostly stayed on the sandflat in the vicinity of the beach. Back in some of the work Larry Hurd and I did (Chapter 4) there was an indication that *Hq*-infected snails tend to hang out close to the beach. We found a near-shore concentration of snails infected with this species during two separate years. Now here again we see *Hq*-bearing snails choosing to remain along the shore near the base of the beach. Of course, *Hq*-infected snails do not all reside on the sandflat along the base of the beach. A look at the 1993 drawing shows that snails bearing *Hq* can be found almost anywhere on the sandflat. It is just that they are somewhat more likely to be found in close to shore. One, including me, wonders why this might be. There is an attempt at an answer in your future if you continue reading. Net moves of *Ls*- and *Zr*-infected snails were different from *Hq*-infected snails. They tended to be in off shore directions. Again, these snails are not only found away from the base of the beach, but that is the direction in which most net moves were made. There were not enough net moves by *HZ*-infected snails (only 13 by 9 snails) to do a statistical test. For what it's worth, they mostly moved to the northeast and northwest (offshore). That there appear to be differences in directions moved by snails infected by different trematode species is immensely interesting. This was my first solid indication that something was going on in this vein. As we have seen, it is not a simple matter to discern where snails, infected or not, go in the field. This study does not well make the point that infecting trematodes influence their host's dispersal. There is only an indication. This needs to be addressed with more work and further looked into -- next chapter.

Some snails in this work were native to the Cape Henlopen habitat and that is where the movements of snails were followed. Many more were transplanted from the Canary Creek saltmarsh. The reason for the transplantation was that I wanted to track snails that were pretty surely uninfected to measure their rate of colonization by trematodes (Chapter 9) and snails from Henlopen were too likely already infected. Cercarial release tests would be unlikely to weed out infected snails with surety. A reasonable question is: do snails from the saltmarsh, that had been living in an essentially one dimensional tidal ditch, bring with them a fixed movement behavior that would affect the way they move around on a broad sandflat? For one thing, inspection of the 1993 drawing shows that uninfected snails from the marsh site and native snails from Cape Henlopen dispersed similarly widely. There was certainly no pronounced difference. Also, it would be decidedly maladaptive for snails to have a genetically fixed movement behavior associated with a particular habitat. It's that planktonic larva (the veliger) in the snail's life cycle again. Suppose, for the sake of argument, that the movement behavior of snails in the saltmarsh ditch was genetically fixed and different. Snails from there move about in a certain way. When snails in the ditch reproduce they would pass the genes that control how they move on to their offspring. The veliger larvae produced would carry the parent's genes and be released free into the plankton. They would drift around for about two weeks and then, if lucky, find a place to settle and metamorphose into a juvenile snail. There is no predicting into what sort of environment a larva might have to settle. It might be another ditch, but just as likely it would be on a sand- or mudflat. There is too much chance that a larva would settle in a habitat to which it is not adapted. If a genetically fixed movement behavior suited it to a saltmarsh ditch, it might be at a disadvantage if it settled on a sandflat. It would be better to have a flexible array of movement patterns, the one adopted being dictated by the environment into which the larva happens to settle. *Ilyanassa* inhabits habitats all up and down the Atlantic seaboard of North America. Not much has been done examining the genetic makeup of local snail populations along its geographic range. One study done in the early 1970's did examine this and the indication was that there is little genetic difference from one snail population to another all along the Atlantic seaboard. The planktonic veliger larva, which would lead to the genetic intermixing of populations, was cited by the authors as the probable reason why. Across the generations larvae drift from one population to another and carry genes with them. The effect is that there is much flow of genes among snail populations and they all wind up having similar genetic compositions. The odds on having a local population, say in a saltmarsh ditch, or on a sandflat, with a unique genetic makeup and consequently a fixed movement behavior are exceedingly slight. Some workers in Australia looked at movement patterns in another snail with a planktonic larva in its life cycle. They found that when transplanted to a new habitat snails did not retain the movement pattern they had in their native habitat. They quickly took on the movement pattern of the snails in the new habitat. All in all it is very unlikely that *Ilyanassa*, transplanted from one habitat to another, carry genetically fixed and different movement patterns with them.

As noted above many snails were sighted more than once. A fair concern is that when the same snail is sighted more than once the data from those sightings may not be independent of one another. When you do statistical tests using these possibly non-independent data, the necessary assumptions under which the statistical calculations are made may be violated. I have used statistics on my data and it is important to examine as far as possible whether sightings of the same snail are independent. Let's be specific. Suppose a snail habitually tends to make northwest moves for a distance of 10 m. If that were true, two sightings of that snail would not be independent. The second sighting is predetermined to be northwest of the first and 10 m distant. Does this sort of thing go on and make sighting data on the same snail interdependent? I make three points. (1) Of all 904 snails sighted, 48% were sighted only once and this reduces the potential problem substantially. (2) The average time between sightings of a same snail was about three weeks. This would decrease the likelihood of a connection between sighting data. (3) There is a branch of statistics called circular statistics that can be used to analyze data that occurs on a circle, such as time or direction data. It can be used to determine if snails took up preferred directions for their net moves. Using uninfected snails what I first did was remove all the net moves during the initial dispersal, which you recall was to the northwest in both years. Including these net moves would bias the test toward that direction and that wasn't the question I was asking. I was asking: what do they do after initial dispersal? Only net moves made after the initial dispersal were used in the analysis. Each net move of each snail included has a direction. Collectively, in both years, there was a total of 109 next net moves (directions taken) by 36 snails in the analysis. Imagine a circle with a central point. Maybe a snail's first net move was in a direction of 360°, a mark at due north (360°) goes on our circle. Let's say its second net move was toward the west, a mark goes at 270° on the circle. Maybe a third net move by another snail was in a direction of 95°, another mark goes on the circle at an easterly direction (95°). This proceeds until all 109 net moves have been plotted on our circle. Where were most of the points on our circle located? It turned out that directions of these 109 net moves were uniformly distributed around the circle. In other words, net moves went every which way, some north, some southwest, some southeast, some west. I used a tool of circular statistics, something called a Rayleigh test, on these data and it told me that there was indeed no preferred direction in which net moves were made. Snails choose directions for the next net move randomly. What this means is that sightings of the same snail are mostly independent of one another.

Since the 1950's based on observations made around Woods Hole it has been known that *Ilyanassa* exhibits schooling, groups of snails appear to stay together. I have noted this phenomenon many times on Cape Henlopen. In the late 1970's a worker asked the question: do the same individuals consistently make up a school? Her answer to this question was no, individuals enter and leave a school. She also determined that individual snails make random walks and my results support that conclusion. The situation seems to be that, following initial dispersal, snails just mill about in any area of suitable habitat. They move in a variety of directions and sometimes over considerable distances, but they mill about randomly. The life of a snail during a summer season is that of a nomad.

Some studies on what influences where a snail happens to go have been done? One influence, suggested by this and some of my other studies, is trematode parasites. Importantly, infected snails do not move less than uninfected snails do. Also, the parasite species infecting a snail can affect the direction it disperses, thus they influence where host snails are likely to be found on the sandflat. Except for the effects of parasites on snail movement, a direct study of what affects snail movement was not the object of my work. Studies done in the late 1960's showed that snails are drawn to sources of nutrition for feeding purposes. That they are drawn to carrion was well known before that, but these studies showed that snails are also attracted to areas rich in bacteria and diatoms.

Jeff Kinley at Savages Ditch in 1996, view from release stake SE across Funnel to Devil Island

As we have covered before (Chapter 2) this snail does have a crystalline style and can make use of "plant" sources of nutrition. The snail also follows water currents draining downhill, which may explain why they initially dispersed offshore (gently downhill from the release stake). The phenomenon of schooling is frequently observed. As mentioned above snails do not necessarily stay in a school, but when in a school snails are influenced by the direction of movement of snails around them. A study done in the 1990's demonstrated that water current direction and speed also affect movement of this snail. In strong currents (greater than 0.2 m/second) snails tend to bury. In lesser currents they remain unburied and tend to move with the current. They collect downstream where currents are less. That study was done in the laboratory where currents could be controlled and also in a saltmarsh stream. During high tides on Cape Henlopen strong currents on the sandflat would not develop in the deep water, but at and around low tide strong currents in gullies and runnels might develop for a short time. As noted above, snails are sometimes tumbled along in such currents. At other times I have noted snails buried when the current is strong. I would say that strong currents are not a widespread, consistently encountered factor on Cape Henlopen. In the final analysis we can list parasites, food sources, and currents as the factors that influence snail movement. On Cape Henlopen parasites and food sources would be the most important.

Chapter 16

PARASITES AND WHERE HOST SNAILS GO AND WHAT THEY DO

As you might suspect I am not done with the travels of snails in their natural habitats. One problem with doing this sort of investigation on Cape Henlopen, as I described in the previous chapter, is that the sandflat is so vast. Often, marked snails get forever lost. In particular, I was only able to find a few snails in years subsequent to their year of release. Longer term tracking of individual snails therefore would be a problem. On the plus side Cape Henlopen had provided instances of parasites influencing where snails go and what they do. There was the case of *Lepocreadium setiferoides* (*Ls)* enhancing the host snail's interest in carrion (Chapter 5) and of course the parasite-altered behavior of snails hosting *Gynaecotyla adunca (Ga)* (Chapter 6). Also of note, in the previous chapter there was the clustering of snails infected with *Himasthla quissetensis* (*Hq)* along shore. In Chapter 11 I recounted moving of my research to a new field site at Savages Ditch in southeast Rehoboth Bay. That began about 1995 as I was heading into a major debilitated foot episode. By this time Cape Henlopen was getting too crowded and, in any case, Savages Ditch had those fascinatingly huge snails to figure out.

That move was facilitated by Nathan Tanner and Jeff Kinley. They each had a project. Nathan (Chapter 11) wanted to follow the accumulation of trematodes by uninfected snails at the site and to measure the frequency of changes in infecting trematode species in already infected snails. Jeff (Chapter 12) wanted to determine the growth rate and age of the snails in the area. In this chapter's frontispiece, he's there and in action. These projects entailed releasing marked snails at Savages Ditch in June of 1996. We released 249 uninfected snails transplanted from outside sites where infected snails were scarce and 231 native snails from the Funnel area that were virtually all infected. I also had a project that I wanted to execute. It was to follow the movements of marked snails in the Savages Ditch habitat. Specifically, I wanted to see if parasites influenced where their host snails distributed themselves. My interest in this was peaked by that first collection of snails from that habitat in 1995. Snails were collected along a shoreline and 85% were infected with *Hq*. What was going on? Were these snails drawn to this shoreline for some reason or were *Hq*-infected snails just very abundant in the whole habitat? The Savages Ditch habitat is smaller by far than the Cape Henlopen sandflat and the odds on being able to find marked snails after release were potentially better. As told before (Chapter 12), some snails were found for up to six subsequent summers. I have already described the circumstances and findings of most of the work done at Savages Ditch. In this chapter I want to specifically treat what I found out by following marked snails around at that site.

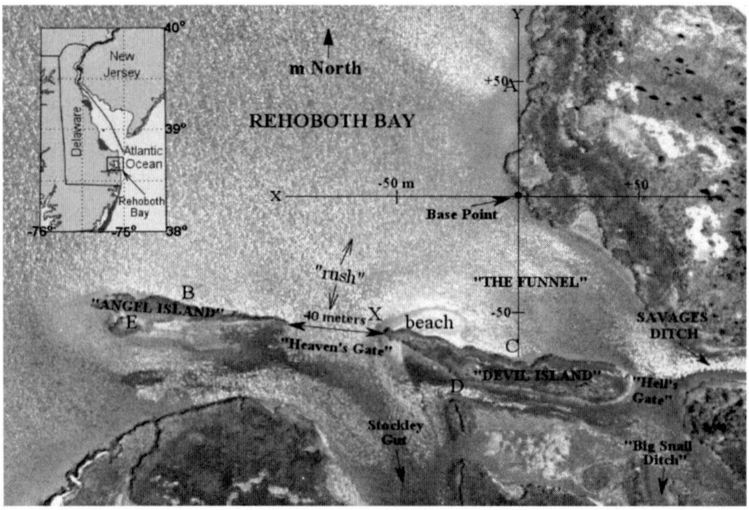

An aerial photograph of the Savages Ditch site in Rehoboth Bay, Delaware (Reprinted from Journal of Experimental Marine Biology and Ecology, 352, Lawrence A. Curtis, Spatial heterogeneity in size and parasitism: How it arises in an estuarine snail population, 317-330, 2007, with permission from Elsevier.)

Previously I described in words the features of the Savages Ditch habitat. Now, considering the work to be described and for purposes of visualizing it, I think an areal photograph of the site will be useful. One very similar to the one I included in the published paper on this work (Curtis, 2007a) is presented above.

This particular photograph was actually published in another paper on Savages Ditch snails (Curtis, 2007b). It works equally well here and in the next chapter where I describe that particular work. You might want to place a marker of some sort at this page. If you put a $20 bill here, you won't forget about the book and will be motivated to return. As you go through this chapter and the next, it will be useful to be able to refer back to this page quickly. The inset in the photograph indicates the positions of New Jersey, Delaware, and between them Delaware Bay on the American east coast. In the rectangle south of Delaware Bay is Rehoboth Bay. This represents the features of the Savages Ditch habitat after we had finished naming the ones that had no names. Features given in quotes are the result of our naming efforts. There are some features in this photograph that do not pertain to this chapter. We'll get to those in the next. Of prime importance here are the Base Point and the axes that pass through it. The Base Point is where all marked snails were released in June of 1996. The axes (lines) passing through that point provide a scale for the whole photograph. For example, across the "Funnel" about 70 meters (m) due south of the Base Point is what we called "Devil Island." On a low tide the emerged and shallow water areas could be searched for from one to four hours, depending on extent of the tide. Recall that the water is about one meter (3-4 feet) deep at high tide. On a decent low tide the searchable area included the Funnel, the areas immediately around Angel and Devil Islands, the area north of Heaven's Gate, and the area along the shoreline that extends north of the Base Point. Between June 1996 and June 2002 the site was searched for marked snails 213 times. I and my two colleagues spent a lot of time traveling back and forth to Savages Ditch. If the traffic wasn't too heavy, the women and other beings around Dewey Beach made the trips interesting. Most sightings (74.2%) of snails occurred in 1996 during the first 100 days after release, but smaller numbers of marked snails were sighted in each of the following summers through 2001. Keeping track of the same individual snails for up to six summers is an unusual feat.

Among the snails tracked were five infection categories, (1) uninfected (NI), and those infected with (2) *Hq*, (3) *Ls*, (4) *Zoogonus rubellus (Zr)*, or with (5) both *Hq* and *Zr (HZ)*. Except for lacking *Austrobilharzia variglandis (Av)*-infected snails, these categories are the same ones I earlier tracked on Cape Henlopen. I need to present some basic information about the number of snails, the number of times they were sighted, and the distances traveled for each of these categories. The most efficient way to do this is to use a table. It is just below. A couple of points of interest are: the average distance moved per day by all categories is in the same range as that determined on Cape Henlopen, someplace between 1 and 3 m/day; for another, all categories had a maximum detected distance moved per day in the same range as on Cape Henlopen. Amazingly, one *Zr*-infected snail was detected moving 57 m in one day, which tops the maximum (46 m) by an uninfected snail on Cape Henlopen. You go ambitious little *Zr*-infected snail! The snail found most distant from the Base Point was an uninfected snail. It was sighted but once and it was 135 m away to the west southwest (259°), down near the west end of Angel Island, 20 days after release. A further point that should be made is that uninfected snails were smaller (average shell height about 21 mm) than all infected categories (average shell height more like 28 mm).

Some basic data for snail dispersal at Savages Ditch

Category	**NI**	***Hq***	***Ls***	***Zr***	**HZ**
Number of snails	139	75	17	43	15
Number of times sighted	381	484	101	192	65
Average meters moved per day	2.3	2.0	1.8	2.8	2.1
Maximum meters moved per day	36	20	13	57	18

The first concern is that I needed a new way to analyze snail sighting data. In the previous chapter, where I analyzed movement of snails on Cape Henlopen, I treated the distance of snails from release separately from their direction from release. Most studies of organism dispersal that I had read were limited to that sort of analysis. This style of analysis works to a degree, but is somewhat disjointed. What I needed was a method of analysis that considered distance and direction from release both at once, in one analysis. I will need you to bear with me a bit because the analysis I came up with to handle Savages Ditch snail sighting data will require some explanation. Why did I not use this sort of analysis on data from Cape Henlopen? Well, we learn as we go along and I hadn't come up with it at that time. I will first describe the approach by using a graph based on hypothetical data, data made up to illustrate the analysis.

The illustrative graph is presented just below. On the horizontal (x) axis is plotted salinity, while on the vertical (y) axis is plotted temperature. Let's suppose that you have been searching for two kinds of organisms and each time one or the other was found you noted the salinity and temperature of the water where the sighting was made. On the graph there are 20 (made-up) sightings represented. Take a minute to understand what the graph is suggesting to you: in salinity and temperature "space" the organisms seem to be of two types, those that are found where temperature is higher and salinity is lower (Group 1) and those found where temperature is lower and salinity is higher (Group 2). Now, this is certainly how things appear and we could probably make this point by raising our voice a little and stating assertively that it is so. We might wave our arms some to re-enforce the assertion. However, in science we can't get away with that, certainly not on points that are basic to whatever claim we are making. We need to back up what appears to be true, our beliefs, with statistical analysis. I made up these particular data so there would be an obvious difference between the two groups in terms of salinities and temperatures where they are found. But, suppose salinity and temperature preferences for the two groups were closer, even to the degree that some of the points overlapped. Then it might not be so obvious that the two groups are different.

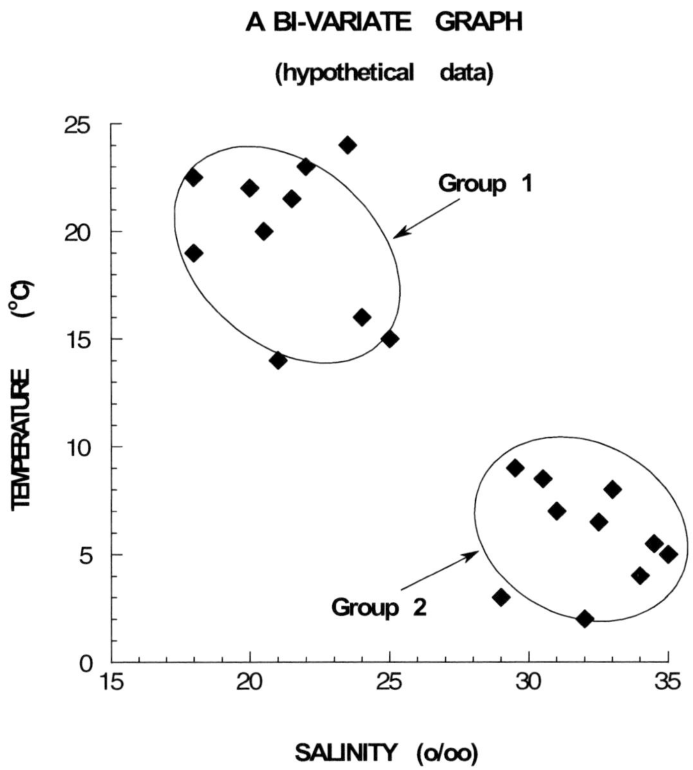

A graph to illustrate the analysis of bi-variate data.

How would it be possible to demonstrate statistically that the scatters of points for the two groups are (or are not) statistically (significantly) different from one another? Most statistical tests operate on a single variable. We might, less correctly, do one test on salinity preferences of the two groups and another test on temperature preferences of the two groups. These would be uni-variate tests (analyzing one variable at a time, salinity or temperature). Usually, it is better to analyze data comprehensively with a single statistical test rather than several. Because we have to consider simultaneously two variables, salinity and temperature, we need what is called a multi-variate (many variables) test to do this. Multi-variate statistics can handle more than two variables at one time, but we don't need to get that complicated here. We start by posing the no difference hypothesis (the so-called null hypothesis). That is, we specifically want to test the hypothesis that the two clusters of points are actually in the same place on the graph, i.e., not in significantly different places. If we wind up accepting this null hypothesis, then the salinity and temperature preferences of the two groups can be considered the same. If we wind up rejecting the null hypothesis, then the preferences of the two groups can be considered different. We get to go on with our reasoning under one outcome or the other.

An aid to interpreting bi-variate (two variable) data clusters, such as those in the graph, is something called a standard ellipse. A standard ellipse is calculated to enclose approximately the central 40% of the points in a scatter of points. I have drawn on the figure what the standard ellipses for Groups 1 and 2 would sort of look like. I have not actually calculated these standard ellipses for the scatters of points plotted, a complicated procedure with which I don't need to clutter this up, but the figure roughly illustrates how the ellipses would appear. It is important to realize that even if the standard ellipses overlapped each other somewhat, the points summarized by each could still be statistically different. That being the case, how might we determine whether the two groups are actually different with regard to salinity and temperature preferences?

We have spoken of uni-variate tests before. The Kruskal-Wallis test is one. Another one already mentioned is the ANOVA, the analysis of variance. To handle the multi-variate situation a multi-variate analysis of variance, a MANOVA, is a means to compare two clusters of points on a bi-variate graph to see if they are statistically in different places. What this test would do for each of the 20 points on the made-up graph above is to consider which group (1 or 2) a point belongs to and its position simultaneously on the salinity and temperature axes. It would be a bi-variate test. The calculations are long, tricky, and complicated. Anyone wishing to guard their sanity would be wise to let a quality statistical software package run the calculations. That's what I would do and did. When the program is done, its output would tell you whether the two clusters of points are indeed in significantly different places on the graph. I have not actually done a MANOVA on data in the graph. After all, data were made up to be different and undoubtedly they are. Besides, why bother? Phony data make no biological point. That the two clusters of points are in different places on the graph (in salinity and temperature "space") may be obvious to the eye, but now, assuming I had run the test, we could say so based on statistical analysis rather than just on eyeball reckoning and arm waving. If you want to make a point in ecological science, whenever possible it is best to do it in this manner.

You need to conceptually transfer the bi-variate graph just described over to the situation according to which Savages Ditch snail sighting data might be analyzed. The two variables to be considered for an infection category of snails, rather than salinity and temperature as in the foregoing graph, are positions of snails on the horizontal (x-) axis and simultaneously positions on the vertical (y-) axis. If you refer back to the photograph of the Savages Ditch area presented earlier, you can see that the x-axis runs east to west. Positions on that axis are reckoned in meters and are positive east from the Base Point and negative west from the Base Point. The y-axis runs north and south. To the north of the Base Point are positive meters and to the south are negative meters.

We originally sighted snail positions in the field using polar coordinates. For example, suppose a snail is sighted at a distance of 26 m from the Base Point in a direction of 165°. This would be out in the Funnel approximately southeast of the Base Point. Now, if you take a pencil, a piece of graph paper, a ruler, and a protractor, you can draw a little graph with axes similar to those in the photograph. You will have two perpendicularly crossed lines (axes) as in the photograph with meters marked out on each. If you have done this on graph paper, let each division along an axis represent one meter. If you have used plain paper, use the ruler to mark out evenly spaced meters on the axes. In either case mark at least 26 m along each axis. Remember, on the south (downward) part of the y-axis meters are given a negative sign. Now, use the ruler and the protractor to accurately plot a point at 26 m and 165° from the vertical axis on your graph. You will see that the point is positioned at +6.7 m on the x-axis and -25.1 m on the y-axis. In other words, any snail position in polar coordinates can be converted to x-y coordinates on a graph. This works the other way too. For example, a point plotted at -73 m on the x-axis and +61m on the y-axis would have polar coordinates of 95 m at 310° (about northwest) from the Base Point. One of my first jobs in this form of analysis was, therefore, to convert the 1,223 total snail sightings gained over six summers from polar to x-y coordinates. This can be done with a physical graph as suggested in the exercise above or with right triangle trigonometry calculations. Maybe you did some of that in high school. I know I did, but by this time it was long forgotten and I had to look it up anew. Once I had the method resurrected, I used a computer spreadsheet to make the calculations. A conversion from polar to x-y coordinates does not alter the physical positions where snails were sighted in the field. All you are doing is representing the same physical place by two different methods. Instead of representing a point as 26 m at 165°, you represent it as +6.7 m on the x-axis and -25.1 m on the y-axis. Why go to all this conversion trouble? It's because the calculations of a MANOVA operate with x-y coordinates rather than polar coordinates.

All statistical tests operate under certain assumptions about the data being tested. The most important assumption of MANOVA calculations is that the data analyzed must be independent of one another. They can't be tied to, or dependent on one another. In the paper describing this work I included a graph showing that individual snails were often sighted multiple times. Many individuals were sighted only once, but a few were sighted over 25 times in six years. As in the last chapter, it is possible that data from two (or more) sightings of the same snail could be dependent on each other. That is, where a snail is sighted today could depend, in some way, on where it was sighted the previous time. If this sort of sighting-data dependence were commonplace, it could totally invalidate my use of the MANOVA test. I needed to be as sure as possible that sighting data for the same snail did not depend on each other, i.e., that where it was sighted once did not predetermine where it would be sighted next. I examined this possibility with respect to Cape Henlopen data discussed in the last chapter and concluded that sighting data for the same snails were independent. So, thus far all indications were that sightings of the same snail were independent of one another, but I needed to check this carefully for Savages Ditch data. I make four points: (1) of the 289 snails sighted, 34% were sighted only once and this reduces the overall problem; (2) snails can move many meters in a single day and of all the same-snail sightings 74% were more than five days apart, which makes data dependence less likely; (3) it was a complicated statistical test that I won't drag you through, but of 42 multiply sighted snails in various infection categories 38 demonstrated random variation in number of meters moved per day (i.e., there's no predicting just how far a snail will travel in a day); and (4) all infection categories took up random directions from net move to net move (i.e., there's no predicting which way a snail will go). In sum, there is no predicting where a snail might be sighted in the future based on where it was sighted today. If there are occasions when where a snail is at one time influences where it will be at a future time, these occasions are few and far between. Such instances would not be pervasive among the data. Therefore, I concluded it is fair to use a MANOVA test to analyze sighting data.

As indicated above the total snail sightings in this study amounted to 1,223. So I ended up with all these sightings plotted on a graph of a form similar to the Savages Ditch photograph provided earlier in this chapter. Recall that there were five infection categories of sighted snails, uninfected, and infected with *Hq, Ls, Zr,* or *HZ*. If an uninfected snail was sighted, its position at that sighting might be represented on the graph by an open triangle symbol. If an *Hq*-infected snail was sighted, its position might be represented by an open circle, and so forth. So the 1,223 points on the graph were of five different types [open triangles (NI snails), open circles (*Hq*-infected), open squares (*Ls*-infected), open lower half circles (*Zr*-infected), and plus signs (*HZ*-infected)]. As you can imagine a graph with 1,223 points on it of five different types is very confusing. Two snails in two infection categories may tend to be sighted in different areas of the habitat, but at another time they can easily be sighted right next to each other. The main thing I was trying to discover was whether snails in the five infection categories tended to be sighted (to cluster) in distinct areas of the graph, much as the two groups did in the previous hypothetical graph. I tested the null hypothesis that snails in all five categories had the same sighting distributions on the graph (actually in the field). If that were the case, if the null hypothesis were accepted, the study could stop right there. It would mean that if a snail were uninfected, or were infected with some trematode, it would make no difference to where that snail goes and what it does. Using a MANOVA test, I tested whether there was any difference in the locations of the sighting distributions of the five infection categories. The result of the test said that indeed there was. That is, at least one category tended to be sighted in a different area of the habitat.

The output from a MANOVA test is pretty complex. From a uni-variate test you usually get one test statistic to tell you whether you have a significant difference among the groups tested, but from a MANOVA test you get four. Without describing them, suffice it to say that all four were highly significant. The null hypothesis that all categories had the same sighting distribution (dispersions in habitat space) was decidedly rejected. Now, there is always a chance that a rejected null hypothesis is in reality true, in this case that the sighting distributions were actually the same. The four test statistics indicated that there was a vanishingly small chance that the null hypothesis was actually true. I could be very confident in the rejection of the null hypothesis. There was indeed some difference in sighting distributions among the categories. [As an aside, if you read much about using statistics to test hypotheses, you will run into something called the P value. This value indicates the chance that a rejected hypothesis is actually correct. By convention, if for a test statistic the P value is less that 0.05, the hypothesis is considered significantly rejected. You can be confident in the rejection: there is less than a 5 % chance that the hypothesis is actually true. You'll take the chance and believe in the rejection. All four of the test statistics from the test we are talking about here were significant at a P value less than

0.001. That is, there is less than one chance in a thousand that snails in all five infection categories in reality had the same distributions in the Savages Ditch environment. We'll confidently take the gamble that there is at least one difference among sighting distributions.] So, now we can proceed with our thinking under the well-founded assumption that all infection categories did not disperse the same.

Once the hypothesis that all categories dispersed the same is confidently rejected it must be accepted that at least one category dispersed differently. The question immediately arises: which category or categories were different and how did it or they differ? The MANOVA test was significant. If, and only if, that is true, it becomes statistically legal to push the analysis further by comparing the distributions of the five infection categories separately on the x-axis and also on the y-axis. Because we are asking a question about one variable at a time (positions on the x- or y-axis) now we use a uni-variate test. Again, legally we can only do these if the multi-variate test was significant. For example, we might want to know whether *Zr*-infected snails tended to be sighted east (offshore) of snails in other categories and what those categories were. The uni-variate test I used for this purpose was the Kruskal-Wallis test, which has been earlier mentioned several times. Here, this test compares the median positions of the distributions of the five categories for example on the x-axis. The median is that point on an axis above which we find half the points of a scatter of points and below which we find the other half. On the x-axis the Kuskal-Wallis test will tell you which infection categories of snails dispersed significantly east (or west) of other categories. On the y-axis the test will tell you which categories dispersed significantly north (or south) of other categories. This level of analysis will tell us the things we really want to know. Which categories tended to be sighted nearer shore? Which categories were likely found farther out from shore? Did any category disperse to the north or south? Once we get answers to these questions we can begin to think effectively about why snails in different infection categories might distribute differently.

When I compared (using the Kruskal-Wallis test) where snails in the five categories were sighted along the east-west (x-) axis there were highly significant (P value less than 0.001) differences among them. What I found basically was that uninfected, and *Zr*-infected, and *HZ*-infected snails tended to be sighted farther east (offshore) than snails infected by *Hq* or *Ls*. Uninfected and *Zr*-infected snails tended to be sighted farthest offshore. The distribution of sightings for a category were naturally scattered above (50% of sightings) and below (the other 50%) the median, but the positions of medians on the east-west axis were -5.8 m for uninfected snails, -6.9 m for snails carrying *Zr*, and -2 m for *HZ* snails. Median positions for sightings of snails infected with *Hq* or *Ls* were respectively -1.5 and -1.8 m, clearly closer to shore than uninfected snails or snails with *Zr* infections. Snails with *HZ* infections were an intermediate case between the offshore and inshore groups.

When I examined sighting distributions of the five infection categories on the north-south (y-) axis there were also highly significant (P less than 0.001) differences among them. From north to south, *HZ*-infected snails had a median sighting position of -10.6 m, uninfected snails a median of -20 m, *Zr*-infected snail's median was -20.8 m, while *Hq*-infected snails (median -24.6 m) and especially snails bearing *Ls* (median -34.9) were farther south. Statistically, there were essentially two groups on the north-south axis, uninfected, *Zr*, and *HZ* snails to the north and *Hq* and *Ls* snails to the south. I readily admit that sighting distributions along the onshore-offshore axis are the more interesting.

When we stop to think that only a limited area of the whole habitat could be searched on even the lowest of low tides, it is amazing that these groups could sort out differently in the space available. As one example consider uninfected and *Zr*-infected snails. Both snail categories tended to be sighted farther offshore. At Savages Ditch, unlike on Cape Henlopen, snails exist in the subtidal zone below the lowest low tide. Had we been able to regularly search this subtidal area it is likely that uninfected and *Zr* snails would have been sighted there often and the median distances of these categories would have been even farther offshore. That is, their sighting distributions would have been even more offshore than I determined. The presence of significant differences in the sighting distributions of these infection categories of snails, even within the limited range of searchable habitat, speaks well for the fact that parasites have a lot to do with where their host snails hang out in the habitat. It may be that ultimately we do not yet know enough to be able to surely assign a reason why snails go where they do. Hypotheses about that will be discussed below. However, just the fact that infection status (whether or not a snail is infected and with what species of parasite) influences snail travels is an important thing of which to be aware. This is seldom recognized when studying snail ecology, but it is indeed important.

It is pretty clear that uninfected snails tended to be sighted away from shore. Based on other statistical analyses, they made significantly fewer than expected net moves toward shore and they were sighted within 10 m of a shore significantly less often than expected. Based on collecting snails, and either testing them for infections by cercarial release or doing dissections, one aspect of the differences in space (spatial heterogeneity) in the area is that the region just north of Heaven's Gate is mainly populated by uninfected snails. This away from shore area was given the name the "rush" because tidal currents there are quite strong, especially when the tides ebb from Rehoboth Bay through Heaven's Gate. In contrast snails from the Funnel, very close to the "rush," are virtually always infected and there currents are much less strong. Now, why uninfected snails tend to hang out away from shore, often in the "rush," at Savages Ditch is a mystery. I can come up with a number of tentative explanations for this, but in the end I must admit that I don't know. It probably has to do with their being younger, parasite-free, or smaller in size.

Now to discuss snails infected with *Hq*. This is probably the most interesting case of a parasite influencing host snail dispersal emerging from this study. Recall that one impetus for the study was that heavy concentration of *Hq*-infected snails Buck Cornell and I found at Savages Ditch in 1995. If you look back at the photograph, those snails were collected along the shoreline that extends north from the Base Point. Since that time many snails from Savages Ditch have been examined for parasites and by now it was clear that *Hq* was not everywhere so abundant as along that shore. What were they doing specifically there? Data from this study indicated that *Hq*-infected snails remained pretty close to shore. On the east-west axis the median of their sighting distribution was only 1.5 m west of the Base Point. Additionally, I made a count of the number of times snails in the five infection categories were sighted within 10 m of a shore and tested the data to see what categories were most likely to be present there. The statistical test I used (a contingency analysis) said that all infection categories were not equally likely to be found within 10 m of shore. It turned out that *Hq*- and *HZ*-infected snails were found there significantly more often than expected. As indicated above, uninfected snails were found there less often than expected. Also, *Hq*-infected snails made a lot of northeast net moves, which would move them toward shore. Statistically, this favoring of northeast moves was not significant, but it was very nearly so. Another indication of *Hq*-infected snails hanging out close to shore is the fact that the snails most often sighted in the five years after 1996 were those infected with *Hq*. The same *Hq*-infected snails were sighted in each of the six summers of the study. What this means is that year after year, after moving offshore for winter, in spring the same *Hq*-infected snails returned to near-shore areas for the summer. They existed close to shore where searches could routinely be made. That's why they were the most often sighted in the years subsequent to release. Based on all these observations, it can safely be said that at Savages Ditch *Hq*-infected snails demonstrably cluster near to shore.

It is now clear why so many snails infected with *Hq* were present in that 1995 along-shore collection. They were there to collect *because* they were infected with *Hq*. Uninfected snails and otherwise infected snails decidedly do not behave that way. Why might *Hq*-infected snails do this? What possible selective advantage might be gained and does it go to the host snail or to the parasite? We haven't been talking about natural selection much in recent chapters. Recall that it operates across generations. The host snail has been sterilized by the parasite and its reproduction shut down. There is no way a snail parasitized by *Hq*, or any other trematode, could pass advantageous genes on to offspring. There are no offspring. Therefore infected snails cannot gain selective advantage from anything they possess or do. The only reproducing entity in the host-parasite system is the parasite. Only it can pass genetic advantages on to its descendants. If there is a selective advantage to hanging out close to shore, it must therefore go to the parasite. Now, what might that advantage be? Let's think about what the most important thing an infection in a snail has to accomplish. It is true that it must be able to exist within the snail's tissues without harming the snail too much. All these parasites are apparently adapted to do this. It is hard to see how distributing close to shore would aid in this endeavor. Anyway, if it did help in some way, why wouldn't all infecting species promote this behavior? No, we shall have to look elsewhere for a selective advantage. However, our reasoning thus far provides a clue. Why do these parasites not harm the host snail too much? It's because they need to emit cercariae from that host for a very long time (years on end). If the host were too debilitated, say to the point of not being able to move about and find adequate nutrition, this could not be accomplished. Production of cercariae would be cut short. And, why is there a need to put out so many cercariae? It's so enough second intermediate hosts will end up infected by them and the life cycle of that parasite genetic strain can potentially be completed. A reasonable selective advantage to explore, and one that is pervasively important, is enhanced transmission of cercariae to a second intermediate host.

The structure of populations of these parasites is necessary to have in mind if you would appreciate the value of successful host-to-host transmission. This has been done in other contexts before, but let me quickly review the structure of these populations. Trematode populations are segmented. Parasite stages exist within hosts in their life cycle and also outside hosts (usually briefly) as they transmit from host to host. This gives the populations special features that make them more complex than populations of non-parasitic organisms. An infection in an individual *Ilyanassa* can be thought of as an infrapopulation. It consists of all the sporocysts, or rediae, and cercariae of a species in the snail. Because probably only one miracidium originally infected the snail, and all these stages are descended by asexual reproduction from that miracidium, they are all genetically identical, a clone. If we consider all the infrapopulations of a trematode species (e.g., *Hq*) that exist in a snail population, we are talking about a parasite component population. Of prime importance here is the suprapopulation level of structure. It consists of all the parasite's stages in all the hosts, plus stages in transit between hosts, in a ecosystem. To account for this population level for *Hq* we would have to consider (1) the miracidia that infect snails (the first intermediate host), (2) the rediae and cercariae in snails that develop from the miracidia, (3) the cercariae that transmit from the snail to second intermediate hosts, (4) the metacercariae in second intermediate hosts (e.g., clams) that develop from cercariae, and finally (5) the adult worms that reside in definitive hosts (shorebirds) and develop from eaten metacercariae that were in infected second intermediate hosts. It is the adult worms that produce miracidia. If they manage to infect a snail, the cycle gets completed. My point in all of this exposition is that if an infection in a snail is to be transmitted from the snail on into the overall suprapopulation, it will have to do it by way of cercariae successfully reaching a second intermediate host. Parasite genetic strains that are equipped to manage this will be selected for and preserved in the suprapopulation. Genetic strains that fail to do this will be lost, weeded out of the suprapopulation.

There are only two known second intermediate hosts that carry *Hq* metacercariae out there in nature. One is a polychaete worm species (*Scoloplos fragilis*) and the other is a clam species, something called the New England soft-shelled clam (*Mya arenaria*). Both occur in the intertidal zone where *Ilyanassa* occurs. In addition to these natural second intermediate hosts there are a number of other species, experimental hosts, that under laboratory conditions can be infected by *Hq* cercariae and develop *Hq* metacercariae in their tissues. These experimental second intermediate hosts are all clams of one sort or another. Included are ribbed mussels, horse mussels, blue mussels, scallops, razor clams, and hard-shell clams. The one of these potentially natural hosts that I want to stress in connection with Savages Ditch is the ribbed mussel, *Geukensia demissa*. The saltmarsh grass shorelines at Savages Ditch, indeed in much of Rehoboth Bay, are lined with populations of the ribbed mussel. Mussel populations are very dense there, in an almost unbroken array along shore. If a parasite "wanted" to get cercariae to a ribbed mussel, it would behoove that parasite to influence its host snail to hang out in the near-shore area. To get to a second intermediate host an *Hq* cercaria must be emitted from its infrapopulation in its snail and swim in the outside world. It is not a foregone conclusion that a cercaria will find a second intermediate host to infect. A cercaria does not feed and is microscopic. It only has a life span of hours (maybe a day) before its energy is used up and it sinks to the bottom and perishes. The closer a host snail is to a second intermediate host when cercariae are shed the more likely they will successfully reach and infect that host. Successfully making it to and infecting a second host is an utter necessity for a successful trematode infecting *Ilyanassa*. So, the tentative conclusion (hypothesis) is that *Hq* infections modify the behavior of their host snails so they hang out close to shorelines and make it more likely that emitted cercariae will be able to get to and infect ribbed mussels.

There have been a number of cases described in the literature wherein a parasite is thought to change the behavior of a host. Pretty clearly this is in fact what is happening in some cases. In others it is less sure. It is usually thought to have something to do with transmission to a next host, usually by making the prior host more susceptible to being preyed upon by the next host. The idea that host behavior can be modified to enhance cercarial transmission is relatively new. One of the earliest cases in the literature was that of the modified behavior of snails when infected by *Gynaecotyla adunca (Ga)* (Chapter 6). Trematodes virtually always transmit from *Ilyanassa* to a next host by way of cercariae. Now here again I am using this reasoning to explain why *Hq*-infected snails would hang out near the shore. There is not much question that they do distribute close to shore. The only question is why they do. That is what calls for reasoning and speculation.

One of my favorite studies of parasite-altered host behavior was that done by one Janice Moore back in the 1980's. She was studying a parasite of cockroaches that needed to transmit to birds, specifically to starlings. Cockroaches, as we all know, tend to remain hidden in the dark. But, cockroaches infected with this parasite would come out into the light and expose themselves to predation by starlings more than uninfected cockroaches would. To prove (support) the efficacy of the behavior modification what she needed to show was that more infected cockroaches were consumed by starlings than uninfected ones. Cleverly, in the laboratory starlings were fitted with neck ligatures. The birds could prey upon and swallow cockroaches, but the cockroaches were caught above the ligature, prevented from passing into the stomach. The cockroaches could be recovered for examination. Indeed, most of the consumed cockroaches were infected, proving beyond any reasonable doubt that the behavior modification led to greater host-to-host transmission.

That is the sort of evidence that has to be accumulated to make the case that a parasite is modifying a host's behavior to effect greater transmission. Ribbed mussels were only an experimental host for *Hq*. Mussels had taken on infections in the laboratory. A basic question about Savages Ditch is: are ribbed mussels in the field there naturally infected by *Hq* metacercariae? You may recall a former undergraduate student of mine named Ann Barse (Chapter 7), now a professor of biology at a nearby university. She and one her students began looking at the bodies of ribbed mussels from Savages Ditch to see if *Hq* metacercariae could be found. Finding metacercariae in mussel tissues is quite challenging, as they are usually well camouflaged and inconspicuous. Anyway, metacercariae were indeed found by them, and by me. They all but positively belonged to *Hq*. The slight caution in that statement is that to be sure it's an *Hq* metacercaria you need to count 31 so-called collar spines on it. Microscopically this has proved difficult, but it has all but been accomplished. We have counted 29 spines and no more than two more can be hidden from view. Besides that, *Hq* is the only candidate for that type of metacercaria at that site. There is virtually zero chance that the metacercariae found in mussel tissue are other than *Hq*. There are a raft of different studies that might be done on this, but for starters the basic point is made: ribbed mussels along the shores of Savages Ditch are infected with *Hq*. It can't be any other metacercariae. This becomes a new natural second intermediate host. If we can count those last two spines, this by itself is worthy of a short publication.

Proving beyond reasonable doubt that a parasite alters a host's behavior, to the parasite's advantage, is notoriously difficult. For one thing we must usually make the case without knowing the mechanism (probably chemical) by which the parasite changes the host's behavior. For another it needs to be shown, as with the starling example, that the behavior change leads to greater transmission. But, we are well on the way. Not only that, but once we publish the finding of the ribbed mussel as a new natural 2nd host for *Hq*, I will have a "grandchild" undergraduate student involved in a publication. That's cool.

So, at Savages Ditch snails infected by *Hq* seem to have their dispersal affected such that they tend to hang out in near-shore positions, but what about in other habitats? Recall in the last chapter that there was pretty strong indication that *Hq*-infected snails on Cape Henlopen also tended to disperse along the shore. The problem is: on Cape Henlopen, *Hq* has to be depending on second intermediate hosts other than ribbed mussels. The mussel is found in saltmarsh grass environments and does not exist along the shores of Cape Henlopen. Therefore it becomes more difficult to explain why *Hq*-infected *Ilyanassa* might cluster along-shore there. Perhaps some other second intermediate host is concentrated along-shore. There is also the possibility that *Hq* has become universally adapted through the process of natural selection to cluster near-shore without regard to the habitat where it finds itself and what second intermediate hosts are available. This parasite is spread among snail habitats by shorebird definitive hosts. Dispersal of parasites among snail populations by birds is analogous to the dispersal of snail larvae in the plankton. Both result in genetically similar populations. For example, a seagull might eat a ribbed mussel that is infected with *Hq* metacercariae in one habitat and then fly to another habitat, perhaps lacking ribbed mussels, where *Hq* miracidia liberated from the bird's fecal material infect snails there. If that strain of *Hq* has genes that make it hang out near-shore in the first habitat, those genes would be transferred to the second habitat and *Hq* would make any snails infected there cluster along shore too. If *Hq* frequently finds itself in habitats, like Rehoboth Bay, where it is advantageous to stay close to shore for proximity to ribbed mussels, it is likely that the genes that promote that behavior will often be transferred into *Hq* (component) populations in other kinds of habitats. In short, this parasite could be universally adapted to make its snail hosts cluster near to shore regardless of what second intermediate hosts are available. This of course is all a product of speculation, but it is reasonable to think it a real possibility.

The 17 snails infected with *Ls* were equally shoreward with *Hq*-infected snails, but their 101 sightings were distributed well south of all other infection categories. What this means is that snails that bore *Ls* were mostly sighted in and around the opening of the Funnel into Rehoboth Bay, usually well away from a shore. There is only one known natural second intermediate host for *Ls*, another polychaete worm. It is less clear-cut than for *Hq*. Actually it's out and out speculation, but the sighting distribution of *Ls*-infected snails can also be related to the distribution of second intermediate hosts. This worm second intermediate host would probably be more numerous out and away from shorelines. In fact, another study by other authors has noted that where many snails are infected by *Ls*, many specimens of this worm were infected by *Ls* metacercariae. In the Mediterranean Sea a congener of *Ls* (same genus, different species) in another snail species, similar to *Ilyanassa*, was studied. The second intermediate host in that system was a clam that lived in the low intertidal zone and that's where snails infected by that parasite tended to congregate. The authors thought maybe that the parasite altered the behavior of infected snails to congregate them there and put them in proximity of that second intermediate host. They did not have data to test that hypothesis, but I would like to think that is the case. There is another possibility. Way back in Chapter 5 we talked about *Ls*-infected snails having an enhanced attraction to carrion. Perhaps carrion is more abundant in the opening of the Funnel than elsewhere. That would have to be demonstrated, but if true that could be the reason why *Ls*-infected snails tended to be sighted there. We must always entertain alternative hypotheses.

What can be said about *Zr*-infected snails? These snails had a sighting distribution as offshore as uninfected snails. There is only one known, natural second intermediate host, yet another polychaete worm. In the marine environment polychaetes are sort of like earthworms are on land. The odds on a *Zr* cercaria finding and infecting one of these worms would be better away from shore. To get cercariae to this worm it would be adaptive for the parasite to induce host snails to remain away from shore. It has been shown by other authors that where many *Zr*-infected snails occur in the field, this worm is often found infected with many *Zr* metacercariae. As a statement based on evidence, that's pretty flimsy, but it's about all that can be said at present. However, *HZ*-infected snails present a potentially interesting case. If transmitting cercariae to a next host is an important issue, within one doubly infected snail is one parasite (*Hq*) that would do well to remain close to shore and another (*Zr*) that probably should stay away from shore. The stage is set for conflict. In this study there were 15 *HZ*-infected snails and they were sighted 65 times. These snails were sighted within 10 m of shore more often than expected. If there is a conflict, it appears that *Hq* won out.

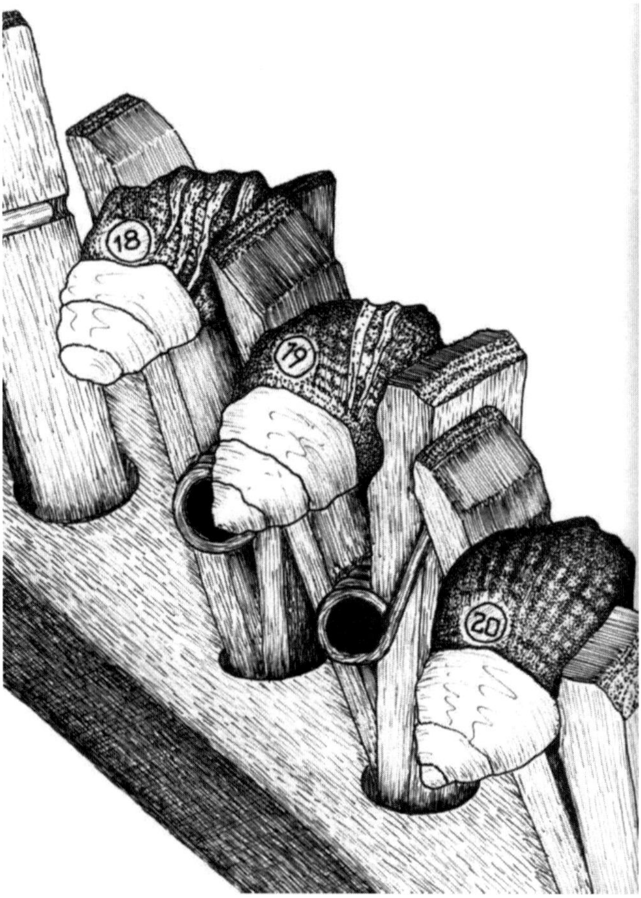

Three newly marked *Ilyanassa obsoleta* in clothespin clamps waiting for their marks to dry.

I used a quote in the final paragraph of the paper published on this work from a 2001 book by Claude Combes, an accomplished and influential French parasitologist: "... when one sees trematodes manipulate their molluscan hosts' growth, fecundity, and behavior, one may wonder whether this situation is exceptional or whether it merely illustrates a more general phenomenon that simply is not well documented." Pretty clearly, with my research history, I am predisposed to believe that it is a common, yet largely undocumented, phenomenon. That means I should guard against accepting too readily such propositions. Do I think that I have demonstrated beyond doubt that these parasites influence the distribution in space of host snails? Well, yes I do. Do I think I have demonstrated beyond doubt that parasites in snails at Savages Ditch are modifying snail behavior for their own adaptive gain? I've made some strong arguments that this goes on, but across the board I'm not so sure. The previous case for *Ga* is strong and the case for *Hq* is pretty strong, but for the other parasites, *Ls* and *Zr*, the case is far more circumstantial. Science will always ask more questions than it answers. That's the nature of the beast and I for one am glad. It keeps things interesting. Some of the things that I have discovered in this (and previous) work have spread to at least one other university. Ann Barse and some of her students have had their interest stimulated and are pursuing certain aspects of the work. That is a satisfying thing. What more can a humble researcher ask?

Chapter 17

How Snails here and there Get to Have Different Features

I mentioned spatial heterogeneity for the first time way back in Chapter 4. In that chapter I was describing some work that Larry Hurd and I (L^2) did on the distribution of age, sex and parasites in the *Ilyanassa* population on Cape Henlopen (Curtis and Hurd, 1983). If you are interested in taking representative samples that successfully characterize a population of snails, spatial heterogeneity is an important thing to take into account. What is "heterogeneity?" Well, it's the opposite of homogeneity. Milk is homogenized to make it the same throughout, no lumps of cream anywhere in the bottle. (Does milk anywhere still come unhomogenized in returnable bottles? In my youth it used to and the cream was separated out on top.) If a snail population were correspondingly homogeneous, then snails collected anywhere would be of the same size. Likewise, proportions of males and females and proportions of snails parasitized by various trematodes would be the same. In general, whatever you looked for, snails everywhere would be the same. Of course, in Larry's and my work that was definitely not found to be the case. When we stood and looked at the snails on the broad, unbroken sandflat they gave the overall impression of being everywhere alike, but when we sampled them for age, sex, and parasites we were surprised at how different things were from place to place. It was an object lesson that proved to be generally true of other snail populations in other habitats. Even at quite small sampling scales (sometimes only a few meters), spatial heterogeneity is a feature of almost any snail population. This made accurately sampling them a challenge. The question to be addressed in this chapter is basically: how does an *Ilyanassa* population develop spatial heterogeneity? A different concern, why heterogeneity comes to be, will also be an issue.

At the time of L²'s work on Cape Henlopen we were unaware of what caused the spatial heterogeneity, but I later was able to shed some light on that. I found (Chapter 5) that female snails and *Lepocreadium setiferoides (Ls)*-infected snails apparently need and were preferentially drawn to sites where carrion was present. There was also the later finding that snails infected with *Gynaecotyla adunca (Ga)* clustered on beaches and sandbars. But, the sandflat habitat was too big, and my grasp on the specific features of the spatial heterogeneity too rudimentary, to do much more about generally addressing the question there. However, the Savages Ditch habitat is smaller and it was at least possible to try. There was one other advantage. There were no sandbars to disrupt snail dispersal. As much as I knew they moved around, with no sandbars to block their way snails from anywhere in the habitat could potentially travel to anywhere else in the habitat. In Chapter 12 I described the release of 500 marked snails at Savages Ditch in 1999. (In the frontispiece for this chapter three of the newly marked snails are shown.) I had this potential for snails to move throughout the habitat in mind when I released them. In that chapter and previous ones, I used these and other snails to explore how long individual snails live, how rapidly snails are colonized by trematodes, and how long infections in individual snails persist. Snails live for decades, rate of colonization by trematodes is low, and infections were observed to at least persist in a snail over six summers. In fact, they certainly persist much longer than that. The 1999 experiment was pointedly designed to examine the spatial heterogeneity seen in the snail population in that habitat. My initial question was quite simple: given that snails could potentially move to anywhere in the habitat, if I take snails from different areas of the Savages Ditch habitat and mark and release them in their own and different habitat areas, how likely is it that they will intermingle within the habitat? Answering that question was anything but simple. I originally called this my "mingling" experiment. Later on I got more officious. The title of the paper ultimately published on this aspect of the work (Curtis, 2007b) was "Spatial heterogeneity in size and parasitism: How it arises in an estuarine snail population."

This was my last piece of field work and I consider myself fortunate to have been able to do it. The work was begun in the spring of 1999. Without going into it again, you may recall that beginning in spring 1996 I had a major faulty-foot episode requiring hospitalization. In early 1997 I had a major operation and the effects of that continued through the summer of that year. During that time my "vampire" disease (diabetes) really sucked the goods out of me, pretty much keeping me from doing anything by myself in the way of field work. However, in the spring of 1999 I was heading into a summer to be followed by a sabbatical leave in the fall. I was feeling pretty good by then. My feet were healed over and I could get around well enough. Moving with grace was not mine to experience, but I believed I could get this experiment up and running. At this time there was no student available to recruit as help with this project. If this was going to be done, I was on my own. Below I will do the necessary thing and describe fully how the experiment was set up. For now let me just say that it involved collecting 500 snails from several sites at Savages Ditch, bringing them back to the laboratory at Cape Henlopen, individually marking them, measuring them for size, testing them for the presence of trematode infections, returning them to their natural habitat for release, and then doing 104 searches of the habitat for marked snails from 10 June 1999 to 2 April 2003. I searched after that, but no more marked snails could be found. Anyway, this all represented a lot of trekking around on questionable feet. As said above, I feel fortunate that I was able to bring all this off without a medical collapse of some sort intervening. Not only that, but I am very pleased with the effort. I probably have discovered more exciting things during my field research, but this is not a bad capstone for it all. I discovered some very interesting and thought-provoking things in this piece of work that will potentially make people think more effectively when doing research on *Ilyanassa* and other snails. As I write this, it is now about 10 years after the fact. My feet are still doing pretty well. But, now in my sixty-seventh year, I'm no longer getting around with anything even so good as my former unsure surety. Could I presently do this experiment again? Probably not, but given a need I would try. In any case, it is good to have done it.

Let me explain how I set up this experiment in the field. I need to be as clear as possible about this. If you don't understand how the experiment was designed, the results and conclusions from it will seem totally mysterious. Back in the previous chapter, where maybe you stowed that $20 bill, there is an aerial photograph of the Savages Ditch habitat. Refer to it now. You will see a letter A about 50 m north of the Base Point. That will be referred to as position A. You will also see an area marked X on the right side of Heaven's Gate and just to the north of it. Heaven's Gate experiences relatively strong tidal currents. I know that sometimes it was a challenge keeping my footing, such as it is, when walking across there. Area X is in what I call the "rush" because of the tidal currents there. I chose this area as a source of snails for release elsewhere in the habitat because there were both large and small snails present. I also knew that most snails from there would likely be uninfected. There are other lettered positions in the photograph, B on the north side of Angel Island, C on the north side of Devil Island in the Funnel, D on the south side of Devil Island, and E on the south side of Angel Island.

What I am going to do is use position A as an example to illustrate how I set things up at all five of these positions. On 3 June 1999 I went down to Savages Ditch. I had the place to myself. It was a pretty day and no one else was there. I took a set of vernier calipers with me and went to the area designated by the X in the photograph. I preset the calipers at 23.5 mm. I collected 149 live snails from the X area that would not pass through the jaws of the calipers, meaning they were more than 23.5 mm shell height. I also gathered 151 live snails that did pass the caliper jaws, meaning they were less than or equal to 23.5 mm in shell height. This was only an approximate measurement of snail size, but it effectively separated them into two categories. Snails collected in area X would ultimately be referred to as "outlander" snails. Why? Well, it's because groups of them would ultimately be released at each of the lettered positions noted above (A-E) and they did not come from any of those positions. They were from another place in the habitat, i.e., outlanders. There would be two kinds of outlanders released, large and small. In addition, on that day, I measured my way to the A position and drove in a metal stake. The polar coordinates of that stake from the Base Point were 49 m at 11°. If you prefer x-y coordinates, it was at -9 m on the x-axis and +49 m on the y-axis. The A in the photograph is about there, but I moved it left a little so it would not confusingly sit right on the y-axis or on the shoreline in the photograph. It's not that the exact location of position A, or any other position, is something that you need to keep in mind. It is, however, important that you know that I knew where it was. Once the stake was in I proceeded to collect 50 snails from the immediate area around the stake. Size did not matter in this collection. If the snail was there and alive, that was all that mattered. I refer to the snails collected from around the A position as A natives.

These three groups of snails, 50 A natives, 149 large outlanders, and 151 small outlanders were returned to the laboratory. I had set up a screen-divided and aerated 76 liter aquarium filled with 28 ppt salinity water from Delaware Bay to hold outlander snails. That's a pretty big aquarium (around 20 gallons) and divided by the screen I could maintain all the large outlanders in one half and all the small outlanders in the other. These would be pools of large and small outlanders from which I could draw specimens for release at each of the five lettered positions.

Now I could get snails ready for release at position A. First I got all the large outlanders out of their half of the storage aquarium and put them into the individual compartments of a plastic grid on a laboratory table. Then using computer-generated random numbers I selected a random 25 to be marked and released. I did the same with the small outlanders. It is important to select experimental subjects (snails in this case) randomly. In the case of large or small outlanders, had I just stuck my hand in the aquarium and picked up 25 snails, I might easily have gotten a biased 25. Maybe I would have unconsciously gotten larger (or smaller) snails, or especially clean (or dirty) snails, or, or, and or. You never know what shenanigans your unconscious mind might be up to. Once I had my 50 A natives, and the random selections of 25 large and 25 small outlanders, the next job was to mark them. Earlier I described all the pleasures of individually marking snails. You know, the buffing in the orgasmitron to get a dry, clean patch of shell to attach the tag, the handling of those pesky little bee tags, the attaching of them to the shell by embedding them in clear fingernail varnish, the tamping down and re-gluing of raised tag edges, and the final sitting back to admire your work. What fun! Snails released at position A got yellow tags. Numbers 1-50 went on natives, 51-75 on large outlanders, and 76-100 on small outlanders. To aid sighting snails in the field shell apexes on all were painted with a blue fingernail varnish. It was Revlon Ultraviolet. Red would have been better as that color was easier to see on a snail when searching, but at this time I was still tracking snails released in 1996 with red on their "tails." I didn't want to confuse snails from 1996 and 1999 groups. So blue apexes it was.

With all that done (a long day's work) and the markings dry, the next operation was to test each snail for the presence of trematode infections. I got out 100 plastic cups and filled them all about three quarters full with filtered Delaware Bay water (28 ppt) and then put a single marked snail in each. Some lightly weighted porch screening kept snails in their individual cups. They will crawl out and you could end up with two snails in a cup, or you might have several snails out on the table. Either way you couldn't be sure which snail released which cercariae and that must be avoided. After a snail was alone in its cup for at least 24 hours the snail was removed and the water examined for any cercariae released. I had set up a computer spreadsheet with snail numbers already entered and the species of cercariae released by a snail (if any) could be recorded. Each snail was measured again. The first measurement had been done in the field with set calipers. It was possible that I had held the snail a bit crooked when I tried to pass it through the jaws of the calipers. It might have passed when it should not or not passed when it should. A laboratory measurement would catch any such errors. In fact a few "small" outlanders had to be reclassified as large. In any case, whatever measurement was gotten this time was that snail's official starting measurement. When all this was done, I was ready to make the trip back down to Savages Ditch to release these 100 snails. In the end for these 100 snails I would know what the numbers on snails were, where they came from, what infection (if any) they had demonstrated, the sizes of snails, and where and when they were released. All this information would be recorded in the spreadsheet.

On 7 June I released these snails at the position A stake. That's a pretty quick turnaround. Four days after initial collection they were marked, tested for infections, measured, and released to roam freely where they would in their home world. When the tide flooded in over them they dispersed away from the stake within minutes. After releasing snails at position A, I went to position B north of Angel Island. I drove in a stake, measured its location, and collected 50 B natives. They were taken back to the laboratory. I randomly selected another 25 large outlander snails and another 25 small outlanders. These 100 snails were to be released at position B. They went through the same procedures as position A snails except they got white tags (B natives 1-50, large outlanders 51-75, and small outlanders 76-100). They were ready for release on 10 June. On that day, after releasing snails at position B, I began searching for and sighting snails released at position A. On 15 June I collected natives from position C in the Funnel north of Devil Island. These guys got blue tags and snails were released at position C on 19 June. Natives from position D were collected on 22 June. Natives and outlanders got green tags and were released on 24 June. Native snails from the fifth position (E) were collected on 28 June. Position E snails got red tags and were released on 1 July. What with 500 snails processed and released, that June was a full, busy month for me. I remember being worn out when it was over. Happy note though, my feet were holding up.

So there we go. The field experiment was in place and running. Now of course there was the searching and recording of locations where snails were sighted. I kept measurements of individual snail sizes as time proceeded, though I never officially analyzed them for growth. That was because no dissections were done. All snails were native to Savages Ditch and, especially for snails that tested as putatively uninfected, and I could not be super sure whether they were actually uninfected or not. Some snails were brought back and subjected to cercarial release tests to see if they had changed infection status since release. After the test they were returned to the field where they were found and re-released. Part of Chapter 12 treats what happened in that research alley. In all 273 of the 500 snails released were sighted 645 times after release in searches from summer 1999 to spring 2002. Of all sightings, 92 % were made in the first half year after release, mostly in the summer of 1999. In the past I had always released snails from one place, but this time as already described they were released from five different positions that were separated by more than 100 m, or an island. These snails became spread out over a wider area and many, nearly half, were not seen at all after release.

The approach to analyzing the data from this study was similar to that described in the last chapter. I won't go through it all again, but I do want to be sure that the analysis here can be understood. In the previous chapter all snails were set free at one place, the Base Point in the photograph, and only one MANOVA (multi-variate analysis of variance) was required for the initial analysis of those data. The basic idea there was that the sighting distributions of the five snail infection categories [NI (uninfected), *Himashla quissetensis*-infected (*Hq*), *Ls*-infected, *Zoogonus rubellus*-infected (*Zr*), and *HZ*-infected) were compared to see if any of them was in a significantly different place in the habitat. If you recall, each category had a collection of points where snails in that category were sighted (a sighting distribution). The MANOVA test was used to compare the five sighting distributions. Here, snails were released from five different positions (A-E) and I analyzed the sighting distributions of the three groups of snails released at each position. That is, I used five separate MANOVA's to compare the sighting distributions of native snails, large outlander snails, and small outlanders snails from each position. In the paper (Curtis, 2007b), I presented five graphs with the sighting distributions of the three groups from each position, each with their characterizing standard ellipse. I won't include those five graphs here, but further on I will provide the summary graph from the published paper. I know, I know you can't wait, but you'll just have to. I don't mean to defer your curiosity, but that graph doesn't exactly fit here. What the MANOVA test for a position would tell me was whether natives, or large outlanders, or small outlanders dispersed differently from that release position. That is, did at least one of these groups have a sighting distribution different from the other two. If the MANOVA test demonstrated some significant difference in the three distributions, as before it became legal to use a uni-variate test (the Kuskal-Wallis again) to examine the distributions of the three groups separately on the east-west and north-south axes. Based on these tests it could be established which snail group if any distributed east or west (or north or south) of which others. I could then begin to identify and think effectively about what really happened and why.

An important assumption of a MANOVA test, that data must be independent of one another, also must be considered here. In the last two chapters I believe I made a convincing case that data on locations where the same snail was sighted were independent of one another. Basically, the case was that snails can move extensively and there was no predicting how far, or in what direction, a snail would go between sightings. Therefore, where a snail was on a former sighting did not predict where it would be on the next sighting, i.e., successive sighting data for a snail would be independent. This was important to establish if a MANOVA test was to be used.

The data considered here were less extensive and not all the tests I used previously could be employed here. Let me summarize what could be done. (1)

Individual snails were sighted up to 10 times, but 40% of snail sightings were of snails sighted only once. This reduces substantially the number of same snail sightings among data and therefore the overall potential problem. (2) Snails can move 10's of meters in a day. Considering release time as the first sighting, I calculated the average time to the 645 subsequent snail sightings and it came out at 116 days (range 3-1,383). Almost always (92% of the time) sightings were more than five days apart. These facts too suggest that sighting data would be independent because they are mostly so far apart in time. (3) From each release position I analyzed all net next moves by snails in the three groups that were sighted three or more times (i.e., that had at least one net next move to put in the analysis) to see if they took up random directions for next moves. There is a table in the published paper providing data associated with these analyses. You may remember me doing something like this in the last two chapters. The Rayleigh test was used to analyze the distribution of points (i.e., directions of net next moves) around a circle. If the points were uniformly distributed, snails had taken up random directions for next moves. The three groups from each of five positions comes to 15 groups of snails to analyze. In all I could analyze 13 groups of snails from the five release positions. The other two groups (large and small outlanders from Position E) did not provide enough data to analyze in this way. Of the 13 groups, analysis showed that snails in 10 chose random directions for next moves. Groups of snails that did not take up random directions were small outlanders from position A (preferred southerly moves), small outlanders from position B (preferred easterly moves), and natives from position C (preferred northerly moves). If you look in the photograph (that $20 bill again) at the locations of positions A, B, and C, you can see why these groups might not show random next move directions. There were other reasons involved, but basically in certain directions there were landmasses in the way and snails couldn't much move in those directions. From position A snails could not found to the north of the stake there (subtidal most of the time) and they could not easily make easterly net moves because of the shoreline in the way there. From position B southerly net moves were unlikely (Angel Island in the way), and this was also the case from position C (Devil Island in the way there). However, most groups did take up random directions. Even the three groups that did not had net moves well distributed among available directions. In other words, all groups of snails from all five positions tended to choose random directions for next moves. With all this, and the analyses in previous chapters, it seems pretty clear that sightings of the same snail are independent of one another. This is the most important assumption of MANOVA. It having been satisfied makes it fair to use this as a tool of analysis. When you use a test of any sort to analyze data, recognizing and checking up on the assumptions that the test requires is a necessary thing to do. There is a chance that unmet assumptions could negate the analysis. It is always best to investigate whether or not they are met as thoroughly as possible and avoid this sort of mistake.

Without loading this (and you) down with every last bit of data with respect to parasitism (they are available in the published paper), let me try to summarize the characteristics of the various groups of snails in this experiment. The parasitism categories we have to deal with are uninfected snails (NI), and snails infected singly with *Hq*, *Ls*, *Zr*, or *Stephanostomum tenue* (*St*). There were also snails with two different double infections, *HZ* and *LZ*. So there were seven infection categories among the 500 snails initially released. Among 112 small outlanders released, only four (about 4%) were infected with *Zr* or *Ls*. Released large outlanders numbered 138. There were more of them because at the second size measurement in the laboratory I found that several "small" outlanders were actually 23.5 mm or larger. Most released large outlanders were uninfected, but 21% were infected, mainly with *Zr* and *Hq*. In the proportions of infected native snails released, there was lots of variability. Recall that at each of the five release positions 50 native snails were released. Natives from position A had a parasite prevalence of 68%, mostly *Hq* or *Zr*. Given the last chapter, the presence of many *Hq*-infected snails along that shore should not come as a surprise. Position B natives were 42% infected, also mostly with *Hq* or *Zr*. The release site in the Funnel, position C, had native snails that were 58% definitely infected, almost entirely with *Hq* or *Zr*. Only three natives (6%) from position D were infected. Again, it was with *Hq* or *Zr*. From position E 34% of released natives were infected, mostly with *Zr*, but also with *Hq* and *St*. Bear with me just a moment longer. Of the total 500 snails released, 72.3% were NI, 11.6% were infected by *Hq*, 1.4% by *Ls*, 12.0% by *Zr*, 0.6% by *St*, 1.4% by *HZ*, and 0.4% by *LZ*. So, most snails were uninfected, or *Zr*- or *Hq*-infected. Let me again say one more thing. Small outlanders were decidedly smaller (average shell height just under 22 mm) than any other released group. All other groups had considerably greater size (average shell height 24 to 29 mm). As we try to figure out why what happened in this experiment happened, these factors will have to be considered.

Many released snails were not later sighted, but the ones that were had similar proportions parasitized to those released. To restate this in a couple of ways just to make it very clear, the percentages were nearly the same, indicating that sighted snails were not for some reason an unusual subset of the released snails. Put alternatively, released snails infected by certain parasites were not lost or inordinately more or less often sighted. All infection categories were sighted (came back) in about the proportions they were available, i.e., sighted snails were representative of released snails.

Results were pretty striking. It would have been reasonable to expect that the three groups would disperse similarly from each release position. That would be the no difference (null) expectation or hypothesis. This was clearly not what happened and those hypotheses can be mostly rejected. Recall that I did a separate MANOVA test on snails from each of the release positions. With the exception of position D, all release positions had a significant MANOVA result. That is, at least one of the three groups, natives, large outlanders, or small outlanders, dispersed differently from the others. So in terms of dispersal it mattered whether snails were native to the position, or large or small outlanders. Now it's time for that eagerly awaited graph I promised earlier in this chapter. It is presented below. Let me take a moment to explain what is in the graph, something you should always do whenever a graph is presented. It is an approximate drawing of the Savages Ditch habitat. The Base Point with x- and y-axes intersecting at that point and showing distances in meters east-west and north-south are shown. The amount of detail is reduced to make things less confusing. Approximate shorelines are indicated as is Heaven's Gate in the gap between the simple lines representing Angel and Devil Islands. Also shown are the five release positions (A-E). From each emanate three lines, each terminating in an arrowhead pointing to a solid dot with an associated 1, 2, or 3. Native snails released from a site are signaled by 1, large outlanders by 2, and small outlanders by 3. Let's use dispersal from position A as an example. The solid dot next to the 1 at the terminus of the arrow coming from the A position represents the center of the sighting distribution for natives released at that position. It is determined by the median positions on the east-west and the north- south axes. Thus it comes to sit just where it is. A median, once again, is that point above which half the points in a scatter of points fall and below which the other half fall. So, for example, on the east-west axis half the sightings of A natives were east of the 1 dot while the other half were west of it. Clearly natives tended to be sighted sort of southeast of the release position, pretty close to shore and not too far south. A number of them were sighted along the northern shore of the Funnel. Apparently, over time there is an interchange of snails between the A shoreline and the north shore of the Funnel. On the other hand, both large and small outlanders from the A position tended to disperse more to the west than natives and farther south, out in the "rush." Of course, I did some statistics (as before, separate uni-variate Kruskal-Wallis tests on the x- and y-axes) to back up observations such as these, but there is no good reason to clutter things up explaining those tests in detail here. Just take my word for it that what I say here about dispersions of groups from positions is backed up by statistical analyses.

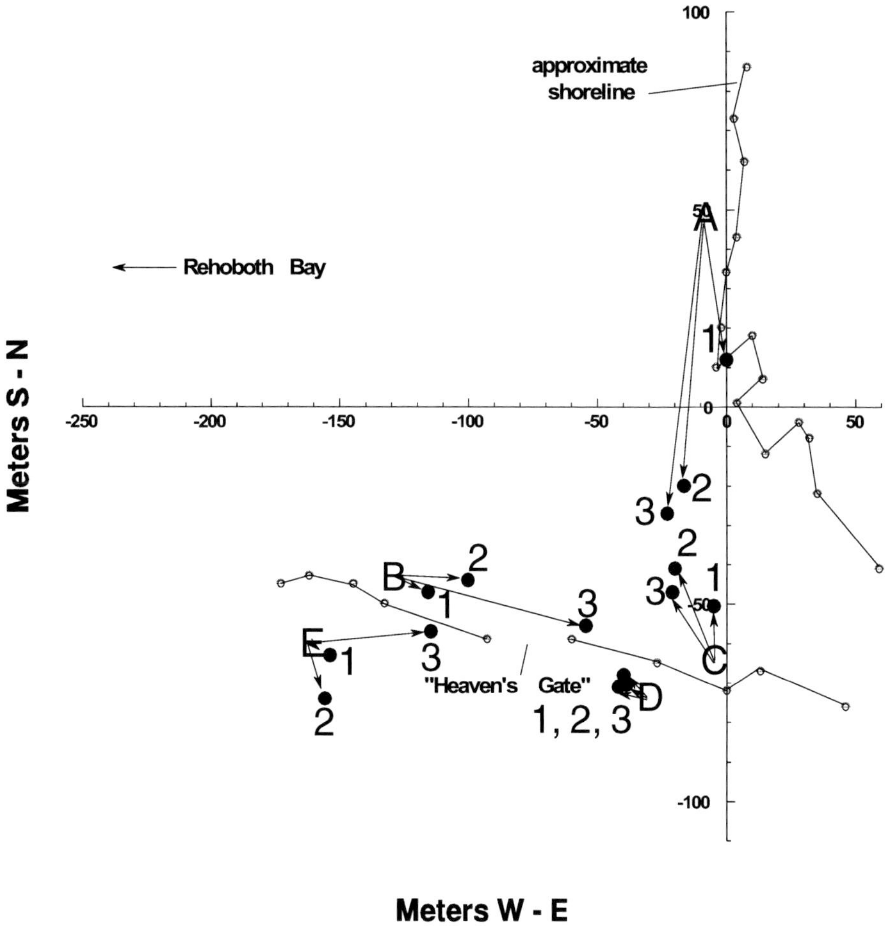

A summary graph showing the basic results of the 1999 Savages Ditch experiment. (Reprinted from Journal of Experimental Marine Biology and Ecology, 352, Lawrence A. Curtis, Spatial heterogeneity in size and parasitism: How it arises in an estuarine snail population, 317-330, 2007, with permission from Elsevier.)

Let's stick with position A. Recall that the native snails were the most frequently infected with *Hq* of any set of natives. I sighted 17 of the 50 snails natives released and over half (59%) of those definitely carried *Hq*. Their sighting distribution is confirmation of a major conclusion from the previous chapter: *Hq*-infected snails hang out close to shore. There are now four strong indications that this goes on, (1) the near-shore distribution of *Hq*-infected snails on Cape Henlopen, (2) that original 1995 along-shore snail collection at Savages Ditch that had 85% *Hq*-infected snails in it, (3) the *Hq*-infected snails released in 1996 that stayed close to shore, (4) and now the largely *Hq*-infected snails released from position A in 1999 that also stayed close to shore. Taken together this makes a convincing case for this phenomenon. Once again, the probable reason why *Hq*-bearing snails do this is because the parasite modifies the behavior of the host snail so it spends most of its time in proximity to a next host (at Savages Ditch, the ribbed mussel) in the parasite's life cycle to which cercariae can be transmitted. Another striking feature of dispersal from this site is the fact that both outlander groups (large and small) tended to disperse back toward where they were originally collected. They came from the "rush' just north of Heaven's Gate and when released from position A they went south and west, back toward their place of origin. Curious.

From position B on the north side of Angel Island the group that dispersed most differently was the small outlanders. Their sighting distribution was east and a little south of the other two groups. Most sighted natives and large outlanders were (putatively) uninfected, though some were infected with *Hq* or *Zr*. They were also quite large, with average shell height about 27 mm. Sighted small outlanders were 15 in number and only one of them demonstrated an infection. Because prevalence was low parasites probably did not much influence dispersion from this position. Basically, natives and large outlanders stayed close to where released, but small outlanders dispersed east back toward Heaven's Gate where they were originally collected. This is the second case where small outlanders did this.

Position C was north of Devil Island in the Funnel. The Funnel has many infected snails in it and not surprisingly many released natives from this site were infected with *Hq* or *Zr*. Of the 50, I was able to sight 37. Fourteen of them did not release cercariae and were putatively uninfected. I suspect that many of these were actually infected, but we'll never know for sure because I dissected none of the snails in this experiment. The rest of C natives demonstrated mostly *Hq* or *Zr* infections. Again natives stayed pretty close to where released. Their sighting distribution indicated they dispersed a bit to the north and decidedly to the east of the other two groups, i.e., they stayed in the Funnel, but not especially close to shore. You would not expect a close-to-shore dispersion for these natives because there were as many *Zr*-infected snails as *Hq*-infected snails involved in the sighting distribution. The two outlander groups tended to go west from the C release site and farther north than natives. That is, both outlander groups tended to move out of the Funnel and back toward the "rush." This is the third time small outlanders showed this tendency.

Position D was the one release site from which all three groups distributed essentially the same (i.e., there was an insignificant MANOVA). This meant that no further analysis, separately on the x- and y-axes, was allowed or needed. If you look back at the photograph, this site was on the south side of Devil Island and basically all three groups dispersed a few meters to the west from release and came to reside on that small strip of intertidal shoreline located there. That strip is only about 2 m wide at low tide. There were very few infected snails among natives or outlanders released from position D. So infections won't help us explain why they went where they went. I believe they were mostly trapped there. Had they moved east from release along the south Devil shoreline they would have encountered very soupy mud. I have laboriously ventured along that shoreline several times and found no marked snails. Indeed, there were very few snails of any sort. *Ilyanassa* seem to avoid excessively watery mud, perhaps because in it they stand to sink to their doom. There is a deep tidal gully just to the south of the release point and a small sandbar to the west, between position D and Heaven's Gate. The tidal gully held about a meter of water at low tide and had very soupy mud on the bottom. I know that because I stupidly got down into it on a couple of occasions and had to struggle my way back out, really soupy. I guess I'll never learn. Tidal currents in that gully are sometimes strong. Snails in all groups, smarter than me, did not wander into to this gully. They stayed close to shore on that little strip of intertidal ground south of Devil Island. Further, for the most part they seemed unable to mount and cross the little sandbar to the west between them and Heaven's Gate. It appears that all three groups were basically trapped (maybe even "happy?") just where they were. They basically stayed there for the rest of 1999 when all sightings of snails from D were made. As might be expected they moved away from that shore as winter came on. Oddly, snails (marked or not) did not inhabit that place for the next fours years.

The last release position (E) on the south side of Angel Island is particularly difficult to assess. Few snails released from there were sighted. Of native snails, 12 that were putatively uninfected were sighted. Five other natives that were infected were also sighted, two had *Hq* and three had *Zr*. All sighted outlanders (eight large and seven small) were putatively uninfected. Only one small outlander from E was sighted after 1999. It was RB90. This guy was sighted twice in 2000 on 26 May and 17 June, around the north side of the eastern tip of Angel Island. It had survived, but it had lost shell in the intervening year, being 23.1 mm when released and 22 mm when seen in 2000. So sighted snails from E were few. On the other hand, these 32 snails were sighted 53 times and when I ran the

MANOVA on these sightings it did come out significant, thus opening the way for further analysis on the two axes. Small outlanders had a sighting distribution to the east of both natives and large outlanders. So once again small outlanders tended to move back toward where they were originally collected near Heaven's Gate. Sighted native snails stayed very close to where released, but large outlanders distributed south of the other two groups. They mostly were found in a shallow gully south of Angel Island. Where did the rest of the released snails from E go? There is a large almost entirely subtidal area west of this release position. You can see the beginnings of it in the photograph. On very low tides this area could be searched, but it's a huge area. Time was always short and searches incomplete. Unmarked snails were abundant out in there. No marked snails from positions other than E were found there, and only a few snails from position E were sighted. It seems reasonable to conclude that many snails from E wandered out into this area and were never seen again. Probably they are still there somewhere, their tags likely long eroded away. Here it is some 10 years later. Could one of those 100 released at position E still be running around with a red tag in place? It is unlikely, but if so and it could be found it would have much to tell us. I can only dream.

To summarize what this experiment revealed one would have to include at least two major things. First, from all release positions the center of native snail dispersal was close to where they originally lived. Of course, natives did not just stay in one place. They roamed about pretty much at random, but they did it in the vicinity where they were originally collected. This alone explains why there is so much spatial heterogeneity in the habitat's population of snails. Snails may move more widely, but they tend to stay in certain broad areas of the habitat. Second, small outlanders were collected just north of Heaven's Gate in the "rush." They dispersed back toward the "rush" almost from wherever released. Compared to other areas of the habitat, tidal currents in the "rush" are certainly stronger and this seems to be involved. Indeed, it's in the "rush" that uninfected snails among the 1996 releases tended to be sighted. Not only that, but most of the smaller snails in the habitat naturally occur there. What about position D, from which dispersal of small outlanders back to Heaven's Gate did not occur? If strong currents are the key attractive factor, the argument can be made for the D site that small outlanders from that site did not need to disperse away. This might be said as well about natives and large outlanders, as both these groups were smaller in size and mostly uninfected. The gully south of Devil Island has quite strong tidal currents, which could have made that patch of ground where they stayed attractive.

Nothing so general can be said about the dispersal of large outlanders from the various release positions. From A and C, north of the islands and from the east, they did tend to disperse west back toward the "rush," but from other release positions there was no such tendency. Most of large outlander snails were uninfected and their only unifying feature was being greater than 23.5 mm in shell height at release. We might speculate that larger, uninfected snails can tolerate almost any area of the habitat (except for the Funnel), while smaller, uninfected snails need the "rush" environment for some reason.

I wish to address two issues here: one is the likelihood that snails from one release position would make it to the vicinity of another; and the other is the likelihood that snails released north of Angel and Devil Islands would make it south of those islands and those released south would make it north. Because most snail sightings were made in the first half year after release the following estimates apply mostly to that year. To get at the first of these issues I arbitrarily envisioned a 20 m radius circle around each release point. That encompasses quite a large area, over 1,200 m^2. I then counted the number of snails from other release positions that showed up within each position's area. This would give an idea of the amount of snail intermingling among release positions that went on. Quite a few small outlanders (19% of them) released at position B made it into the area of position D. Given their general dispersal (look back at the summary graph) this is not surprising. However, overall travel among release positions was much less frequent than that. Only 3.6% of snails made it from one release position to the area of another, the 19% of small outlanders from B to D not withstanding. For a couple of specifics, almost no snails from other positions made it to position A and zero snails from other positions made it to position E. So commingling of snails among release positions was quite minimal. This would maintain spatial heterogeneity in size and parasitism in the snail population in the Savages Ditch habitat.

For the second issue, how likely is it that a snail would pass north or south past Angel and Devil Islands? They could make this passage by going through Heaven's Gate or by going around the west end of Angel or the east end of Devil. I often saw marked snails in Heaven's Gate, but never did I see any passing around the other ends of the islands. So most, if not all, transits were by way of Heaven's Gate. As indicated in the photograph that passageway is about 40 m wide and you might expect much passage of snails through it. Quite a few (47.6%) small outlanders from release position B did pass through from north to south, but other than that transits through Heaven's Gate were more modest than I expected. In overall terms, i.e., all positions combined, about 10% of snails released north of the islands passed to the south and likewise about 10% of snails released south of the islands made it north. Because most of this movement was in 1999, this becomes a yearly estimate of snails moving through Heaven's Gate (about 10% each way per year). Most of that 10% each way was done by small outlanders, which would follow from where they distributed themselves, in the "rush." So we get the idea that passage of a snail through Heaven's Gate is a surprisingly rare event. This was also true of snails released in 1996. Then all snails were released north of the islands and only a couple of them were sighted south of the islands. It appears that islands like Angel and Devil, and the somewhat restricted passageways past them, tend to separate batches of snails on one side or the other. That surprised me. I had expected that snails that moved around randomly and possibly 10 or 20 m per day would easily pass such obstacles. But, they don't. Surprise, surprise.

The major features of snail spatial heterogeneity in this habitat are now reasonably well identified. One of those features is the existence of almost exclusively large and parasitized snails in the Funnel. They were mostly in excess of 24 mm and almost all were infected by *Hq, Zr*, or possibly with *Ls*. Why they stay in the Funnel is something requiring explanation. We can't explain this concentration of snails with universal proximity to a shoreline because not all snails bear *Hq* and there are many areas of the Funnel quite removed from shorelines. Also, there are annual migrations wherein virtually all Funnel snails leave for winter, but many of the same marked snails return each spring. Not only that, but during a summer marked snails wander in and out of the Funnel, even as they tend to stay within its confines most of the time. With all this, at least from 1996 through about 2004, a concentration of large, infected snails was maintained in the Funnel. So this is a hard to explain aspect of the spatial heterogeneity in the habitat. Why should large, infected make up the Funnel population? In the end I can't explain it, but there it is. Also contributing to spatial heterogeneity is the concentration of *Hq*-infected snails along shorelines. We might add the tendency of *Zr*-infected snails to seek offshore locations. The concentration of smaller, uninfected snails away from shore in the "rush" is yet another contributor to spatial heterogeneity. This may all seem a case of T. M. I., too much information. You undoubtedly never aspired to know all this about populations of *Ilyanassa*. But, to folks interested in studying this or other snails, and believe it or not there is a current and future population of such individuals on the planet, these are just the sorts of things they need to know or suspect. It will help inform their own work.

Let's speculate a bit about why smaller, uninfected snails prefer to hang out in the "rush." As I have noted before the stronger currents there probably have something to do with it. But, what? I can come up with a number of untested hypotheses that may singly or in combination explain this phenomenon. (1) Smaller snails are more at risk from predators (e.g., blue crabs, saltmarsh terrapins) than larger snails. This is because larger snails have decidedly more robust shells that can't be so easily broken by predators. Maybe predators don't much occur in higher current areas. I can say that I have mostly seen blue crabs in lesser current situations. So this at least accords with casual observations that I have made in the area. Thus, smaller snails may escape predation by hanging out in the "rush." (2) It is also possible that currents routinely carry some nutrient that is needed by younger snails. (3) Another possibility is that in stronger current environments snail shells stay cleaner. As far as it goes, compared to other habitat areas snails in the "rush" are noticeably freer of crud (organisms and mud) on their shells. Maybe young snails with less crud on their shells get to grow faster. This has been shown to be the case with at least one other snail. (4) My favorite untested hypothesis, and one I wish I could test, is that in places with stronger currents snails are less likely to get infected by trematodes and they stand to remain uninfected for longer. In such environments the weak-swimming miracidia that infect snails are likely to be swept along and unable to stop long enough at a snail to do the job (see Chapter 13). Is any of these hypotheses actually at play in the preference of smaller, uninfected snails for the "rush" environment? They are all untested. If there is an interested student out there somewhere seeking a research problem, feel free to go after this.

Why does anyone need to appreciate spatial heterogeneity? Let me present just one specific example of a reason why. Suppose you were a researcher and interested in predicting the expected proportion of snails at Savages Ditch doubly infected with *Hq* and *Zr*. You want to make this prediction assuming that the two parasites combine together in individual snails at random. A reasonable way to approach making this prediction, which I first introduced to you back in Chapter 10, would be to sample snails from the habitat and find out what proportions were singly infected with *Hq* or *Zr*. My point here is that spatial heterogeneity matters, but let's suppose you know nothing of it in this habitat. For the sake of description let's say you wander out in the habitat one day and happen to collect snails in the vicinity of position A. Your plan is to use your snail collection as a basis for a prediction of the proportion of doubly infected snails. For illustration, let's use the native snails that were collected in this study for release at position A: 38% were *Hq*-infected and 20% were *Zr*-infected. Based on these prevalences and the product law of probability, you could then multiply the two proportions together (0.38 X 0.20 = 0.076) and predict that 7.6% of snails in the habitat would be doubly infected (*HZ*). Specifically, that prediction applies to the proportion of snails that would be expected to be doubly infected if parasites combine in individual snails at random. That is, if a miracidium of one species approaches a snail to infect it, it will have the same chance of successfully colonizing it whether that snail is presently uninfected or is already infected by the other species. (As an aside here, in actuality 1 % of released A natives were *HZ*-infected, considerably less than 7.6%. Below I will cover what this might, probably incorrectly, be taken to mean.) So, based on a collection of site A natives, you predict that 7.6% of snails in the whole habitat should be doubly infected if the parasites combine in snails at random. The appropriateness of this prediction depends upon how representative of snails in the general habitat your original estimates of proportions were. If your original estimates (0.38 and 0.20) of snails singly infected with each parasite are correct for the whole habitat, then it is likely that your 7.6% prediction of randomly associated parasites existing together in snails is also accurate. If not, then your prediction is in error.

Whether 7.6% *HZ* doubles is a good prediction depends on where you collected your original samples. If you happened to collect them in the Funnel or along that north-extending shoreline, the proportions of singly infected snails you obtained were probably higher than they would be for the whole habitat. Therefore, when you multiply those proportions together, in terms of the whole habitat your prediction (expectation) would be too high. You would predict too many doubly infected snails in the overall population. If you happened to collect the initial samples in the "rush" or maybe on the south side of Devil Island, then you undoubtedly would have arrived at unduly low prevalences of *Hq*- and *Zr*-infected snails. Your prediction for the proportion of doubly infected snails in the general habitat would then too be low because you are multiplying together two proportions that are lower than they should be.

Suppose you wanted to take another step and compare the actually observed proportion of doubly infected snails with the predicted proportion of doubly infected snails. Then, you have potential error in both observed and predicted proportions. Your observed proportion of doubles depends on where you collected snails, i.e., from the Funnel you would have found a greater proportion of double infections and from the "rush" a smaller proportion. The predicted proportion of double infections would have the same potential errors described in the previous paragraph. Any conclusions you might draw about whether *HZ*-infected snails are more or less common than expected (predicted) will depend on how well you understand spatial heterogeneity in the habitat.

By the time of this study I understood heterogeneity in this habitat pretty well and could sample the snail population quite representatively and comprehensively. I was able to get good estimates of the overall proportions of singly infected snails at Savages Ditch. Therefore, assuming the two parasites combine in snails at random, a well-founded expected proportion of doubly infected (*HZ*) snails could be made. Let's assume you did this with snails collected in this study. Among the 500 released snails 11.6% were singly infected with *Hq* and 12.0% were singly infected with *Zr*. Based on these values, if the parasites combine in snails at random, the predicted (expected) proportion of *HZ*-infected snails in the general habitat is 1.4% (0.116 X 0.120 = 0.014). Among the 500 released snails, the actual proportion of *HZ*-infected snails is just that, 1.4%. So, the observed proportion equals the expected, suggesting that the parasites combine at random in snails.

But, suppose it came out otherwise. If doubly infected snails were more common than expected, you might hypothesize that the two parasites somehow aided one another in infecting the same snails. If they were less common than expected, you might hypothesize that they somehow "fought" (amensalism, competition, predation) when co-infecting the same snail. (Note that in this last case we deal with exactly the situation above where snails were collected only from around position A. Based on those snails, doubly infected snails were expected to be 7.6% of the population, but only 1% were observed. So, according to those figures, *HZ*-infected snails were quite a bit less common than expected. The ill-founded conclusion would be that the two parasites "fought" somehow and could not infect individual snails at random.) Had you sampled the whole population unrepresentatively you might have been led to accept one or the other of these ill-based hypotheses. Depending on which conclusion you happened on, doubles being more or less common than predicted, your research would take on one direction or the other and be on the wrong track. But, we're assuming that you accounted for spatial heterogeneity and took samples representative of the whole habitat. Because predicted and observed proportions came out equal, you are led to accept the hypothesis that these two parasites do indeed combine randomly in individual snails. They do not "fight" or facilitate one another when infecting the same snail. This is in keeping with other results obtained by former students (Karen Hubbard, Mike Hendrickson, and Nathan Tanner) and myself. It is most likely just what goes on. The message is: lest your research be led astray, when you are studying an *Ilyanassa* population (or any other) you had better come to understand the spatial heterogeneity in it.

And so we come to the end of my last chapter dedicated to specific cases of what I found out and the circumstances under which I found it out. Whew, you say! As I have indicated, this was my last piece of field work. In recent years, after retiring some four years ago, I have kept involved with this my life's interest. I can't travel very far so I have been writing papers and giving a few seminars (talks) at nearby universities. And, of course, there is this production before you.

Lonesome searcher in 1999, the author searches for marked snails in the Savages Ditch habitat.

The subject matter of this chapter is perhaps not especially of pointed interest to people not involved with biological field work. I mean who among the general populace worries about spatial heterogeneity in snail populations? But, it is another example of how the lore, the facts, of ecology and natural history get into the literature so people can read them and use them, or not, as they see fit. It is this process into which I have been trying to give you an insight. So it belongs here. Some of the stuff I have been lucky enough to find out may have been of more TV-documentary type interest, even if it does all happens at a snail's pace. Maybe those other findings have been easier to understand. I guess I'm thinking about the work on *Ilyanassa's* altered behavior under the influence of the parasite *Gynaecotyla adunca*, but this investigation, while on different sorts of problems more of direct interest to practitioners of marine biology, nevertheless is for me a good way to finish up. It has fleshed out my understanding of the beast *Ilyanassa*, and its trematode parasites. That is no small deal. It's what I began striving to do over 35 years ago, back in the middle 1970's. And, considering my intended audience for all this, if you followed this chapter and the previous ones on snail movements in the field all the way through, you are to be congratulated. It's not easy stuff.

Chapter 18

THE ENDING

What did we learn that we did not know? In keeping with the maxim that the best made point is the one made most often, it is my intent in this last chapter to briefly review what was learned from all this work on *Ilyanassa* and estuarine ecology. The job is sort of like writing a very long abstract. Those things always give me trouble and I confess that I dread it, but here we go. Science is a never ending process. As already noted, it always asks more questions than it answers. So, as I go along, I will identify some potentially fruitful and available avenues for further research.

The first discovery revolved around the diet of *Ilyanassa*. This snail was known to consume sand and mud grains, it was a deposit feeder. It had long been known that the snail possessed a crystalline style. Because of the enzyme(s) produced by it plant material can be digested. It was also well known that the snail fed upon carrion. Among neogastropods, the group to which *Ilyanassa* belongs, this is unusual. Almost all neogastropods lack a crystalline style and are strictly carnivorous. This is because the enzymes of a carnivore that digest meat would also digest a style. Consequently, neogastropods cannot normally digest both plant and animal material as food, i.e., be both herbivore and carnivore. But, here was *Ilyanassa* able to mingle both modes of nutrition, digesting both plant and animal material. What was not known was what snails required in their diet. The question that Larry Hurd and I initially asked was: can they get along (survive and grow) on plant or animal sources of nutrition alone, or do they need both. The results of our laboratory experiments (Chapter 2) led us to conclude that the snail is an "obligate omnivore." To survive and grow it requires both vegetable *and* animal nutrition in its diet. Our analysis of the gut content of snails from the field showed that they consume mostly sediments, but also a wide variety of other foods (vegetable and animal) from the environment. So, what we added to the knowledge with this work was that the snail needed, not just consumed, a "balanced" diet. It would be possible to do other experiments similar to ours, but using a variety of other foods. Does the kind of carrion or vegetable used make any difference to the outcome? There have been several studies of feeding in *Ilyanassa*, mostly envisioning the snail as either a carnivore or an herbivore. In future studies it will be important to recognize that the snail must most likely be both.

An important finding was that out on the Cape Henlopen sandflat snails do not simply have a crystalline style all the time. Whether the style is present in the stomach of a snail does not depend on what type of food (animal or vegetable) had been recently eaten. Rather, it forms and dissolves (waxes and wanes) according to a pretty definite schedule. In simplified overview, the style is lost in the morning after sunrise, regained during the day, lost again around dusk, and then redeveloped during the night. This schedule is pretty much kept as long as low salinity events (under 20 ppt) do not interrupt it. My proposal is that the snail's deposit feeding behavior, that of a "biological vacuum cleaner," keeps a more or less constant stream of sediments and other food types coming in, as if on a conveyor belt. This stream contains both plant and animal sources of nutrition. According to the schedule, when the style is present plant materials are digested, when absent animal materials are digested. This way the snail is able to digest a mixed diet and thus remains healthy, grows, and reproduces. I believe that this proposal (hypothesis) covers what is happening, the adaptive strategy of the species, but there are some open questions. What causes the style to dissolve and disappear at times? It is likely to be associated with the appearance in a snail's system of an enzyme that digests protein. That would cause dissolution of the protein-containing style, but it would be good to demonstrate definitively that this is what causes it. What enzyme or enzymes is or are involved? On what schedule are it or they produced? Is something else involved? The style is likely key to the success of *Ilyanassa* as a species and it would be beneficial to understand it's waxing and waning better. Another avenue that could be followed has to do with the effect of low salinity on the style. The style disappears when salinity falls below about 20 ppt and Larry Hurd and I suggested that this is why the snail occurs only in the higher salinity regions of estuaries. The style does not seem to dissolve faster when removed from the stomach and placed in low salinity water. So it is not a direct effect of low salt content on the style. What is the mechanism by which low salinity causes style loss? An answer to this would help to explain why this snail cannot take advantage of lower salinity habitats in an estuary. If it could, it might be an even more abundant and successful species.

The initial study on trematode parasitism done by Larry Hurd and me was an important one. It gave us and the broader scientific community a view into how important parasites could be in a population of marine snails and asked many of the questions that I would address in my future research. The work was done in the late 1970's and very early 1980's when there was not as much interest in marine ecological parasitology as there is today. We were trying to characterize the Cape Henlopen snail population with respect to age, sex, and parasites. From a size-frequency analysis we got our first idea of the relationship between snail size and age. *Ilyanassa* larvae (veligers) settle from the plankton in summer and metamorphose into juvenile snails. By the end of that first summer they are about 6 mm in shell height. They do not grow in winter. By the end of the second summer they reach about 12 mm. In the third summer, at about 14 mm shell height, they become sexually mature, i.e., begin producing sperm or eggs, and by the end of that summer may reach about 18 mm. Snails as large as 26-28 mm are rare on Cape Henlopen, but do occur. Clearly, they are more than three years old. How much older they get to be would have to wait for a future investigation. In any case, on the sandflat there would be three groups (cohorts) of snails present at any given time, first year snails, second year snails, and snails three years and older. If you want to understand any population, this is one of the basic things you need to know.

Almost as soon as snails settle on the sandflat, trematode parasites begin accumulating in them and sterilizing them for life. A size-prevalence analysis of over 3,000 examined snails told us that as snails get older (bigger) there is increased chance that they will have become infected. Viewed from afar the population appeared to be homogeneous throughout, but on closer examination we found it to be rife with spatial heterogeneity.

On Cape Henlopen the *Ilyanassa* population exists in a band along shore about 2,000 meters long and 25 to 200 meters wide. The shore extends essentially from southwest to northeast. On both ends of the sandflat snails larger than 23 mm were almost all infected, and snails under 19 mm were largely uninfected. As noted above, snails begin reproducing in their third year at about 14 mm. A major finding was that snails of middle shell heights, 19 to 23 mm, were more likely to be infected if they lived on the northeast half of the sandflat. Because there are no obvious physical barriers present to cause this difference in parasite prevalence, the basis for it was and still is a mystery. In the years immediately following sexual maturity, especially on the northeast end of the sandflat, parasites begin accumulating in snails in earnest. Because infected snails become sterile, this means that the window for snail reproduction is opened longer on the southwest end and closes more quickly on the northeast end of the sandflat. That is, reproduction among Cape Henlopen snails would be concentrated in the southwest end of the population because of parasitism. In a major way that affected the way this population has to function and was a big surprise to us and to others interested in these things.

Even within the southwest and northeast sections of the Cape Henlopen sandflat there was much spatial heterogeneity. Within each area male and female snails and snails of different sizes (ages) were scattered about heterogeneously. Uninfected and infected snails also occurred in scattered pockets. Further, there were areas where snails were mostly infected by certain parasite species. This all means that representative samples of snails, that would characterize the whole snail population, would be very difficult to come by. We needed to be cautious about saying anything in general about the whole population based on a few samples of snails. That was a major lesson and instructed much of my future work.

Another concept we picked up upon as a result of this work was that, while many hosts were involved in the collective life cycles of the parasites we observed in snails, they all had to use *Ilyanassa* as a first intermediate host. The central and necessary position of *Ilyanassa* in all these parasite life cycles led us to propose that the snail was a keystone species, central to the whole community of species in the estuary, because of all the parasites that had to cycle through it. Were the snail absent many species would be affected and the whole estuarine system would be very different. That idea, though still needing examination, testing, and substantiation, has been with me ever since.

One of the more fun discoveries was that female and *Lepocreadium setiferoides (Ls)*-infected snails had an enhanced positive response to carrion. This was a pretty cool finding in its own right and beyond that it contributed to explaining some of the spatial heterogeneity seen in the *Ilyanassa* population on Cape Henlopen. Clusters of females and snails bearing *Ls* infections might be explained by their attraction to carrion food sources. It is probable that females need the extra protein delivered in carrion because they have to do something that males do not, they have to make eggs and their packaging capsules. A new line of investigation might demonstrate that females do indeed use the extra protein for that purpose. It would also be most interesting to find out just how the *Ls*-induced attraction to carrion is caused by the parasite. This is not in the bailiwick of a field biologist, but what I am thinking about here is a proximate cause. There is work showing that snails infected with *Ls* have a unique protein metabolism. So the attraction of these snails to carrion is likely caused by some sort of chemical alteration of the snail's internal environment brought on by the parasite. Maybe the parasite's rediae or cercariae produce a chemical that causes the enhanced carrion response, or maybe the parasite just draws on the snail's internal protein resources and the result is greater host interest in carrion. Whatever the mechanism (cause) it would be good to nail it down.

Imagine it is low tide in the middle of the afternoon on a summer day. A snail infected with *Gynaecotyla adunca (Ga)* is located at the base of a beach. Such a snail, and only such a snail, has a strikingly interesting agenda to keep. The coming high tide will be followed by a low tide that occurs when it is dark. As the late-afternoon flood tide begins the parasite signals its host, its vehicle, to follow the rising water up the beach. It will be high tide at around sunset. After high-water, as the parasite and its snail are left stranded by the receding tide in the dark, the snail crawls along on the surface of the wet sand. As it does it leaves a mucus trail behind. In that trail on the sand surface are deposited hundreds or thousands of the parasite's cercariae. These infective stages will remain alive in the trail, exposed on the beach, for several hours. This is when crustacean next hosts in the parasite's life cycle (nocturnal, semiterrestrial beach-hoppers) are active on the beach face. Thus, live cercariae are placed on the beach just when the next host is present there and active. It is proposed that the parasite caused its snail vehicle to make the trip up the beach during the flood tide so just that situation would occur and there would be an opportunity for the parasite's cercariae to get from snail host to beach-hopper host. After a length of mucous trail with cercariae in it is made on the sand surface the snail buries in the sand for the rest of low-water. About six hours later (it's sunrise about now) the beach would again be covered by a rising tide. The trail left behind, and the cercariae in it (probably by now dead), would be scattered. The buried snail would emerge from the sand on this high-water and crawl back down the beach.

Beach-hoppers are only active on the beach when the tide is out. They would drown if they remained on the beach during high-water. During the day and during high tides at any time they retire to a refuge, above the water, in algae stranded along the high tide line. Had that rising tide in the mid-afternoon, that I started the previous paragraph with, occurred earlier in the day, say about 9:00 AM, the snail (the host-parasite) would "decide" not to move up the beach with the rising water. Presumably the parasite would give no signal that its snail vehicle should follow the water up the beach into the higher intertidal zone. Were such a signal given, it would mean that snail and parasite would be stranded in the high intertidal zone during a daytime low tide. Even if a host-parasite did make the trip, and they sometimes do, no cercariae would be put in the post-emergence snail trail. Such a trip and any deposited cercariae would be a waste. Beach-hoppers are nocturnal and would not be active, hopping around, on the beach in the daytime. So there is no point to undertaking an upward migration on an early-day flood tide. So, on this tide the host-parasite would probably stay near the base of the beach.

In sort of idealized rendition, that is what *Ga*-infected snails do during the summer. Uninfected snails or snails infected with other parasites do not undertake such migrations. It is nothing short of remarkable that a parasite can somehow make a snail host do something it would not normally do and to boot something as complex as that outlined in the preceding paragraphs. It took much effort to find out that *Ga*-infected snails behave this way and we now know quite a bit, but there are lots of questions remaining. Let me list some. (1) The parasite must alter snail behavior by producing a chemical(s) that affects its host snail. What is that chemical(s) and how does it affect the snail's physiological state? That is perhaps the most important unanswered question. (2) This has to be an adaptation of the parasite and not of the snail. So, in past parasite generations when, why, and how did *Ga* take to using select semiterrestrial crustaceans as second intermediate hosts? However it happened it seems like an improbable event. There must have been an important, critical reason for developing this adaptation. (3) Infections with *Ga* persist in snails for years, but it seems that the altered host behavior is only associated with the summer season, between the equinoxes, when active cercariae are present. There is some connection between the presence of cercariae and altered host behavior. What is it? (4) Most *Ga*-infected snails migrate to the peaks of sandbars and the beach face on flood tides that occur late in the day. This is so they will be stranded in the upper intertidal zone when it is dark and next hosts will be active there. Fewer host snails migrate up when they will be stranded on low tide in daylight. How is the timing of migrations controlled? What signals from the environment are read and does the parasite or the host read them? If the parasite, how could a being tucked away in snail tissues inside a dark snail shell read a reliable environmental cue? I could go on, but I'm sure you get the idea. There is much that might be further done.

There is also the case of the "sexually confused snail." As far as imposex in female snails goes, what Ann Barse and I found out in the 1980's, what I found out by the early 1990's, and what Jeff Kinley and I found out in the middle 1990's was not startlingly new. True, the initial connection between TBT and snail imposex was made (not by me) around 1980 using *Ilyanassa*, but by the middle and late 1980's imposex had been observed in a variety of snails around the world. Here it is more than 10 years since my last publication on this issue, but articles on the effects of organotins (e.g., TBT) on organisms other than snails and imposex in snails still often show up in scientific journals. What the collective work by Ann, Jeff, and me did was to pretty thoroughly document the occurrence of imposex in *Ilyanassa* around Lewes, Delaware. This was a useful contribution.

It broadened out our understanding of how widespread the problem was, but imposex was hardly something new we had discovered. The key new thing that Jeff and I showed was that, even after the ban on TBT in boat bottom paints in 1988, imposex females were still of frequent occurrence in 1995, though the degree of expression of imposex had declined. We also documented for the first time its frequent occurrence along the whole southern length of the Lewes and Rehoboth Canal. Based on a 1992 survey, together with snail samples collected throughout the 1980's, I was able to show that imposex frequency had increased dramatically (some 30 fold) on Cape Henlopen since about 1980. In retrospect, probably the most novel discovery was made during Ann's and my work. That conditions promoting penis development (imposex) in female snails also had penis-generating effects on males was not part of the knowledge back then. Males in the Canal, where imposex females were frequent, retained a penis even when they were parasitized. Parasitized males normally have lost their penises. Also, an uninfected male normally loses the penis after the reproductive season. In the Canal they tend not to. So, where females express imposex, males abnormally retain a penis. I like this finding because, over 20 years later, work by others (E. Oberdorster and colleagues) showed that under the influence of elevated TBT concentrations (as in the Canal) snails produce more of a normally produced chemical compound (to bug you a little, called APGWamide) that leads to penis development. In the absence of TBT, females don't produce much APGWamide and don't normally develop a penis, but this nicely explains why both male and female snails in the Canal would have a penis when they should not.

In Chapter 8 I discussed Karen Hubbard's and my work. This work is kept forever front and center in my mind whenever I think about this whole snail-trematode system. In biological systems, "who" can be adapted to "whom" is a fundamental thing to understand. Our basic question in this work was: when more than one parasite species infects the same individual snail, how do those species get along? What interactions occur and what is the basis for them? Based upon my previous work around Cape Henlopen sandbars, we knew that a large number of snails infected with a variety of multi-species infections could be obtained there. This was because *Ga*-infected snails clustered there and that parasite is often combined with other trematode species in snails. In 379 snails, six parasite species were found in 21 different combinations, including snails with one, two, and three species infecting them. We found that the larval stages (sporocysts, or rediae, and cercariae) of no parasite species were differently distributed in snail tissues just because another parasite species was present in the same snail. For all species, larval stages were mostly distributed in all snail sections regardless of what other species were present. We also showed that, when parasite species co-occurred within a snail, no species shut down the cercarial production of any other in any section of a snail. Further, cercarial production of no trematode species was consistently (statistically) reduced when co-occurring with other species. For the most part trematode infrapopulations in *Ilyanassa* seemed to ignore one another when they co-occurred in the same snail.

When we thought about the relationships of the various species in this system, we reasoned our way to the idea that the parasites must be adapted to deal with the snail host, but that is the only adaptive relationship in this multi-species system. The host cannot be adapted to deal with (resist) the parasites and one species of parasite cannot be adapted to deal with another parasite species. Whatever interactions occur between parasite species when they co-infect a snail, they are accidental. All are ultimately based upon how the parasites are adapted to deal with the host, not adaptation to the other parasite. We emphasized adaptive relationships (interactions) among species because they would be the product of natural selection. As such they would be non-accidental and therefore the most consistent. We believed, and still believe, that it is very important to understand that in this system the only adaptive relationships are those of parasites-to-host. Adaptive relationships of host-to-parasite or parasite-to-parasite should be demonstrated to be operating rather than assumed to be.

There is much more to be investigated. One area would be to know more about how multiple infections come into being. Later chapters, after the one I am discussing here, deal with some of this in a field setting. For example, at least one case where a snail became doubly infected out in the field was recorded, but this sort of thing is a surely an infrequent phenomenon, like the infection of snails generally. Field work cannot reveal much more about this. What is needed is the ability to infect *Ilyanassa* with these trematodes in the laboratory. With the exception of the infection of two snails with *Zr* in H. W. Stunkard's laboratory at Woods Hole back in the 1930's, this has never been done. The problem is that fertilized parasite eggs in their capsules are very slow to hatch out into active snail-infecting miracidia in the laboratory. Other than *Zr* and *Austrobilharzia variglandis (Av)*, miracidia that infect *Ilyanassa* have never even been observed in the laboratory. Were it to become possible to get miracidia and expose uninfected snails to them in the laboratory a whole panoply of avenues of investigation would open up. As a beginning, there are nine trematodes infecting this snail in Delaware and it would become possible to try to infect snails with different pairs of species. If this could be done, with even one pair, it would tell us a lot. That would be a difficult enough undertaking. Another area to investigate would be the adapted state of the species involved in this snail-trematode system. The difficulty with any such investigation is that a species becomes adapted to another across generations and you would have to look into the history of the species you have chosen to investigate. A gene or genes might be preserved and promoted in a population by the favored survival and reproduction of individuals that possess it. It might be possible to identify such genes in a population that help it to deal with another species in the system. The prediction from our work is that such genes would be found in parasite populations that make it possible for them to deal with *Ilyanassa* as a host. The further prediction is that no genes that outfit the snail to deal with parasites and no genes that enable one parasite species to deal with another will be found. It is outside my capabilities to pursue this line, but it would be quite a step forward to find out if it came out as we predicted. A geneticist might be able to do it. I suspect that even for such a scientist any investigation along these lines would be quite a journey.

A truly major discovery was a better appreciation of how old *Ilyanassa* individuals can get to be. When I began all this work in the 1970's and 1980's the maximum age for individuals of this species was thought to be something like seven to 10 years. As outlined earlier in this chapter, on Cape Henlopen there are usually three groups of snails present in the population. The smallest cohort is in their first year, a middle cohort is in its second year, and the largest group (snails 18 mm and over) is made up of individuals three or more years old. The unknown was how old those snails in the largest group got to be. In the early 1990's I began individually marking and following those largest snails in the field to measure their growth rates. I kept track of how fast they added shell height. I found that there was much variability in growth, some snails would add maybe two or even three millimeters in a summer, others would not grow at all. I used data to put brackets on snail growth. Minimal and maximal growth rates were determined. If a snail grew at the minimal rate, it would add shell height very slowly and a snail of a given size would therefore be older. If one grew at the maximal rate, it would add shell height more rapidly and a snail of a given size would be younger. The upshot of the analysis of snails on Cape Henlopen was that an uninfected snail could reasonably be expected to reach 40 years of age. Rather than living about one decade they can live for several decades. It was also discovered that an infected snail (no matter by which species) grows about five times more slowly than an uninfected snail. On average a large size-class, unparasitized snail puts on about 1 mm per year, a parasitized one puts on about 0.2 mm per year. We cannot guess the age of an infected snail because we do not know how long it has been living in the infected condition and growing slowly. All we can be sure of is that it is older than its size suggests. Virtually all 24 mm snails on Cape Henlopen are infected. If it is assumed that the snail grew at a minimal rate while uninfected, got infected when 24 mm, and then grew at the rate of 0.2 mm per year after that,

I determined that an infected snail of 27-28 mm on Cape Henlopen could be as much as 57 years old. That's a truly astounding age for a little black snail about an inch long. It is older than most humans wandering around on the sandflat and happening to look down on it. There are decades of historical content in that snail population that we don't think much about. To push our knowledge further, what is needed is an alternate method, besides monitoring shell height changes in the field, of determining the age of snails so that comparisons can be made. What might that method be? It will probably be some chemical signature reflecting the age of the shell. I make no claim to presently understand all the ins and outs of it, but one idea is that radionuclides (breakdown elements formed from pre-existing elements under the influence of radiation) from nuclear weapons tests run above ground in the atmosphere back in the 1950's might provide such a signature. Perhaps a radionuclide produced then was incorporated into snail shells at that time and would show that snails alive now were in existence back then.

The other surprise generated in the early to mid 1990's was that snails in a population accumulate trematode infections very slowly. In an area of Cape Henlopen where prevalence of already infected snails was very high (80-90%) 1,400 marked and putatively uninfected snails were released. Only 1.6% of them picked up infections between June and late October. I was expecting them to be outright assaulted by trematodes when I put them out, but I was wrong. What this meant was that, even though many snails in the population were already infected, in any given year relatively few snails got infected. When I took a sample of snails and found say 80% infected, I was looking at infections that had accumulated over a very long time, years or decades. The field of ecological parasitology could use more colonization rate estimates of this type to substantiate (or refute) this estimate, but it is what it is. I, perhaps presumptuously, refer to this finding as my Darwin moment. Just as Darwin came to recognize that species in general have evolved and become adapted and diversified over millions as opposed to thousands of years, I came to recognize that *Ilyanassa* individuals live long and populations accumulate infections slowly over decades of time as opposed to just a year or two. The present situation is the product of a long history. If you would understand this snail, this is an important idea to grasp. The first half of the 1990's was definitely a time of stepping forward.

The latter half of the 1990's was a time of creating a capstone for previous efforts and beginning efforts in a new habitat, Savages Ditch. I had worn out the sandflat on Cape Henlopen as far as new investigations there were concerned. It was getting too crowded. My capstone was a matter of gathering together data from 11,774 snails that I had dissected from 1981 through 1995. These were all collected after initially looking into parasites with Larry Hurd around 1980. The idea was to try and sum up what I had learned about *Ilyanassa* as a host for trematodes in Delaware. I have taken to thinking of this snail as the most infected snail on the planet and this is probably true. Of all the snails examined 51% were infected with one or more larval trematode species. There are local "hot spots" where this and other snails have been found so often infected, but in broad surveys of other snails, seldom do they come close to the prevalence of infections in *Ilyanassa.* Nine different species infect the snail in Delaware. There were of course snails singly infected with all nine species. In addition there were many multiple infections. There were 16 of 36 conceivable double (two species) combinations encountered, seven of 84 conceivable triple infections, and one of 126 conceivable quadruple infections. It is very rare to find four trematode species infecting one snail. I found it in two out of 11,774. As far as I know this is the maximum number of trematode species found infecting any snail. More multiple infections (12.6% of 11,774) have been found in *Ilyanassa* than in most other snails. If you wanted to guess whether a particular snail would be infected and with what species, the accuracy of your guess would be influenced by which species you guess might be there [*Zoogonus rubellus (Zr)* would be a good guess, as that is the most frequent, widespread species in snails] and by how big the snail in question is (larger snails are more likely to be infected). On the other hand, knowing the sex of the snail would not help. Females are as likely as males to be infected.

The other topic emphasized in this capping study was that, if you wish to study the way trematode species get along when infecting the same snails, look directly at multiply infected snails. Assemblages of parasite species (communities) can be found at different levels of organization. A parasite infracommunity would be found in an individual snail. A parasite component community (all infections by all species in all snails) would be found inhabiting a snail population. Trying to infer infracommunity dynamics from the state of the component community is correct enough in principal, but fraught with sampling problems. Suppose you sample some snails and use them to determine the prevalences in the snail population of two species of parasites, A and B. From those snails you also determine an observed prevalence of doubly infected (AB) snails in the population. You then multiply the prevalence of A times the prevalence of B to estimate what the prevalence of AB snails should be if they combine in snails at random. Lastly, you then compare the observed prevalence of AB snails with what it should be according to your combine-at-random calculation. As an example, suppose AB-infected snails are much less common than predicted. The likely conclusion would be that there is some sort of barrier to co-infection, the parasites "fight" when they try to co-occur, or that for some reason the two miracidia seldom get to the same snail. This conclusion is dependent on at least two things. (1) Trematode miracidia function like colliding billiard balls bouncing around randomly on a billiard table. I have my doubts, but let's tentatively accept that proposition. (2) Your original sample of snails was representative of the whole snail population. If your observed prevalences were inaccurate, then the whole structure of the argument collapses and with it whatever conclusion you reached. I was later able to learn enough about a population of snails in a smaller habitat (Savages Ditch) to sample representatively and make a reasonable calculation (Chapter 17). My point is that seldom is enough known about a snail population to do this. Spatial heterogeneity can be very complicated. Therefore, we should be very guarded about drawing conclusions about what goes on in parasite infracommunities based on data from the component community. It would be better to look at infracommunities directly, but there are only few studies so far that have been done this way. A useful future direction would be to pursue this approach.

Nathan Tanner's (and my) work began in 1996 and quite a bit of new information was obtained. When we started it appeared that trematodes in the snails at Savages Ditch were everywhere very prevalent. They are very prevalent there, but it turned out to be more complicated than that. A simple "everywhere" was not a good descriptor. Anyway, we released sentinel snails as I had done earlier on Cape Henlopen. The idea was to see if trematodes colonized uninfected snails more rapidly in another habitat where trematodes appeared to be even more prevalent than on Cape Henlopen. We released 249 sentinel snails, but in the end in 1996, 1997, and 1998 we could ultimately only recover 16 of them for dissection. That is a poor showing. We would have liked to recollect more, but most sentinels dispersed from the area where they were released and disappeared. Few could be recollected. Anyway, two of the 16 carried newly acquired infections. It happened over a couple of years, so that's a colonization rate of 6.25% per year. This rate was a higher than what I got on Cape Henlopen (1.6%). It would have to be cautiously used because it was based on so few sentinel snails. In any case, here was another estimate of trematode colonization rate of snails, of which there are not many, and it was again quite low. Apparently, there is about a 5% or less chance that an uninfected snail will get infected in any given year.

Nathan and I were also interested in the chances that an already trematode-infected snail might change the species that infected it in any given year. At Savages Ditch, along with sentinels, we released 231 native snails. These came from the Funnel area of that habitat and virtually all were already infected. We recollected some of these at the end of 1996 and dissected them to see what had happened in terms of infecting species. Some of the ones that escaped dissection in 1996 could be found in 1997 and 1998. These were retested by cercarial release for possibly changed infection status. We also made use of 300 native snails that Karen Hubbard and I had released on Cape Henlopen back in 1993. She and I had found and recollected for dissection 61 of these later in 1993 and one was found in 1994. In summary, Nathan and I concluded that six of 123 native snails, that had been free for one summer or more at Savages Ditch or on Cape Henlopen, had changed infection status. These native snails were mostly infected when released and over one or more years 4.9% had actually changed the species composition of trematodes infecting them. This could happen by adding, subtracting, or trading infecting species. So, in general, the chance of this happening in any given year also appears to be under 5% per year.

There is a question afoot regarding whether trematode infracommunities in snails are interactive or isolationist in nature. The key distinction would hinge on whether the trematode colonization potential is high or low. How likely is it that a trematode miracidium will attempt to infect anew an uninfected or already infected snail? Our contribution to answering this question was that, at least in the *Ilyanassa* system, the most infected snail on the planet, it appears to be very low. As Nathan usefully phrased it, if chances of changing infecting species status are say 10% per year, then on average snails would go some 10 years without experiencing a change. Such a community of species would not be very interactive. An interactive parasite infracommunity would require that new parasite species often come and go in a host. Our measured rate of something like 5% per year or less (on average 20 or more years to experience a new parasite species) suggests strongly that larval trematode infracommunities in *Ilyanassa* are isolationist in nature. It takes a lot of time and effort to get field estimates for colonization potential of trematodes in snails and I do not foresee a raft of new such estimates coming down the pike. Such estimates are a key to understanding the dynamics of parasites in nature and I hope I am wrong in that expectation. I am pleased that through my work with Nathan and Karen (and Chris Morrisey) we could contibute some solid new data to the interactive-isolationist parasite infracommunity debate.

Jeff Kinley was responsible for adding a lot to the knowledge about *Ilyanassa*. The work that he, Nathan, and I did on the potential age attainable by individual snails is a necessity for understanding how this snail functions ecologically. I had found that snails could live about 40 years based on Cape Henlopen work. When the Savages Ditch work was done we could make a strong case that snails lived for 60 or 70 years, or possibly longer were the habitat older. Even if one is reticent to buy longevity estimates this long, it is surely true that snails can live for several decades. As I had found on Cape Henlopen, we confirmed that smaller unparasitized snails in the largest size-class (18 mm and up) might grow a millimeter or two each year. A growth model for snails in this group, growing at Savages Ditch, showed that growth slowed as they got older and stopped when they attained about 30 mm (1.2"). The model indicated that an uninfected snail would be about 40-45 years old at this size.

Our work showed that snails are often infected at smaller sizes, but at Savages Ditch few snails get past 30 mm without becoming trematode-infected. Previous work done by me and work done by Nathan and I had shown that the probability of a snail gaining an infection in any given year is quite low (5% or under) and once an infection is established in it a snail carries that infection for years, probably decades. By the time a snail has lived 40-45 years it is a virtual certainty that in one year or another it has become infected. After infection, and in line with Cape Henlopen results, we estimated a growth rate of 0.2 mm per year for snails at Savages Ditch. This estimate may be a little high, but it accords pretty well with observations made on infected snails growing over several years. Many snails over 30 mm shell height have been found at savages Ditch, including one slightly over 39 mm. Even if that 39 mm snail was still unparasitized (it had *Ls*) it would have put on those 9 mm after 30 mm at an extremely slow rate. If we figure that the snail lived uninfected until it attained 30 mm, that's about 40 years right there. To grow 9 mm more as an infected snail (at five years to the millimeter) would take it some 45 years, making the snail about 85 years old according to our estimates. Estimates are just that, estimates. For example, maybe on average the snail grew a little faster than estimated and was actually a little younger. There was a major disruption between 1925 and 1928 when the Indian River Inlet closed and stayed closed. Rehoboth and Indian River Bays were shut off from the ocean for those three years and ecological conditions in the system, including the Savages Ditch habitat, became seriously deteriorated. We concluded that no snail could have lived through that. Snail larvae would have come in with ocean water when the Inlet reopened in 1928. So in the year 2003, when I found that largest snail (and others just a little smaller), the oldest any snail in that habitat could have been was about 75 years. Given all the assumptions that go into estimating a life span that long, that's close enough. Very conservatively we can safely say that *Ilyanassa* individuals stand to live for many decades. Therefore, a snail population such as on Cape Henlopen or at Savages Ditch is a treasure trove of historical content.

Where might we go from here on the longevity thing? As I remarked on my Cape Henlopen results, for comparison, refutation, or support we need an alternate method to estimate snail age. Another question deserving of further investigation is: do parasites perhaps lengthen the life of a snail that they have infected? It makes sense that they would. Think of the enhanced output of cercariae that would accrue to a trematode capable of lengthening the life of its host. But can they? This would be difficult to decide with either a field or laboratory approach. Rhetorically, how do you experiment with longevity on organisms (the snail and the parasites) that potentially have such long life spans? I surely don't have a workable answer. Finally as regards this topic, it might be possible to gage the ecological effect of the snail's (and parasite's) extended longevity by curtailing it. It wouldn't be easy or quick, but it might be possible to maintain a field population of snails where individuals are not allowed to get more than say five years old. (For comparison, you would also have to maintain a control population in which snails more than five years old are included.) You could simply harvest (perhaps sieve) snails from the experimental population when they hit that age (size). Very few snails would get infected in that amount of time and there would be no larger individuals in the population. The situation would be far from usual. Could the experimental snail population survive? What does the absence or near absence of parasites do to the snail population, to other organisms associated with it? Would the snail population over reproduce? Would it reproduce later? I suppose I could go on, but lots of questions might be asked, investigated, and answered. Maybe you can come up with a few.

There is a whole area of saltmarsh islands adjacent to and south of the Savages Ditch site in Rehoboth Bay. Having canoed around amongst them I knew that there were *Ilyanassa* all over the place down in there. Here was about 1.5 km^2 of area with a large snail population about which I knew next to nothing, a situation that nagged me. In summer 1997 I decided to remedy that. With the necessary and appreciated aid of an undergraduate student (Robb Harman) I sampled the snails in the whole habitat and learned quite a bit. Questions were basically two: what were the characteristics of the host snail population (a parasite resource); and how do the set of trematodes infecting snails in the studied area make use of that resource? Snails to infect were abundant (millions upon millions of them in the area); numbers and sizes of snails were very different from place to place; and the population appeared to recruit new juveniles from the plankton regularly. This should be a heavenly place for parasites, but for some reason trematode infections in snails were not very common. Over many years, probably decades, of accumulation only about 14% of snails in the area had become infected. Because the parasites must live in snails, one would expect that snail distribution in space would affect parasite distribution in space. The spatial distribution (dispersion in space) of uninfected snails (the resource) was clumped or aggregated. That is, there were larger numbers of them in some patches of ground, smaller numbers in others. Not surprisingly the spatial distributions of all categories of infected snails were also clumped. The interesting thing was that, with one exception (*Zr*-infected snails), the distributions in space of all categories of infected snails were two or more times as clumped as uninfected snails. This told me that there were more factors governing the distribution in space of parasite infections than simply the distribution of snails to infect. Did species of parasites infecting snails show any association (positive or negative) in where snails were colonized by them? Basically, no. On the other hand, nestedness analysis of parasite species assemblages at locations in the study area revealed some order. Miracidia of three species of trematodes [especially *Zr*, but also *Himasthla quissetensis (Hq)*, and *Ls*] were more likely to colonize snails at locations than other species. Two questions now arise that are related to success of miracidia: (1) why are so few snails in this population infected; and (2) why are *Zr, Hq,* and *Ls* better able to colonize snails at locations? My speculation in answer to (1) is that the habitat's strong tidal currents make it difficult for miracidia to infect snails. They would be swept along and it would be difficult to stop long enough to bring off an infection. An answer to (2) leads us into the depths of our scientific ignorance. We have almost no idea of how the infection of *Ilyanassa* by trematode miracidia proceeds in the field. We don't even know what features or behaviors would contribute to a species of miracidia being more or less successful. So why some species are better able to colonize snails can't be answered. Indeed, research now needs to be focused on miracidia. The miracidia of most of the nine species that infect snails in Delaware have never been seen, even in the laboratory. This is where work should begin. Once we can get miracidia to work with in the laboratory, maybe investigations can shift to how miracidia operate in the field. This is sort of the great ominous unknown regarding this snail-trematode system.

A particularly tedious undertaking was looking into the way a specific pair of parasites (*Hq* and *Zr*) got along when they infected the same snail. Most of the tedium I readily admit was weathered by my co-author in this work, Mike Hendrickson. I guess that's what anybody gets when they decide to undertake a scientific project. After Mike had become expert at collecting snails and identifying parasite stages we examined a total of 45 snails (15 *Hq*-infected, 15 *Zr*-infected, and 15 *HZ*-infected). The exacting business of counting all the *Hq* rediae and/or *Zr* sporocysts was performed on each snail. Based on the infrapopulation sizes of each species (numbers of rediae and/or sporocysts) in singly and doubly infected snails we had data to assess the way each responded when they co-infected the same snail. It turned out that *Hq* infrapopulations are unaffected by coexistence with *Zr*, but *Zr* infrapopulations are reduced when coexisting with *Hq*. So on co-occurrence (when interacting) in a snail there is a 0 -

relationship. When the species are in separate snails (not interacting) there is a 0 + relationship. In ecology, amensalism is the term used for a pair of species that relate this way. Usually in such a relationship one of the species (the amensal) produces a chemical that inhibits growth of the other's population. In keeping with other results on this species pair we concluded that the two species co-occurred in snails too seldom for *Hq* to develop, through natural selection, a chemical specifically for the purpose of inhibiting *Zr*. So it looks like amensalism, but it is not in all probability based on an adaptive relationship between the parasite species. Rather, the relationship is more likely a result of the way these species are adapted to make use of *Ilyanassa* as a host. That is, if there is a chemical involved that inhibits *Zr*, it is likely made as part of *Hq's* method of dealing with host snails. It would be accidental with regard to how *Hq* gets along with *Zr* when they infect the same snail. It's derived from *Hq's* adaptation to use the host snail. A couple of possible future research projects along these lines come to mind. We could extend the knowledge by examining other pairs of parasites in a similar way. Also, what does *Hq* do (metabolically, chemically) that results in *Zr* being inhibited on co-occurrence in the same snail? If we are correct in our conclusions, whatever is done it should be found in snails singly infected by *Hq* and also in snails doubly infected with both species (*HZ*).

The final three chapters all have to do with *Ilyanassa* travels in the field. When I started on this almost nothing was known because no one had previously marked and followed individual snails in the field. My first experience with this general topic was around a sandbar in 1985 on Cape Henlopen, but that did not tell me much about how far or where snails were likely to move on the broader sandflat. To get an idea of this Karen Hubbard and I followed the movements of individually marked snails released on Cape Henlopen in 1991 and 1993. We had pretty good luck with this part of that field experiment. In 1991 we released 1,200 snails that had tested as uninfected. For another purpose (determining trematode colonization rate) all were transplanted from a saltmarsh where likelihood of being infected in the first place was quite small. Of those 1,200 we found 554 snails 971 times in 1991. In 1993 we released 500 more snails, 200 uninfected from the same saltmarsh site and 300, probably infected snails, native to Cape Henlopen. We were able to find after release 350 of these snails 949 times. Snails moved around quite a bit. Often they moved 10 or 20 m per day. One marathoner moved 46 m in a single day. The snail that moved the farthest was found 180 m offshore from the release point. As around that sandbar in 1985, parasitized snails moved around as much as uninfected snails did. Snails did not mount sandbars in the area where released in either year. So higher sandbars qualify as barriers to dispersal. This was demonstrated best in 1991 when snails moved into an area between two large sandbars and basically stayed there all summer. In 1993 sandbars in the release area were smaller and routes for dispersal more wide open. Snails quickly dispersed farther and more widely in 1993 than in 1991. At any moment snails with any species of trematode infection may be found anywhere on the sandflat, but *Hq*-infected snails tended to disperse along shore close to the beach, while snails carrying *Ls* or *Zr* gave indication of dispersing offshore. This was my first indication that other parasites, besides *Ga*, could affect snail dispersal. This field experiment provided much basic information. We now have an idea about how much snails are likely to move around in a habitat. Over the shorter term of one season they move a lot. Over their long lives they stand to move much greater distances. So we learned a lot, but considering the effort required I'm not sure that more experiments along these general lines on Cape Henlopen would tell us much more. You would have to design more specialized experiments investigating particular aspects of snail movement for the knowledge to increase. To begin to learn more a switch in research locations to the smaller Savages Ditch site and more manipulative experiments had to take place.

The first of the Savages Ditch experiments was based on the movements of the 480 marked snails released there in June of 1996 by Jeff Kinley, Nathan Tanner, and me. Among these snails individuals were included that were uninfected, *Hq*-infected, *Ls*-infected, *Zr*-infected, and *HZ*-infected. Most snail sightings were made in 1996, but some of these snails were followed until 2001. The amount of movement demonstrated by snails in all these categories was similar to what they had shown on Cape Henlopen. Moves of 10 or 20 m per day were common. Infected *Ilyanassa* moved around as much as uninfected snails did, which is contrary to what is usually thought about infected snails. One moving individual, a *Zr*-infected snail, traveled 57 m in one day. That's an incredible distance for a little snail to go in so short a time. An uninfected snail got farthest away from release, 135 m southwest 20 days after release. In the end I had a plot, representing the habitat and similar to the photograph in Chapter 16. There were 1,223 snail sightings scattered about on it that stretched over six years of observation. The question was: did snails, based on the parasite species they were infected with, tend to be sighted in different areas of the habitat? I used what for me was a new method to analyze the scatters of points on the graph (the sighting distributions of snails in each category). The answer was a resounding yes. The most interesting dimension was east-west, or on-off shore. Uninfected and *Zr*-infected snails tended to be found farther off shore (west), while *Hq*-infected and *Ls*-infected snails were found closer to shore (east). Snails infected with *HZ* were sort of between these two groups on this east-west dimension. One category of infected snails (*Ls*) was located well south of other groups, which put them in the mouth of the Funnel mostly away from shores. So where snails are likely to be located in a habitat is indeed influenced by the trematode infections they carry. As far as why snails in infection categories distributed the way they did, the best answer can probably be given for *Hq*-infected snails. Quite clearly year after year these snails hugged the shoreline. There is a dense population of a potential 2nd intermediate host for this parasite, the ribbed mussel, that exists along shore. The proposal is that the parasite changes the behavior of host snails so they prefer to hang out close to shore. This would make it more likely that *Hq* cercariae emitted from snails and having to swim in the outside environment would actually make it to and infect a ribbed mussel.

This whole area of cercariae getting to appropriate 2nd hosts is fruitful as a topic for investigation. For example, we (that is, Ann Barse, one of her students, and me) have all but surely identified *Hq* metacercariae in ribbed mussels from Savages Ditch. Undoubtedly these are in fact *Hq* metacercariae, but upon confirmation this would become a new natural 2nd host for *Hq*. Lots of questions for future investigations arise. What environmental cues are used (by the host snail, or the parasite?) to indicate a closeness to shore? What are the dynamics of this snail-to-2nd-host transmission? How many ribbed mussels are *Hq*-infected? Are infections only found where there are lots of *Hq*-infected snails in the area? How many metacercariae are likely to be found in a ribbed mussel? Is there a seasonality involved or do metacercariae persist in mussels throughout the year? Do metacercariae harm mussels in any way? More specifically, do metacercariae make the mussels gape so they are more vulnerable to predation by shorebird definitive hosts? Snail infections with *Ls* and *Zr* have to get to appropriate (worm) 2nd hosts by means of cercariae too. What other 2nd hosts can be infected? What dynamics are involved there? When *Hq* and *Zr* occur in the same snail, what about the apparent conflict between their needs as far as getting cercariae to a next host? I suppose it is time to shut me off.

The last piece of work I discussed concerned the mechanics of how spatial heterogeneity in size and parasitism arises and is maintained in an *Ilyanassa* population. That snails here and there have different characteristics was one of the first things I learned about snail populations way back in the late 1970's and early 1980's. I owe much to Larry Hurd for his involvement in that early work. Explaining spatial heterogeneity has been an occupation of mine ever since. Clues to several elements of what causes heterogeneity in space have been uncovered on Cape Henlopen and at Savages Ditch (e.g., favored response to carrion by females and *Ls*-infected snails, aggregation of *Ga*-infected snails on sandbars and beaches, and *Hq*-infected snails clustering near to shore). But, until Savages Ditch and 1999 came along, I never got the chance to run a manipulative field experiment on a whole snail population to study what sorts of things go on to bring spatial heterogeneity about. That experiment begun in 1999 is hardly the answer to it all, but it is an interesting start. The more we understand spatial variability in populations of this beast, the more representatively we will be able to sample them, and the more accurately we will be able to characterize populations of this important estuarine organism.

I set up an experiment in 1999 that revolved around release of marked snails at five different positions that were separated by more than 100 m or an island. Native snails from around the position and well as "outlander" snails, from another site within the Savages Ditch habitat, were released at each position. Salient findings were: (1) native snails from each position tended to be found in the general area where they were collected and released; (2) smaller outlander snails tended to move back toward their site of original origin from almost all release positions; (3) larger outlander snails, apparently able to handle a wider array of environmental circumstances, did not similarly disperse back toward their site of origin; and (4) *Hq*-infected snails again clustered near to shore. All these factors would promote spatial heterogeneity. From 1995-96 to about 2004, while I was keeping track of things at Savages Ditch, there were at least two consistent elements to the spatial heterogeneity in the habitat. One was the aggregation of mostly uninfected snails in the strong current area I called the "rush," the area where outlander snails were originally collected. Another was the aggregation of large and infected snails in the Funnel. From the results we can explain how uninfected snails come to reside in the "rush." They actively tended to go back toward there when released elsewhere. It is likely that for some reason they preferred the stronger tidal currents there. On the other hand, I have no explanation for why large, infected snails year after year collect in the Funnel. I wish I could figure that out. Parts of the Funnel are right next to the "rush." Could it have to do with the generally reduced currents in the Funnel compared to the "rush?" Maybe a low current area translates into better success of cercariae emitted from snails at infecting their 2nd hosts. This is a reasonable hypothesis that could be tested initially by exposing cercariae in the laboratory to various current velocities to see how they behave. With that as a background, maybe cercarial success as they infect a 2nd host in the laboratory could be monitored in different current velocities. If laboratory tests indicate that current velocity is important to cercarial success, moving the study to real field conditions would be an intensive test of somebody's ingenuity.

Let me use a little space here to extol the intrigue of a life of science. I have my own predilection for the discipline area, but pick your own. There is no guarantee that an interest in science will lead you to material riches. It might, but given a temperament for research, that's probably not your prime motivation. Nevertheless, we all need sustenance and, if you are any good at it and apply yourself, you stand to earn regular meals and be reasonably comfortable. The one thing I can guarantee is that your life will be damned interesting. Unraveling the workings of nature is absolutely fascinating and you will be mightily entertained. You will have to take on thinking in a structured way. I explained the scientific method (a.k.a., the hypothetico-deductive method) in the first chapter and that is quite a good model for this way of thinking. Keep in mind that it's a model and doesn't cover everything. As explained it consists of (1) make an observation of the natural world, (2) ask a question about that observation, (3) come up with an hypothesis to tentatively answer the question, (4) devise a test of the hypothesis, (5) test it, and (6) use the results of your test to support, modify, or refute your hypothesis. In Chapter 1 some issues are covered about exactly how this method should be applied, mostly having to do with the order in which the steps are used. Even with that qualification the method is powerful.

This way of thinking has been infused into my being and I couldn't be happier with the gift. I deeply owe all my former teachers, especially biology professors, for that infusion. It won't work in all venues. It won't help my enjoyment of a piece of music, or a nifty photograph, or a good workout, but it is pretty strong in most. A life devoted to investigation is rewarding and indeed a pretty good one. In all its complexity *Ilyanassa* has provided me with abundant entertainment for these last 40 years or so. Can we ask more of a passion? This chapter has reviewed a lot of stuff that we now know, but previously did not. I confess to satisfaction in having been a prime mover of that contribution.

Many areas of human endeavor will not lend themselves to this way of thinking, but insofar as it can be applied it will provide a powerful indication of how well-founded any idea or point of view is. Of course, when practicing science you would best try to follow the steps all the way through. I might add to those steps, publish your results and interpretations so others can benefit. Following through on an observation of the natural world and to a degree explaining it is a supremely satisfying experience. The thrill of that moment when understanding comes to you can't be recommended too highly. Probably my cleanest application of the scientific method began with the observation in 1983 of *Ilyanassa* littering that beach in Indian River Bay. I'd never seen that before. What were they doing there? A little further investigation led to the hypothesis that it was caused by snails bearing infections with *Ga*. The test of that hypothesis consisted of collecting snails along that beach. For more than a year collections were made during low tides from below the waterline, at the waterline, and above the waterline. It wasn't long before my moment came. Snails from above the waterline, on the beach face, were nearly all *Ga*-infected, which was not true of snails at or below the waterline. This clearly supported my hypothesis. The unusual presence of snails stranded on the beach in the upper intertidal zone was somehow related to their being infected with *Ga*. I went on with various aspects of this phenomenon for several years, making it a major activity to publish the results and my thoughts about them. So I was a good little scientist. Many of my other investigations did not fit the model scientific method quite so exactly. Such is the way it often goes when you try to explain the natural world. You do whatever you have to do to uncover the truth and it's often not simply a matter of following a formula.

To my avid reader who has stuck with me through all this, it is my hope that you too feel rewarded and satisfied. Maybe you are a reader who never plans to do any science, but would just like to understand the process better. That is a perfectly fine motivation. Perhaps my prime target audience consists of high school students near graduation or college freshmen or sophomores. Why them? Well, they especially are the ones who might be given a different idea of what to do with an undergraduate career. Maybe one of you will look into the possibility of involvement with a faculty member in a research project. Most of the time all you will have to do is ask and then execute. Such an involvement will open up opportunities of all sorts. Maybe it will be that first door opened, your first entry into a whole new way of looking at things. Whether you choose to go on in research does not matter. Just taking part in a serious pursuit of new knowledge is a life-altering thing to do. Whatever your stage of life it will help you do whatever you ultimately do.

I have tried to write this so that someone not steeped in my particular discipline or in science can understand what I was investigating, how I proceeded, why I proceeded that way, the circumstances under which it all happened, and what the results and interpretations were. After all, why would almost anyone already understand or know in detail about snails and parasites. To re-use a phrase from Chapter 1, my intent was to provide a down-in-the-trenches view of the research process. To the degree that I have been successful, a person not involved in science will have been given a glimpse into the process of discovery, how knowledge comes into being. The point is the process of science. There will be no quizzes. It is unimportant whether you remember exactly what the life cycle of a parasite is or some parasite's name. All things must pass, especially details. If you needed details for some reason, I have faith that you could take command of them. As you walk away from this reading leave the details here, but please take any basic insights about the process of science that you have gained with you into your daily life. That's the enduring thing.

For quick reference I present here a listing of the various trematode species (with their abbreviations) that use the snail *Ilyanassa obsoleta* as the 1rst intermediate host in Delaware. I have not written much about all of these, but here they all are. I have also indicated the pronunciation of the genus and general types of other hosts involved in the life cycle of each parasite besides the snail.

Genus				
Trematode species name	Abbreviation	Pronunciation	2nd hosts	Definitive hosts
Himasthla quissetensis	*Hq*	Him-ass-thla	variety of clams	shorebirds
Lepocreadium setiferoides	*Ls*	Lepo-cree-dium	worms	fish
Zoogonus rubellus	*Zr*	Zoo-gon-us	worms	eels
Austrobilharzia variglandis	*Av*	Austro- bill-har-zia	none	shorebirds
Gynaecotyla adunca	*Ga*	Gyn-ay-co-tie-la	crustaceans	mostly shorebirds
Stephanostomum dentatum	*Sd*	Stef-an-o-stom-um	fish	fish
Stephanostomum tenue	*St*	—	fish	fish
Diplostomum nassa	*Dn*	Dip-lo-stom-um	unknown	unknown
Pleurogonius malaclemys	*Pm*	Plur-o-gon-ee-us	*Ilyanassa*	saltmarsh terrapin

BIBLIOGRAPHY

Curtis, L. A. 1975. Distribution of *Sabellaria vulgaris* Verrill (Polychaeta: Sabellariidae) on a sandflat in Delaware Bay. *Chesapeake Science* 16:14-19.

If you are not familiar with this mode of referencing, the above reference is directing you to a library that includes Chesapeake Science *among its serial publications. Once you find that journal go to Volume 16, pages 14 through 19, and there you will have it.*

——— 1978. Aspects of the population dynamics of the polychaete *Sabellaria vulgaris* Verrill, in the Delaware Bay. *Estuaries* 1: 73-84.

——— and L. E. Hurd. 1979. On the broad nutritional requirements of the mud snail, *Ilyanassa (Nassarius) obsoleta* (Say), and its polytrophic role in the food web. *Journal of Experimental Marine Biology and Ecology* 41: 289-297.

——— and L. E. Hurd. 1980. The effect of salinity on crystalline style occurrence in the estuarine snail, *Ilyanassa obsoleta* (Say) (Mollusca: Neogastropoda), and its potential significance with respect to local distribution. *The Veliger* 23: 155-158.

——— 1980. Daily cycling of the crystalline style in the omnivorous, deposit-feeding estuarine snail *Ilyanassa obsoleta. Marine Biology* 59: 137-140.

——— and L. E. Hurd. 1981a. Nutrient procurement strategy of a deposit-feeding estuarine neogastropod, *Ilyanassa obsoleta. Estuarine, Coastal and Shelf Science* 13: 277-285.

——— and L. E. Hurd. 1981b. Crystalline style cycling in *Ilyanassa obsoleta* (Say) (Mollusca: Neogastropoda): further studies. *The Veliger* 24: 91-96.

——— and L. E. Hurd. 1983. Age, sex and parasites: spatial heterogeneity in a sandflat population of *Ilyanassa obsoleta. Ecology* 64: 819-828.

——— 1985. The influence of sex and trematode parasites on carrion response of the estuarine snail *Ilyanassa obsoleta. Biological Bulletin* 169: 377-390.

——— 1987. Vertical distribution of an estuarine snail altered by a parasite. *Science* 235: 1509-1511.

Bryan, G.W., Gibbs, R.J., Huggett, R.J., ———, Baily, D.S. and D.M. Daurer. 1989. Effects of tributyltin pollution on the mud snail, *Ilyanassa obsoleta*, from the York River and Sarah Creek, Chesapeake Bay. *Marine Pollution Bulletin* 20: 458-462.

——— 1989. Raptor. *Perspectives in Biology and Medicine* 33: 154-155.

——— and A. M. Barse*. 1990. Sexual anomalies in the estuarine snail *Ilyanassa obsoleta*: imposex in females and associated phenomena in males. *Oecologia* 84: 371-375.

——— and K. M. Hubbard*. 1990. Trematode infections in a gastropod host misrepresented by observing shed cercariae. *Journal of Experimental Marine Biology and Ecology* 143: 131-137.

——— 1990. Parasitism and the movements of intertidal gastropod individuals. *Biological Bulletin* 179: 105-112.

——— and K. M. K. Hubbard*. 1993. Species relationships in a marine gastropod-trematode ecological system. *Biological Bulletin* 184: 25-35.

——— 1993. Parasite transmission in the intertidal zone: vertical migrations, infective stages, and snail trails. *Journal of Experimental Marine Biology and Ecology* 173: 197-209.

——— 1994. A decade-long perspective on a bioindicator of pollution: imposex in *Ilyanassa obsoleta* on Cape Henlopen in Delaware Bay. *Marine Environmental Research* 38: 291-302.

——— 1995. Growth, trematode parasitism, and longevity of a long-lived marine gastropod (*Ilyanassa obsoleta*). *Journal of the Marine Biological Association United Kingdom* 75: 913- 925.

——— 1996. The probability of a marine gastropod being infected by a trematode. *Journal of Parasitology* 82: 830-833.

——— 1997. *Ilyanassa obsoleta* (Gastropoda) as a host for trematodes in Delaware estuaries. *Journal of Parasitology* 83: 793-803.

——— and J. L. Kinley*. 1998. Imposex in *Ilyanassa obsoleta* still common in a Delaware estuary. *Marine Pollution Bulletin* 36: 97-101.

——— and N. L. Tanner*. 1999. Trematode accumulation by the estuarine gastropod *Ilyanassa obsoleta. Journal of Parasitology* 85: 419-425.

----------, Kinley*, J. L. and N. L. Tanner*. 2000. Longevity of oversized individuals: growth, parasitism, and history in an estuarine snail population. *Journal of the Marine Biological Association United Kingdom* 80: 811-820.

Esch, G. W., —— and M. A. Barger. 2001. A perspective on the ecology of trematode communities in snails. *Parasitology* 123: S57-S75. (The S stands for Supplement.)

—— 2002. Ecology of larval trematodes in three marine gastropods. *Parasitology* 124: S43-S56.

Hendrickson*, M. A. and —— 2002. Infrapopulation sizes of co-occurring trematodes in the snail *Ilyanassa obsoleta*. *Journal of Parasitology* 88: 884-889.

—— 2003. Tenure of individual larval trematode infections in an estuarine gastropod. *Journal of the Marine Biological Association United Kingdom* 83: 1047-1051.

—— 2005. Movements of *Ilyanassa obsoleta* (Gastropoda) on an intertidal sandflat. *Marine Biology* 148: 307-317.

—— 2007. A Small Canoe Adventure. In the *2007 Dan River Anthology*, p. 23-29, Richard S. Danbury, III (ed.), Dan River Press, Thomaston, Maine.

—— 2007a. Larval trematode infections and spatial distributions of snails. *Invertebrate Biology* 126: 235-246.

—— 2007b. Spatial heterogeneity in size and parasitism: How it arises in an estuarine snail population. *Journal of Experimental Marine Biology and Ecology* 352: 317-330.

—— 2009. The gastropod *Ilyanassa obsoleta* as a resource: utilization by larval trematodes in a low-prevalence system. *Journal of Parasitology* 95: 799-807.

Note: * signals a University of Delaware student co-author.

INDEX

B

C

D

E

F

G

H

I

J

K

L

M

N

O

P

Q

R

S

U

V

W

Y

Z